Platonic Wholes and Quantum Ontology

# Dia-Logos

Schriften zu Philosophie und Sozialwissenschaften
Studies in Philosophy and Social Sciences

**Herausgegeben von / Edited by**
**Tadeusz Buksiński / Piotr W. Juchacz**



**Bd. /vol. 20**

PETER LANG
EDITION

Marek Woszczek

# Platonic Wholes and Quantum Ontology

translated by Katarzyna Kretkowska

PETER LANG
EDITION

**Bibliographic Information published by the Deutsche Nationalbibliothek**
The Deutsche Nationalbibliothek lists this publication in the Deutsche Nationalbibliografie; detailed bibliographic data is available in the internet at http://dnb.d-nb.de.

**Library of Congress Cataloging-in-Publication Data**
Woszczek, Marek, 1980-
[Ukryta całość przyrody a mikrofizyka. English]
Platonic wholes and quantum ontology / Marek Woszczek. — 1 [edition].
pages cm. — (Dia-logos, ISSN 1619-005X ; vol. 20)
Translated by Katarzyna Kretkowska.
Includes bibliographical references.
ISBN 978-3-631-66629-6
1. Quantum theory. 2. Bell's theorem. 3. Microphysics. I. Title.
QC174.12.W6713 2015
530.12—dc23

2015001514

This is a revised and expanded edition of the book published in Polish as *Ukryta całość przyrody a mikrofizyka*, copyright © 2010 by Marek Woszczek.

The edition has been prepared and financed in the years 2013–2015 as part of the Polish Ministry of Science and Higher Education's National Programme for the Development of the Humanities (grant No. 31 H12 0439 81).

NARODOWY PROGRAM
ROZWOJU HUMANISTYKI

ISSN 1619-005X
ISBN 978-3-631-66629-6 (Print)
ISBN 978-3-653-05342-5 (E-Book)
DOI 10.3726/978-3-653-05342-5


Peter Lang Edition is an Imprint of Peter Lang GmbH.

Peter Lang – Frankfurt am Main · Berlin · Bruxelles · New York · Oxford · Warszawa · Wien



This publication has been peer reviewed.

www.peterlang.com

# Contents

ἁρμονίη ἀφανὴς φανερῆς κρείττων. (Heraclitus)

φύσις κρύπτεσθαι φιλεῖ. (Heraclitus)

*In omnibus autem partibus relucet totum, cum pars sit pars totius.*
(Nicholas of Cusa)

*Nam non scitur pars nisi toto scito; totum enim mensurat partem.*
(Nicholas of Cusa)

*Müsset im Naturbetrachten / Immer eins wie alles achten…*
(Johann W. von Goethe)

# Introduction

Composition has doubtlessly been one of the intensely discussed problems in the whole history of metaphysics, but the development of fundamental physics has placed it in a completely novel light. Niels Bohr, when developing his refined interpretation of the newly formulated quantum mechanics in the 1920's, placed the category of 'the indivisible wholeness of the quantum phenomenon' at the very centre of the physical theory and thus triggered off another round of disputes within the philosophy of nature, which – though that might seem a bit surprising – have recently been more and more intensely pervading the foundations of physics in the context of quantum cosmology. The next round of speculation was set off by the more or less independent reflections of Heisenberg and Pauli, who strove to find a place for their own accounts of quantum holism in the shadow of Bohr's influential semi-positivistic statements. But then, in the 1960's, John Bell [1987a] opened new perspectives for the discussion of 'quantum nonlocality' and quantum entangled states, and Y. Aharonov, P. Bergmann and J. Lebowitz [1964] (later also [Aharonov, Vaidman, 1990] and [Aharonov *et al.*, 2009]) disclosed a time-symmetrized structure of quantum theory with a timelike entanglement, relational states and a deep connection to the Kochen–Specker [1967] theoretical result, causing quantum mechanics to be even more astounding to metaphysicians (which, probably, was later stimulated by David Bohm's daring ontological ideas subtly indebted to 'Copenhagen' holism). Quantum wholes became even more non-classical than in the pre-Bell era, the latter not having bothered itself too much with the then elusive 'nonseparability' and the physical meaning of entanglement, not to mention the geometry of quantum histories and the exotic quantum temporal-causal 'paradoxes'. Mara Beller [1999] argued, perhaps rightly, that early 'Copenhagen' efforts concerning holism were at best somewhat foggy and philosophically dilettantish. However, the anything but trivial philosophical question, once provoked by Heisenberg and Pauli, can be boiled down to an attempt at judging how radical a holism we need to construe acceptable *quantum ontologies*, and not merely the pared-down epistemologies for quantum labs and quantum mechanics handbooks.

In the present book, I am going to put forward the view that the category of the hidden 'primary whole' has a fundamental meaning in quantum physics and allows us to speak reasonably about its radical, structural holism, which markedly differentiates it from all the earlier classical theories. The very concept of 'holism' is wide and unequivocal, hence the principal issue is to ascertain what we really mean by quantum mechanics being 'radically holistic'. And further: why do the phenomena of microphysics and the structure of the theory itself force us to speak about a logically and ontologically primary 'hidden whole' at a cosmological level? Such reflections result in the need of a new, deeper understanding of the ontological sense of the quantum theory and such an attempt at its presentation shall be found at the end of the book, as its strictly relational interpretation, proposed in the 1990's by Carlo Rovelli. The relational-informational approach (as a certain programme of formulating quantum theory as an Einsteinian 'principle theory' and at the same time the pediment of the future relational quantum cosmology, at least *in spe*) is sufficiently radical to totally transform an understanding of quantum mechanics as the theory of information transfers in the Universe, whose non-classical, irreducible properties are both relativity *and* the global nonseparability of physical states. I will argue that that global nonseparability is *the* fundamental, structural and irreducible feature of the quantum Universe, and every quantum ontology must acknowledge that fact. Sooner or later, a metaphysician is forced to cope with the nonseparability of quantum actions, which means a stormy divorce with the familiar, post-Aristotelian world of separable objects and their states as 'elements of reality'. But one might ask: what are we supposed to do if those 'elements of reality' vanish into thin air and what about ontology itself? Even some adventurous ontologies of nature we sometimes probe have a frustratingly conservative core consisting of some form of noncontextual realism, preferred presentism with the 'real becoming' and/or thinking of wholes as secondary or derivative as if the lure of metaphysical atomism was literally irresistible. Hence, quantum phenomena in the world of Aristotle's successors are doomed to be 'paradoxes' or 'anomalies', and that is of course why 'commonsensical people' like Karl R. Popper or David Lewis [1986b, p. xi] are inclined to complain about the quantum 'doublethinking deviant logic'. I shall claim that as far as quantum theory is concerned, noncontextual realism is doomed to fail and a new kind of a Platonic-type compositionism is needed, which I call an 'internalistic compositionism', generating a kind of non-classical, internalistic logic and an internalistic (non-Boolean) theory of probability.

In Part I of the book, stress shall be put on the more historical-philosophical aspects of the problems in question. When reflecting on the category of the 'hidden wholeness' of nature, I have found it indispensable to limit the historical material directly concerning microphysics to two selected ontological models, key ones in the whole Western ontological tradition, namely those of Plato and Leibniz. In the case of the former, I shall concentrate on highlighting the model of the relations between the whole and the parts (intensional–internalistic), which has been rather seldom analyzed by historians of philosophy and which is competitive to the mereological (extensional) and Aristotelian ones (the latter being intensional, but at the same time 'externalistic'). 'Aristotelian (externalistic) wholes' – considered not just as Aristotle's concept but more broadly as a residuum of the post-Aristotelian imagery in metaphysics – are such irreducible wholes which presuppose a prior existence of the compositional 'matter' for them, i.e., some 'elements' (possessing definite, even if minimal properties) being combined and recombined, or primitive base individuals with solely some non-relational 'thisness', or any other pre-existing empirical 'material' being structured by some 'forms' and creating further empirical wholes. The key feature of that kind of compositionism is its association with the naïve intuitions concerning becoming and change – the secondary wholes arrange and rearrange ortochronally (along with the 'flowing time'), come into being along history with the objective 'now' at the front, while the holistic structure of thus coming-into-being and dissolving objects is external to their supposed conditioning 'pre-matter'. I shall thus conduct, following Verity Harte's important study [2002a], a more detailed analysis of the *Parmenides* dialogue, to show how Plato attempted to justify the value of radical holism and the primacy of the 'hidden whole' over the phenomenal, apparently 'becoming' ('as in a dream' of the senses/sense perception, *Tim.* 52b) parts. I shall then, against the backdrop of the ancient debate on the relation of the 'One' and the 'many', present the problem of the relationship between contradiction and 'absolute elements' (the absolutes of cognition) in the conceptualization of nature, and propose a working formulation of the Universal Principle of Relativity as a postulate and at the same time a procedure of avoiding those contradictions and eliminating the absolutes from the theory. The Platonic model of holism and the postulate of relativizing empirical 'elements' will next be preliminarily used to modify the popular understanding of the nature of the quantum 'state'. It turns out that it is possible, even at quite a preliminary level, to treat quantum mechanics as a theory of hidden, relational structures (atemporal wholes) of physical micro-processes. 'Quantum states' of the imagined, pre-existing 'material systems', which are usually said to be 'measured' and 'registered' as if they were classical, cannot be treated as 'elements' of mechanics, and, even less so, separable 'elements of reality'; for if they are, a whole array of paradoxes, contradictions and physical deadlocks is produced. Quantum wholes are not composites of pre-existing elements (such as 'states of the subsystems'), even the minimally simple ones, thus they are 'Platonic wholes'.

In Chapter 2, I shall concentrate on the holistic aspect of Leibniz's ontology, usually set aside in favour of his theory of substances ('monadology'). I shall attempt to outline Leibniz's interesting project of the ontology of nature's hidden wholeness, construed upon the category of perspectival 'action', and I shall also show how Leibniz's thesis on the processes of nature being 'micro-actions' corresponds to an important idea of 'action physics', put forward recently by David Finkelstein [1999a]. I shall also place particular stress on the connection between that ontology of the mathematizable process and Leibniz's theory of representation in the context of the 20th century theory of categories, used in recent years to construct relational quantum cosmology. One of the most fascinating facets of the Leibnizian project is its subtle interplay between the notions of the 'process' and complete (atemporal) 'series of actions', which preconditions the consistency of cosmology, hence Leibniz can be seen as a pioneer of the classical blockworld ontology with its built-in internal 'dynamics' and with all the tensions inherent in such a construction. That is the case since the Leibnizian 'series' (internalistic cosmological histories/atemporal functions defining blocks of local events) are an instance of Platonic wholes, and such a universe can be described by a general algebra of the process defined for fundamental, fully relational actions inside it, but nevertheless the condition of the consistency of histories (a predecessor of modern general covariance) enforces an acceptance of the blockworld cosmology. That is why Leibniz's system is also so instructive for modern relational cosmologies as their *classical* toy model. Quantum experimental 'paradoxes' such as that of the delayed choice quantum eraser, seen in a cosmological context and not just as amusing lab effects, are especially interesting in that respect. Quantum wholes manifest themselves *fully locally* in the *apparent* causal-temporal 'paradoxes' forbidden by classical mechanics.

Chapter 3 opens the second part of the book, in which stress will be put not on the historical-philosophical aspects of the radical forms of holism, but on the foundational aspects of the ontology of microphysics as such. I shall begin with a presentation of the basic feature of microphysics – the nonseparability of quantum states which, I claim, is also at the core of almost all the philosophical controversies concerning the quantum theory (even if not recognized as such). I shall start with a concise analysis of the ontological assumptions of the Einstein–Podolsky–Rosen Gedankenexperiment, and I shall stress the fact that a simplistic reading of Einstein's arguments against Bohr – too quickly reducing them to the former's defence of relativistic (causal) locality – misses the point, since Einstein in fact always demanded, as has been rightly argued e.g. by Don Howard, that physical ontology should uphold the *strong principle of separability* and have separable 'elements of reality' as the classical 'states of affairs', thus Bohr was right when he insisted that it was *not* spatiotemporal locality itself which was of prime importance in the foundational dispute between them. Bohr's local contextuality and Kochen–Specker contextuality, as implied in the (Bell)–Kochen–Specker theorem, have a deep connection with quantum nonseparability, which is a much more general and fundamental property and makes quantum entanglement itself more primitive than spatiotemporal relations. The EPR argument does not work, because all the quantum 'states' of the subsystems of the Universe are nonseparable, *hence* inherently contextual. The famous Bell theorem does not indeed list nonseparability among its independent conditions, but the latter is deeply ingrained in it as its general metaphysical underpinning and cannot simply be ignored as a purely philosophical issue, thus the problem of determining which of the Bell conditions is/are being violated, gets even more complex. I shall argue that Einsteinian fully noncontextual, i.e., separable quantum 'states of affairs' indeed cannot exist, and quantum ontology must acknowledge that fact by recognizing nonseparability as *the* ur-phenomenon of mechanics, which obviously constitutes a great challenge for metaphysicians: a new, more general kind of realism must be accepted and it is an immediate consequence of the failure of externalistic compositionism. It turns out that the most natural alternative is the radical, yet still local Contextual Realism, which is – true to Bohr's simple intuitions – relativity working at the fundamental layers of mechanics/dynamics. Quantum 'states' as quantum actions within global histories are purely relational and contextual, thus Einstein's hopes cannot hold and Bell-type inequalities must be violated. The question remains what the global physics/cosmology underlying that is.

The consistent relativization of quantum mechanics has serious consequences, among which the most important one is the full relativization of quantum 'states'/actions and quantum 'observers' themselves. *Quantum wholes do not possess any non-relational base elements*, because their internal structure itself obeys the Universal Principle of Relativity. In Chapter 4, I am going to analyze the consequences of adopting such a model of internalistic compositionism in quantum cosmology if we wish to understand the correlations of the EPR or GHZ-type on cosmological scales. Treating quantum histories as 'Platonic wholes', inaccessible to internal subsystems of the Universe, requires an acceptance of quantum blockworld ontology as a deepening of the block ontology of spacetime in the relativity theory. I am persuaded by David R. Finkelstein's [1996, p. IX] slogan: 'Quantum theory is above all an extended kind of relativity', which eventually makes a generic quantum blockworld inevitable. The argumentation in favour of the block universe is stronger than is usually claimed by conservative analytical metaphysicians (as is clearly shown e.g. by Petkov [2006, 2009]) and it is *not* true that that is an ontology excluded by quantum mechanics, often being falsely depicted as time-asymmetric and pro-presentist. An idea of spacelike (signal) nonlocality or nonlocal 'action-at-a-distance' is the direct outcome of a naïve, quasi-classical interpretation of the quantum 'state' and the 'state function', but seems unnecessary if we allow the mechanical 'elements of reality' together with presentism to disappear and treat quantum correlations between entangled subsystems as the atemporal global structure of the nonseparable blockworld, and not the properties of some imagined, coming-into-being micro-entities. 'Aristotelian wholes' and 'Aristotelian parts' may be desirably intuitive when one is accustomed to think about cats, stones and cells, but are useless when there are no base entities placed in the 'points of space' and no intrinsic properties changing 'in time'. Finally, I shall present the very important idea of quantum frames of reference and of the fundamental connection between the Heisenberg–von Neumann cut and the logical paradoxes of self-reference in the context of quantum measurement. The quantum (action) frame is a fundamental concept for the foundations of quantum information theory and the basic ingredient of the ontology of contextual realism in a quantum blockworld.

In the final chapter, I am going to argue, following Carlo Rovelli, that a conspicuous property of quantum mechanics is its immanently relational structure making it a type of metatheory which ascribes different descriptions to different observers without the loss of an inner consistency. I shall thus systematically present Rovelli's time-symmetric relational interpretation of quantum mechanics and the way in which it solves the EPR paradox without abandoning causally local interactions. A consistent relational (and atemporal) interpretation of the 'quantum state' is much more radical than might seem at first glance and it is to Rovelli's merit that his proposal is ontological, which differentiates it from ones that are similar in spirit, but one-sidedly epistemological or, even more disappointingly, constantly drifting towards the baroque many-worlds ontology. It is obvious that, as in the case of the relativistic blockworld, the principal resistance to the *quantum blockworld* and time-symmetric quantum theory has its roots in presentism and the ideology of objective 'coming-into-being', thus the problem of time (and probabilities) in quantum theory comes to the fore. I shall argue, as in Ch. 4, in favour of quantum 'Parmenidean (or Platonic) physics' without any irreducible time flow which cannot be consistently injected into the fundamental, changeless layer of physics, composed of global, atemporal histories. Quantum mechanics is deeply time-symmetric (which feature manifests itself in almost all its pecularities, strangely invoking a banned spectre of teleology in physics), but in a quite different way than classical mechanics, and that makes any relational quantum cosmology even more non-intuitive in the case when the initial and final boundary conditions of the history of the Universe should be seen as relationally 'conjugated' or coupled. The issue concerning mathematical formalism, which would be able to describe such a generalized, internalistic time-symmetric causality is quite interesting, since it might need purely global notions as in noncommutative geometry and view spacetime histories with local relational events as derivative. In any case, Rovelli's relational paradigm, consistently eliminating time flow from the foundations of physics is perfect for construing any quantum blockworld ontology, even if one does not interpret quantum mechanics itself as the principle theory in Einstein's sense (as e.g. Rovelli does). I shall argue that it is exactly that quantum time-symmetry which makes quantum mechanics *fully relational* and the 'quantum states' of arbitrarily singled-out subsystems of the Universe obey the Universal Principle of Relativity. Quantum wholes are atemporal 'Platonic wholes'.

Philosophy is sometimes unable to keep apace with new developments in physics, but since ancient times it has also in a number of ways created the basic paradigms of ontological thinking realized by theoretical physics itself (for example 'atomism' as a general conceptual scheme). Modern metaphysics, dominated by post-Aristotelian imagery and discussing spatial 'things', the becoming 'objects' and 'bare individuals' as if nothing had changed much since the Middle Ages, seems to be estranged in the world of quantum theory and there are good reasons to assume that it has been destined to that alienation from the very beginning. One can pinpoint many reasons of that but I shall claim that one of the most important ones is not only the conservative, but plainly false model of composition or the whole–parts relation. That is the reason of the stress I shall be placing on the alternative paradigms of holistic thinking (Platonic and Leibnizian) and their non-zero value for theoretical physics, at the cost of a wider evaluation of different, competitive standpoints (particularly the more conservative ones) which, I claim, are disappointingly impotent in the field of quantum ontology. In a sense, the book is to encourage philosophers to experiment with some radical ontologies, which from the perspective of standard discussions appear to be definitely exotic. I have attempted to construe all of my presentation in such a way so as not to weigh it down with physical-theoretical details (excluding Ch. 4 and 5, where I have included more technical subject matter), but instead to analyze the foundational problem of compositionism from the perspective of a metaphysican grappling with the challenges brought on by the development of fundamental physics and the new ways of solving particular theoretical problems (particularly in quantum mechanics and quantum cosmology). It has been my particular concern to present purely metaphysical subject matter in a pronounced way, one which is sometimes overshadowed by technical discussions on the details of physical theories, so important in the philosophy of physics.

The final shape of the book has been influenced by many of the ideas which have not yet been fully elaborated by metaphysicians and traditional philosophers of nature, but what has been of fundamental importance to me are the penetrating studies by Carlo Rovelli [1996; 2002a; 2004], David R. Finkelstein [1996; 1999a;

2003] and Yakir Aharonov and his group (e.g. [Aharonov, Botero, Scully, 2001], [Aharonov, Popescu, Tollaksen, Vaidman, 2009], [Aharonov, Tollaksen, 2011]). The simplicity and insight in Rovelli's argumentation have convinced me to a large extent as to quantum mechanics containing many fewer absolute elements than its common presentations suggest, and when its intrinsically relational nature is taken into account, many of its apparent 'paradoxes' are much more easily understood. Rovelli's physical ideas lie at the basis of the interpretation of quantum mechanics here presented and form the central topic of Ch. 5. A couple of his hypotheses, particularly the one put forward together with Alain Connes concerning the quantum thermodynamics of time, are particularly inspiring and worth a philosophical consideration. I am also convinced, as Rovelli, Aharonov and many others, that quantum mechanics is fully time-symmetric and that the asymmetry of the 'flowing time' is one of the most obstructive idols in contemporary physics. Finkelstein's ideas, in turn, have made me realize the necessity of changing the archaic, substantialist way of thinking in microphysics and making use of the physical category of elementary, *generalized* and *relativized* 'actions'. Modern theoretical physics has been developing autonomously and in its own rhythm, but it is obvious that a significant part of that conceptual work has been difficult to perform without a consistent reflection on the adequacy or inadequacy of such categories as 'whole'/'composite', 'reality', 'thing', 'property' or 'change', being at the centre of the interest of ontology since the ancient Greek period. The main concern of the book is a reconsideration of the ontological problem of composition in order to make it adequate for quantum ontology, but the immediate result is much more general: it turns out that that kind of conceptual accomodation is simply impossible without a radical change in the philosophical construction of 'realism' itself. After all, that is what one can expect when one knows the rich history of metaphysics. Finally, though in post-Kantian ontology that might seem a bit naïve, it all boils down to posing still anew N.D. Mermin's [1998b] question contained in the title of his excellent article: 'What is quantum mechanics trying to tell us?'.

The original Polish edition of the book appeared in 2010 and taking advantage of its present new edition I have corrected some printing mistakes, shortened some parts and upgraded or supplemented some others. I hope that the book will thus be easier to read and more interesting, not only for metaphysicians and philosophers of science.

The book would not have been written without the support of a number of persons, whom I should here like to express my gratitude to. I wish to thank first of all Prof. Antoni Szczuciński for his trust, apt criticism, long-standing support and excellent work conditions; Prof. Michał Tempczyk and Prof. Jan Such for their astute reading, thoughtful discussions and valuable remarks; my colleagues at the Philosophy of Science Seminar at the Department of Philosophy at Adam Mickiewicz University, who often were the first listeners and reviewers of some of the ideas the book contains; Prof. Zbigniew Tworak for reading and offering remarks to the preliminary version of Ch. 1.2; Prof. Danuta Sobczyńska, Prof. Ewa Czerwińska-Schupp and Prof. Paweł Zeidler for their good will and friendly words of support. Naturally, the responsibility for all the daring theses presented here, for the arguments or their lack is mine alone.

I wish to warmly thank my nearest and dearest – my parents and sister, for their love, support and staunch patience.

# PART ONE

# Chapter 1
# *Megiste mousike*: The Hidden *logos* of Nature and Platonic Wholes

## 1.1 The relation of the whole and its parts in Plato's ontology, the problem of the hidden structure and a new reading of the *Parmenides*

At the source of modern mathematical physics, inaugurated since the beginning of the 17th century by Galileo, Kepler, Huygens and Newton, and ever since undergoing a continuous development, we encounter a cognitive impulse of the metaphysical and mathematical research undertaken in Plato's Academy and its environment in the 4th c. B.C. and in the still earlier Pythagorean studies. Historically speaking, the remark is not an exaggeration or a mere compliment to the archaic layers in the history of ideas, but it is a statement of a certain fact of historical importance when attempting to comprehend the genealogy of natural science in the tradition of Western culture. Though Plato himself, as can be inferred from the existing sources, rather did not conduct mathematical research, yet the range of his inspirations and his encouragement of such studies was so vast that we can call the 4th c. B.C. an age when Greek, 'Platonic' (i.e., 'analytic' and 'dianoetic') mathematics was in full bloom, with their peak achievement undoubtedly attained in the *Stoicheīa* by Euclid from Alexandria (4th/3rd c. B.C.) ([Lassere, 1964], [Maziarz, Greenwood, 1995, p. 75*ff.*], [Boyer, 1991, p. 86*ff.*]). The strength and range of that research inspiration in Greek science can hardly be overrated and the Academy itself seemed to play the role of a mathematical research centre of the Western world in the first half of the 4th c. B.C. And yet, in the end, despite that early determining influence, the Western speculative images of nature in physics and cosmology did not retain, as we know, the radicalism of Plato's central solutions and intuitions. What is more, in line with the Aristotelian tradition, they *upturned* them, which – as we shall later observe – has had marked consequences for theoretical physics right up to the present day when one reflects upon the elementary constitution of the physical reality.

Carl F. von Weizsäcker, recalling Heisenberg's opinion in one of his lectures, called Platonic philosophy 'a skeptical theology of the mathematical paradigm' and highlighted the fact that Plato – despite our contemporary impression of his tendency to a mythical imagery – had a certain consistent physics at his disposal based on a deep reflection on the foundations of mathematics [von Weizsäcker, 1987]. We would nevertheless like to know to what extent that ancient proto-physics anticipates contemporary physics, even the most recent one, and to what degree Plato has offered inspiration and the methodology to explore the problem of the hidden whole(s) in nature. It is also worth considering whether Plato's meditations on the internal logic of nature can help us to understand certain paradoxical phenomena which fundamental physics has been having to deal with. Heraclitus's remark concerning the way 'all things' become 'according to *logos*' (κατὰ τὸν λόγον)[1], gains particular importance when we treat it as the starting point of Plato's way of construing physics, also in the context of the latest research on identifying the mathematical structure of elementary physical processes (see Ch. 2.2 and 5). A skeptical interpreter could accuse such attempts of being a sort of wishful reading and ascribe the possible validity of Plato's constructive thinking merely to his abstract ('empty') generality, of no import to the modern natural sciences. Still, when we look more closely at the mathematical and physical subject matter studied at the Academy, with its unwaning involvement in solving the riddle of the whole–parts relation and its connection to the metaphysical problem of the relativity of reality, it becomes clear, as I shall claim, that it reappears in an astonishing manner in 20th century microphysics. The phenomenon can be convincingly explained by the way several constant categories and ways of interpreting nature or the universe reappear in Western thought (cf. especially Ch. 5).

A reflection on the issue of the 'physical system as a whole' and its relation to 'nature as a whole' requires us to ponder the riddle of the relation between 'the One' and 'the many', which particularly the Pythagoreans had been so interested in, as the certain paradox in the statement that '$x$ is one/unity and is made up of $y_1, y_2, \ldots, y_n$' is automatically transposed even to the sphere of fundamental physics, when we speak about the 'global state' (the unity) of a system composed of many subsystems. Verity Harte [2002a], in her study *Plato on Parts and Wholes*, proves that the problem of the relation between the part (μέϱος) and the whole (ὅλον) has constituted one of the central axes of Plato's late ontology: the obligation to seek its constructive solution spurred on the academic research on the real mathematical structures ordering the physical world. (The very vocabulary of the 'whole', 'structure', 'order' and 'harmony' – though originally Pythagorean – has pervaded the natural sciences mainly due to the intellectual mediation of the Academy). The significance and topicality of that issue, in the context of

1 Heraclitus, B 1 (Sextus Empiricus, *Adv. Math.* VII, 132); cf. [Kirk, 1975, p. 32*ff.*].

the philosophy of physics, arises from the fact that the philosophical choice of a certain ontology of composition (compositionism) and its associated strategies forces us to decide what that 'composition' of objects consists of and what exactly are those base 'objects'/entities of which the wholes construed and observed in nature (such as very simple quantum systems) should be made up of. The commonsense approach in mechanics and dynamics, preferred in physics in the early 20th century and still avidly defended by Einstein and many others (before physics came to be completely dominated by the counterintuitive Quantum Field Theory, henceforth QFT), favoured, in a singular manner, both the methodological and the ontological parcelling of physical phenomena into independent objects of a different scale, whose relations are purely external and usually have the character of an action of forces. Such an *externalistic compositionism*[2] has

---

2 Further on, externalistic compositionism/holism will be treated as an ontological standpoint which states that the parts (elements) of a whole are ontically *primary* and *well-defined* regardless of the irreducible whole (composite) they create, while the whole as such is secondary and dependent on the relations those elements enter into. Those composition relations do *not supervene* on the other properties of the parts and can be realized only when *all those parts/elements are actually present*. Ontological extensional (non-intensional) compositionism is much more radical and simple, defined as the view that real elements unanimously determine all the composites (arrangements) they create, i.e., the so-called 'fusions' or 'aggregates', whose properties are completely reducible, and hence they supervene on the elements they are composed of and their properties, whatever they might be. (The term 'extensional' is used here with reference to extensional mereological systems, containing the so-called Supplementation Principle). In such a situation one can either claim that there exist all manner of possible fusions ('pseudo-wholes') of randomly selected elements of the world (the so-called *universalism/conjunctivism*, or related 'explosivism', defending an unlimited composition), or that it is difficult to talk of any 'elements composing the whole' at all, as the 'whole' does not have a special, ontological character or the 'composition' in fact never occurs (the so-called *nihilism* or *compositional eliminativism*). The most well-known representatives of universalism are Donald Baxter [1988], David Lewis [1991] and Theodore Sider [2001], as well as Allan Gibbard and Stephen Yablo; nihilism, in turn, is analytically defended by e.g. Keith Hossack [2000], Cian Dorr [2005] or Jeffrey Grupp [2006]. The popular externalistic compositionism/holism with a limited composition (defended in a neo-Aristotelian spirit by e.g. Kathrin Koslicki [2008] or Crawford L. Elder [2011]), together with radical internalistic compositionism/holism (discussed further on) are the basic types of intensional compositionism (holism), in which the whole/composite/compound object *cannot* be exhaustively determined by the individual characteristics of its parts, and hence it is defined by an *essential, irreducible structure* (of relations or non-relational properties), which does *not*

turned out to be – that is undoubted – an extremely productive research strategy (regarding explanation and prediction) and can be assumed to be responsible, as reductionism in general, for an unprecedented theoretical and technological effectiveness of physics (the suppression of or appriopriate control over secondary factors and relations enables us to effectively steer processes and to study them comprehensively). The fascinating novelty of quantum mechanics (henceforth QM), nevertheless has depended on the fact that it has challenged, in a nontrivial and somewhat surprising manner, the credibility of such a strategy of explanation, hence the accusations of it being counterintuitive or 'deviant', as some analytical metaphysicians might say. In such a case, it seems reasonable to analyze somewhat more thoroughly the metaphysical character of the compositionism we usually believe in, when we speak of the 'whole' of a certain system in physics, as it might turn out that there exist alternative ontological categories, which might effectively remodel our conception of mechanics and fundamental physics in particular. In the background of that issue we find a deeper question about the structure of the physical world and its relations with mathematical structures: if there is any sense in speaking about nature's 'hidden wholeness', its *purely global* (irreducible) structure, then what language could at all model the structure of that whole and what, all in all, is the status of that category in physics? I shall argue that the very notion of 'reality', 'time' and 'change' is at stake here (cf. Ch. 4 and 5).

On account of the problem of the whole–part relation, and also due to the development of Plato's metaphysics as such, the central place amongst the Platonic dialogues is held by the enigmatic *Parmenides*, where the fundamental issue of the One (ἕν) and the 'infinitely many' (ἄπειϱον πλῆθος) defines the principal axis of the narration. Verity Harte [2002a], in her analysis mentioned above, managed to decipher certain semantic intricacies in that narrative in a significant way, locating the Eleatic-Platonic ontological problem of the 'whole' (ὅλον) in a wider context of the history of metaphysics and recognizing in the *Parmenides*'s argumentation a 'mereological undercurrent', generating general paradoxes concerning the very concept of 'composition'[3]. The argumentation can, according to Harte, be ordered into two movements: a negative and a positive

---

*supervene* completely on the properties of the parts, and first-order micro-entities in particular, hence the impossibility of its complete microreduction. Thus, every intensional compositionist must subscribe to the minimal thesis that wholes are not sums.

3 The 'mereological undercurrent' in the second part of the *Parmenides* was also discussed by Marian Przełęcki [2000] in his interpretation of the dialogue.

one, of which the former contains a series of theses–paradoxes (puzzles) arising from identifying the whole with the sum of its parts, while the second enables a conception of the whole as a *real* (not mereological) unity [ibid., p. 49*ff*.]. The whole dialogue begins with a short exchange between Socrates and Zeno, with the former bringing up one of the principal claims of the latter, concerning the impossibility of plurality:

> What do you mean, Zeno? Do you think that if the being things are many (εἰ πολλά ἐστι τὰ ὄντα), they need at the same time to be both similar (ὅμοιά) and dissimilar (ἀνόμοια), and that is impossible, as neither can the dissimilar be similar and the similar dissimilar (οὔτε γὰϱ τὰ ἀνόμοια ὅμοιά οὔτε τὰ ὅμοια ἀνόμοια οἷόν τε εἶναι)? Is that what you claim? [*Parm.* 127e1–4]

The climax of the conversation and at the same time an answer to Zeno's paradox is to be found in a longer speech by Socrates, in which he presents his theory on the idea of 'similarity' and 'dissimilarity' as an attempt at challenging – by the *modus tollendo tollens* – Zeno's pattern of logic from the fragment above:

> Things which participate in similarity become similar in such a way and to such an extent as that to which they participate (τὰ μὲν τῆς ὁμοιότητος μεταλαμβάνοντα ὅμοια γίγνεσθαι ταύτῃ τε καὶ κατὰ τοσοῦτον ὅσον ἂν μεταλαμβάνῃ); whereas when they get a share in dissimilarity – become dissimilar, and both similar and dissimilar when they take part in both. Therefore, if all things participate in those opposing (ἐναντίων) beings and – having them in themselves – are hence both similar and dissimilar at the same time, then should we find that astonishing (τί θαυμαστόν)? [*Parm.* 129a4–b1]

The most interesting fragment from the perspective of the issues of the one and the many is contained in the thought, usually unnoticed by commentators treating the whole passage as a standard presentation or repetition of Socrates' teaching on ideas/forms (*eidē*) or avoiding to literally translate it:

> If someone could demonstrate that the very same kinds and forms (αὐτὰ τὰ γένη τε καὶ εἴδη) possess those opposing properties, I would be astonished. *But if someone argues that I am one and many* (εἰ δ' ἐμὲ ἕν τις ἀποδείξει ὄντα καὶ πολλά), then what is surprising in that? If he would like to demonstrate that I am many (πολλά), then he will say that some things are on my right side, some on the left, some in front, other in the back, and similarly on the top and on the bottom – as I think that I have a share in plurality (πλήθους γὰϱ οἶμαι μετέχω). But if he were to prove that I am one (ἕν), then he would say that I am one of seven people present here, because I have a share of the One (τοῦ ἑνός). In both the instances, his argument is correct. [*Parm.* 129c2–d2]

Socrates's answer is also, as Harte rightly notices [2002a, p. 56], a paradox 'in Zeno's style' and has been construed to directly address Zeno's argument: Socrates defends the thesis that the statement of the similarity and dissimilarity of a given entity is true in many situations, which undermines the second prerequisite of the *modus tollendo tollens* used in the argument. At the root of both the first and second paradox lies a significant presupposition about the function of the parts within the whole, which can be termed the *Pluralizing Parts Principle* (PPP) [ibid., p. 53]: *when a certain entity is made up of parts, then it in itself is many ('plurality').* Socrates, having announced the problems to be discussed in the whole dialogue, introduces the concept of 'forms' (*eidē*), which, as can be seen, are to be completely free of similar difficulties: opposing definitions can be attributed only to phenomenal beings, 'stones, trees and other things' (*Parm.* 129d3–4). The problem does not seem to end, though, as Socrates goes on to use and uphold the PPP, whose benign or solely perspectival-epistemological interpretation could with great probability be excluded [ibid., p. 60]. According to the benign interpretation of that particular fragment of the dialogue (*Parm.* 129c-d), it is assumed that Socrates speaks in it of how a thing or object is presented when we view it from a given perspective: the person 'is one' when we consider his/her individuality (*possesses* a unique identity), whereas it 'is many', when we differentiate his/her distinct parts (for instant he/she *possesses* a multitude of body parts). A very strong ontological (not epistemological) equation of 'to be' = 'to possess' takes place here, though, as in order to answer Zeno in a proper manner, Socrates has to uphold, and does keep in line with a strict understanding of the PPP, according to which: 'I am plurality' means '*I am* many' (= 'the whole of many') in a strictly ontological sense, and not only: '*I have* many parts' [ibid.]. It is only by retaining the PPP that Socrates can undermine Zeno's radical, ontological thesis about the antinomy introduced by plurality, and strongly state: 'I am *one* (ἕν) and *many* (πολλά)', thus proposing a new antinomy – the paradox, to be solved solely with the help of the theory of ideas[4]. The assumption of that strong PPP implies that Socrates' paradox is based on a universal form of extensional compositionism, in which the 'whole' is precisely identical with its parts ('the whole' = 'a sum') – otherwise Socrates could not have said that he '*is* many'. The solution proposed in the theory of *eidē* is an assumption of the existence of 'paradoxical beings' ('participating', 'having a share'), the (indivisible) One and plurality (ἄπειϱον πλῆθος as a pure ontological basis of a multiplication of beings) [ibid., p. 62]. Paradoxical beings are not astonishing (*Parm.* 129c4: τί θαυμαστόν), as they are a part of both the categorically designed *eidē*, but unfortunately – and that is the first serious disappointment – that

4 Harte is right to notice that a rejection of the Eleatic PPP found in the first prerequisite would make Socrates' answer to be inadequate, i.e., deprived of an analogy to Zeno's former reasoning. Socrates's thesis has to be a simple questioning of the second prerequisite in that reasoning, that is it has to have the same form [Harte, 2002a, p. 60].

does not clarify what the actual 'identity' and 'unity' of those wholes is, if they are in the empirical sense *only* composites of parts.

The second of the series of paradoxes concerning the whole and taking part in the dialogue is the 'Parmenidean paradox', formulated in his dialogue with Socrates. It is certainly a step forward with regards to the former, as Plato here moves the structure generating a paradox onto a higher level, into the sphere of the *eidē* themselves and their relation to empirical beings, contrary to Socrates's earlier conviction (the legendary Parmenides himself performs this shift in the plot of the dialogue):

- Does it not seem to you that the whole form (ὅλον τὸ εἶδος), is one and the same, and still, being just one, is to be found in each of the many beings?
- Why should it not be so, Parmenides? [*Parm.* 131a8–10]

Socrates invokes an analogy of the day, which is a fact in many places, but 'nonetheless remains to be one and not divided in itself (αὐτὴ αὑτῆς χωρίς)', but Parmenides takes advantage of that spatial picturing to prove its weak point in the famous example of a 'sail spread over a multitude of people' (*Parm.* 131b). Regardless of whether we assume that an empirical being has a share in all of the idea, or just its part, that leads to an unavoidable paradox together with an additional – recognized by Socrates – assumption that an idea/form is present in the participant thing: in the former case, the same *eidos* is found in many different beings and remains 'separated (χωρίς) from itself', and in the latter case is divisible, and thus as an '*ideal* multitude' it cannot at the same time be Socrates' *ideal* unity (*Parm.* 131c). The conclusion is possible solely on account of Socrates' steadfast assumption of the ontological PPP, which is binding on all the levels of being, and hence including the concept that 'parts pluralize the whole', which makes it impossible for such an *eidos* to still be an authentic *unity* [Harte, 2002a, p. 66]. The 'whole' as a category of composition referred to empirical beings has not caused difficulties, as the unique relations between the beings and *eidē* have not until now been disrupted, but at present that relation has also been threatened by the steadfast upholding of the extensional compositionism given in the PPP formula. There is no need to analyze, at this point, what the nature of that 'participation' or indeed 'presence'/'immanence' of 'forms' in things is, as both of the above problems are independent of such a resolution – their persistence in turn is a direct result of the mereological foundation of the whole reasoning. We are assuming here that the created 'wholes' are mereologically 'innocent' [Lewis, 1991, p. 87], in the sense that they do not contain anything apart from the parts they are composed of, and therefore an *ontologically* understood whole can simply be put together from those parts (whenever there is a given set of elements, there is also a fusion, i.e., the whole – or pseudo-whole – of those elements in the form of their sum, which is *in an equal degree* a multitude and a unity). David Lewis put it most briefly in his famous *Composition-as-Identity Thesis* [ibid., p. 81*ff.*], according to which every mereological fusion of elements is ontologically innocent, i.e., 'the "are" of composition is, so to speak, the plural form of the "is" of identity' (see also [Harte, 2002b, p. 5*ff.*].

The first two paradoxes of the series are crowned in the first two hypotheses (and deductions at the same time) of the *Parmenides*[5], in which the above model of compositionism has been applied to a supreme, abstract *principle* of unity ('the One'), thanks to which, according to Harte [2002a, p. 73*ff.*], its whole aporeticality is exhibited all the more clearly. Parmenides starts off with encouraging Socrates to return to the inquiry, despite all the earlier conclusions seeming so discouraging (*Parm.* 135c-e): the assumption that every empirical object is both the one and the many, contrary to *eidos*, and also that *eidos*, by determining the structure of being, is somehow present in it, has led to the difficulties which seem to suggest that the theory of ideas itself might be questionable. Parmenides's conversation with the youngest participant, Aristotle, begins with the first hypothesis, which directly (as far as terminology goes) refers to the relation between the parts and the whole:

> εἰ ἕν ἐστιν, ἄλλο τι οὐκ ἂν εἴν πολλὰ τὸ ἕν;
> – Πῶς γὰρ ἄν;
> – Οὔτε ἄρα μέρος αὐτοῦ οὔτε ὅλον αὐτὸ δεῖ εἶναι.
> – Τί δή.[6] [*Parm.* 137c4–6]

At the very beginning of that deduction (carried on until *Parm.* 142a) we notice the presence of that same PPP as in the preceding paradoxes, only in that phase of the analysis it is hypothetically used to something that can be called 'unity' in the proper understanding of the term (εἰ ἕν ἐστιν – οὐκ … πολλὰ), hence those hypotheses are often commented to concern the very *principle* of the One. If we agree to acknowledge – says Parmenides – that 'something' (unity) as the One (*Parm.* 137d3: 'εἰ ἓν ἔσται τὸ ἕν') *is* the whole, therefore by definition

---

5 An additional analysis of the *Parmenides* hypotheses in K.M. Sayre's classification (from his work *Plato's Late Ontology: The Riddle Resolved*, Princeton 1983) is to be found, for instance, in S. Blandzi's study [2002, p. 152*ff.*].

6 'If something is One, then the One would not be many? – Of course not. – Then it should not have any part nor be any whole? – Why not?' With regards to the translation of the dubious phrase 'εἰ ἕν ἐστιν' see e.g.: [Bigaj, 2003], [Blandzi, 2002, p. 37*ff.*], Harte [2002a, p. 74*ff.*].

it possesses parts (*Parm.* 137c6: 'τὸ μέρος που ὅλου μέρος ἐστίν'[7]), then it is also many (multitude): 'Ἀμφοτέρως ἂν ἄρα οὕτως τὸ ἓν πολλὰ εἴη ἀλλ' οὐχ ἕν' (*Parm.*137c9–137d1)[8]. The cutting-edge of Eleatic PPP is thus turned against any possibility of predicating a certain multitude or complexity of the One, which results in ruling out the option of using the term 'whole' (ὅλον) to denote the One as a metaphysical principle and foundation of unity in the empirical world: 'Thus, if one is to be the One, then it will neither be a whole (οὔτ' ὅλον ἔσται), nor will it have parts (οὔτε μέρη ἕξει)' (*Parm.* 137d2–3). It is worth noting, at this point, that Plato does not assume here, just as he had not done earlier, a 'benign' understanding of the meaning of the whole (and of PPP), where it merely denotes 'possessing parts', and not '*being* parts'. We can clearly see that that exclusion of a benign interpretation, which we usually tend to apply when speaking of 'wholes' as true 'unities', is also a concealed prerequisite of the whole deduction in the run of the *Parmenides*'s first hypothesis. According to it – and at the same time in tune with the third paradox in the series under discussion – the consequences of such a limitation are far-reaching, as the One may then become a merely *mereological atom*, which answers to the strong version of the PPP on the basis of a lack of any parts [Harte, 2002a, p. 78]. The effect of such a deduction is well-known. The One 'can neither have a beginning, nor an end, nor a centre' (*Parm.* 137d4–5: οὔτ' ἂν ἀρχὴν οὔτε τελευτὴν οὔτε μέσον ἔχοι), 'it is unlimited' (*Parm.* 137d7: ἄπειρον), 'bereft of shape' (*Parm.* 137d8: ἄνευ σχήματος), 'it is to be found nowhere' (*Parm.* 138b5: οὐκ ἄρα ἐστίν που), 'it is neither at rest, nor moves' (*Parm.* 139b3: οὔτε ἕστηκεν οὔτε κινεῖται), and, lastly, is neither identical, nor different from itself nor from anything else (*Parm.* 139b). The mereological atom turns out to be an unknowable unity, in which no effective change can take place, no movement, passing away or pluralization.

With Parmenides's second hypothesis, we finally reach the climax of the whole series of paradoxes and we are at the same time at the focal point of the whole dialogue. Parmenides proposes a return to the assumption of the first deduction, but at the same time, together with the hypothesis that 'the One exists' (*Parm.* 142b: ἓν εἰ ἔστιν), the PPP is finally openly applied in the course of deduction as a principle organizing the way of thinking about the whole and parts: 'And what is divided into parts must of neccessity be as numerous as the parts it possesses' (*Parm.* 144d4–5: καὶ μὴν τό γε μεριστὸν πολλὴ ἀνάγκη εἶναι τοσαῦτα

7 'For a part is after all a part of some whole'.
8 'Hence, in each case, One would be many, not One'.

ὅσαπερ μέρη). Harte [2002a, p. 79] correctly remarks that the Greek μεριστὸν may denote both '*divided* into parts' and '*divisible* into parts', but, as we have already concluded, from the very beginning of the dialogue, Plato's use of the strong PPP introduces an identity of 'being divisible' and 'being a multitude'. In all of that deduction, the role of multiplying is played by being/existence (fully independent of the One), which causes, in accordance with the PPP, a radical pluralization of the principle, which finally results in a conclusion that proves opposite to the one arising from the former deduction: 'The One as such, divided into parts (κεκερματισμένον) by being, is thus many (πολλά), an *infinite plurality* (ἄπειρα τὸ πλῆθός)' (*Parm.* 144e3–5). As we know, that conclusion has allowed Parmenides to deduce a plurality of ideal (*eidetic*) numbers, which are defined by combinations of the simplest, multiplied elements: 'Hence, if there is one (ἕν), there must by necessity be also a number (ἀριθμὸν)' (*Parm.* 144a4). It finally appears that, contrary to the previous paradox, 'that existing One is both one (ἕν) and many (πολλά), as well as an unlimited plurality (ἄπειρον πλήθει)' (*Parm.* 145a2–3). A consistent use of PPP ends the discussion on the problem of the part and the whole with the latter paradox, which is at the same time Parmenides's complete answer to Socrates's theory from the first paradox: the problem which Socrates wished to solve by transcendent 'forms', is transposed onto them and on 'the One as such' (τὸ ἓν αὐτὸ) [Harte, 2002a, p. 82*ff.*]. Parmenides's deductions no longer concern things that are discernible by the senses, but purely abstract unities, though it turns out to be possible to account for them in quasi-sensual categories (movement, finiteness etc.), as the character and source of those categories is supra-sensual, *eidetic* (e.g., an observed physical movement has to have some founding metaphysical ground apart from the fact of its sheer physical reality). We should not feel discouraged at this point by Plato's use of such expressions as κεκερματισμένον ('chopped', 'sliced', 'parted') with reference to the abstract concept of the One (*Parm.* 144e4), as what comes to the fore here is the metaphysical unity of different levels of reality, and the Socratic aporetic dualism of the 'empirical world' and 'world of ideas' recedes into the background. With the help of Parmenides's *eidē*, Plato attempts to prove that the problem of the parts and the whole (referred not only to specific empirical beings or the whole universe) concerns the foundations of human thinking about the relations between beings, leading, with clearly defined assumptions, to identifiable paradoxes.

The reasoning reaches a climax at this point with the suggestion that an adequate theory of the whole should violate the PPP, as the whole cannot be identical to its parts – this has been pointed out by G. Owen [1986, p. 92*ff.*] and, critically, by V. Harte [2002a, p. 89*ff.*]. At *Parm.* 144e–145a, Plato unexpectedly suggests a different understanding of the whole, one arising from the thesis that 'the One as a whole would be limited' (πεπερασμένον εἴη κατὰ τὸ ὅλον τὸ ἕν), 'the parts (τὰ μόρια) are encompassed by the whole (ὑπὸ τοῦ ὅλου)', whereas what is encompassing (τό περιέχον) is a *limit* (πέρας). The conclusion is then incorporated into the above quoted fragment of the *Parm.* 145a2–3 – which has proved to be so tricky to generations of the dialogue's interpreters – in which totally opposite definitions of the One are proposed. The above suggestion implies that the whole would be something decidedly exceeding its parts, irreducible to them, but on the other hand, a solution suggested so turns out to be unacceptable as a good alternative to the PPP. That is so as in that case the whole as a limit and container of the parts has no connection with them, is separated from them [Owen, 1986, p. 92], and the whole idea of the solution can be brought down to an absurdity. In the end, in a decisive part of the *Parmenides* (145e3–5), Plato goes back to the conclusion ending that troublesome passage (*Parm.* 145a2–3) and the PPP is retained in the whole paradox: an identity of the one, the whole, parts and the limit, all made possible solely due to the PPP [Harte, 2002a, p. 92*ff.*]! One should note that those conflicting conclusions are bred in the fact that the *Parmenides* at first presents two separate treatments of the whole (the whole as *a limit* and the whole as *all the parts*), until finally they are dissolved in one judgment on the One as a limited whole *and* of all the parts together (*Parm.* 145e3–5). It is exactly at that point that a climax of the 'negative movement' takes place in the dialogue: the PPP as a prerequisite is seen to end in models of the whole which produce aporias. The 'mereological undercurrent' of thinking about the whole presents all of its problematic nature, as the whole is either a mereological atom, about which nothing can be said ('the One as such'), or it is a mereological sum (fusion) of the elements, in which unity is spurious and ontologically 'barren', ending in a metaphysical extensional compositionism ('the One as many'). Putting it more precisely, from the radical Platonic perspective the very mention of 'compositionism' and 'wholes' in the latter case seems utterly problematic, as the parts are only added and no unity in a non-trivial meaning of the term arises from that (where 'unity' means just 'parts'). We arrive at the same paradox Socrates described at the beginning of the dialogue – only in that case it concerns the One itself. Here the game with wholes is over.

A 'positive movement' in the reflection on the whole in the *Parmenides* is contained in a short passage of the text in the course of investigating the second hypothesis, *Parm.* 146b, and it is from that very fragment onwards that it becomes clear that Plato begins to sketch or suggest a deeper, ontological understanding of the real unity of the whole (composite). As Harte has presented with great precision [ibid., p. 118*ff.*], Plato sketches there a model of four different relations which can be ordered into pairs:

> Everything with relation to everything else is either the same (ἤ ταὐτόν) or different (ἤ ἕτερον); or, if it is neither the same nor different, then it would either be a part (μέρος) of what it is in relation to, or relate to it as a whole (ὅλον) to its part. [*Parm.* 146b2–5]

The above implies that the relations are mutually exclusive – not only within one pair, but also *as* pairs. According to such a differentiation, made by Parmenides, the statement '*a* is different from *b*' excludes the possibility of stating that '*a* is part of *b*' or that '*a* is a whole containing *b*'. Therefore, a differentiation takes place here between the relation of *identity*–difference and that of the *whole*–part. What at the beginning seems like a conceptual pedantry turns out to be useful as a defense against erroneous categorization, as it is impossible to put a sign of equality between the theses that: (1) '*a* is different from *b*', and that: (2) '*a* is not the same as *b*'. For statement (2), '*non-b*', may mean that: i) '*a* is different from *b*'; or ii) '*a* is a part of *b*'; or iii) '*a* is a whole containing *b*' [Harte, 2002a, p. 119]. Hence we cannot say that a part of the whole is identical or different from the whole – such a rule remains to be only a limitation placed on the predication procedure, but the very recognition of that fact is important. *Parmenides*'s 'negative movement' was based on the PPP, which equates the whole as a sum of its parts with identity (unity), nevertheless at present it turns out that those relations have to be categorially separated, as putting/arranging the parts together or summing is not equivalent to attaining any unity. In the course of the dialogue, that idea appears at the climax, but without its importance being stressed, hence we can infer that it has the status of a suggestion or announcement on Plato's part, who as if in passing introduces a deeper dimension of the ontology of the whole. We are dealing in this case with a breakdown of the mereological way of thinking about the one and identity, which is found to be inadequate to capture wholes which are *authentic unities*. We can ask why that takes place. As Harte rightly points out [ibid., p. 121], one of the main reasons of that weakness of mereology is its axiom of extensionality, which states that wholes composed of the same parts are *identical*, and also the mereological definition of the 'improper part' of the whole, which is identical to that whole. Instead of that simplification, which leads to paradoxes and an impoverished understanding of the unity, Plato suggests that the mystery of the latter cannot be explained by referring to an ordinary, commonsense procedure of dividing into parts and of a repeated, extensional construction. It is such a composition procedure in the form of applying the PPP that lies at the root of the *Parmenides*'s series of paradoxical theses. Thus, the game with wholes is not over.

The breaking away from an extensional compositionism becomes most evident in the third hypothesis of the dialogue, when the possibility of a part being part of many, instead of a certain *whole as one*, is rejected (*Parm.* 157c). The reader, accustomed to an earlier assumption of the PPP, present throughout Parmenides's reasoning, encounters an unexpected new formulation, breaching the mereological way of thinking about the wholes:

- If something were to be part of many (πολλῶν), and it would itself be one among the latter, then it would be part of itself, which is impossible[9], and part of each of the many, as it is to be part of them all. And if it were not to be part of one, it would be part of the others apart from the one, thus not being part of each of the many, and not being part of each one, it would not be part of any of the many. And it is not possible for something that is not part of any of the many, to be part of or anything else for those that it is not any part of.
- So it appears.
- Therefore a part is neither part of the many nor part of them all (οὐκ … τῶν πολλῶν οὐδὲ πάντων), but of some single form (idea), some 'one', which is called 'a whole' (ἀλλὰ μιᾶς τινὸς ἰδέας καί ἑνός τινος ὃ καλοῦμεν ὅλον), a perfect 'one', generated from all its parts (ἐξ ἁπάντων ἓν τέλειον γεγονός). That is exactly what a part is part of.
- Surely. [*Parm.* 157c8–e2]

The above deduction follows a key statement: 'Ἀλλὰ μὴν τό γε ὅλον ἓν ἐκ πολλῶν ἀνάγκη εἶναι' (*Parm.* 157c5–6: 'But a whole must be a one composed of many'), where it becomes clear that a whole cannot be 'innocent', taking the form of a simple fusion, but has to constitute *a real unity* (ἑνός τινος). Parmenides therefore stresses clearly that every whole composed of parts participates in unity, as the blithe statement that 'a part is a part of a sum of parts' has no ontological sense (it is at best a mental shortcut). Hence, a distinction is made in that context between a 'false whole' (in the mereological, i.e. collective sense), which is captured by the PPP, and an 'authentic whole', which is a 'perfect one/unity' (ἓν τέλειον), which is impossible to trace with the help of a simple analysis of the parts composing it and which constitutes a certain 'form'. The concept of a 'whole' (ὅλον) gets a new meaning in the above fragment (ὅλον = ἓν τέλειον), earlier absent, and which an inattentive reader could have easily overlooked.

9 We can at this point notice an open rejection of the mereological concept of an 'improper part', and further on also a distributive understanding of the 'many' and not a collective (mereological) one, which is the basis of the whole reasoning. Parmenides requires the set of 'many' and 'all' to be a *pure plurality*, without any real unity, hence a collective understanding is excluded.

There is a positive movement visible here within the dialogue and at the same time there is a progress in the investigation (from the initial Socrates's paradox to the third hypothesis), which has often been unnoticed or ignored. If we are at all to be able to differentiate one part of a whole from the other parts, it is necessary not only to present that whole as a certain unity, but also each of its parts has to participate in 'one' (ἕν). The whole, in turn, if it is to be real, should never – in line with Parmenides's argumentation and contrary to commonsense assumptions – be split into parts (as 'atoms'), through which it can *as a whole* be conceptually or in reality reconstructed (cf. the defence of such essentialism e.g. in [Johnston, 1992] or [Baker, 1997]). Plato's Parmenides, although it originally proclaims the PPP, ceases to be an Eleate here [Harte, 2002a, p. 129], and in the very centre of the reflections there suddenly appears the issue of *a hidden whole* of empirical phenomena as their structure. Plato himself unobtrusively discards the Eleatic assumption of an 'innocent composition', i.e., of a fusion, and it is to Harte's credit that she highlights that fact in the context of the positive movement proposed by Plato.

What we find most interesting in Plato's treatment of the whole is the way he suggests to define it in the above fragment as ἓν τέλειον ('a perfect unity/one') and μία τις ἰδέα ('a certain singular form')[10], where τέλειον can also be understood as 'full', 'utter', 'complete' or 'fulfilled', without suggesting an axiological aspect (although it is always important to Plato). The whole thus understood, otherwise than the whole multiplied by parts on the basis of the PPP, possesses an immanent, internal order, without which it is meaningless to speak of the whole – it is that very order/structure that excludes an 'innocent', i.e. mereological concept of the whole as being inadequate. The fact that that pseudo-compositionist view is rejected by Plato simply implies that the truly radical, *extensional* version (reductionist or anti-holistic) is revoked in the course of the dialogue. What about the moderate form of holism, though, the one we have called 'externalistic'? As we have already noticed, according to the latter it is reasonable to talk of a certain structure of the observable whole (e.g., its physical state), but its elements (e.g., some physical states/properties of its subsystems) enjoy an ontological independence from that structure, which is in a way external to them and characterized by a separate category of concepts (e.g., spacetime or field potentials in physics). What is important for externalistic compositionism is that the relations between the parts cannot be reduced to their intrinsic properties,

10 Cf. also μία τις ἰδέα ἀμέριστος ('some singular, indivisible form') from *Theaetetus* 205c2.

and thus the dynamic or static structure of the whole is indeed of key importance to understand its character. It is in that sense that we usually say – taking a paradigmatic example from pre-quantum mechanics – that the whole, a physical system, *'has' a structure* whose actuality is strictly dependent on the pre-existing parts that compose it, and we can assume that structure to be the system of bulk relations between them (which can change 'in time', whatever that might mean).

The above simple type of holistic compositionism can also be transformed into an Aristotelian holism, provided that we clearly separate the primary category of the monadic 'substance'/'substrate' from the category of 'relation' (in a broad meaning here). Aristotle himself made a clear distinction between 'aggregate'/'totality'/'all' (πᾶν = *omne*), which is always a loose, extensional set of certain free elements, from the 'totality'/'whole' (ὅλον = *totum*), in which the relations between the differentiated parts are critically important and their change means a change in the structure–form of the whole entity, and hence also possibly of its essence (*Metaph.* V.26, 1024a). He was also an evident holist in line with the simplest interpretation of 'holism' – taken from Plato – as a view that the whole can never be fully defined by its parts due to the primary opposition of the one, τὸ ἕν, and many, τὰ πολλά (*Metaph.* X.2, 1054a20): the famous definition from *Metaph.* VIII.6, 1045a8–10 lays down that the whole is something 'above', something 'surpassing its parts', παρὰ τὰ μόρια = *praeter partes*, nevertheless – as Aristotle stresses separately in his *Physics*, analyzing the problem of time – it is 'a part which is a measure of the whole, and the whole has to be composed of its parts' (*Phys.* IV.10, 218a6–7). Aristotle, as every holist, steadfastedly defends the option that every 'whole is some kind of unity' (*Metaph.* V.26, 1023b36), but that unity does not solely depend just on a certain system of relations (it cannot be reduced to that), as it forms a *substantial whole*, and not only a relational one. In the central part of his *Metaphysics*, Aristotle reflects on the key (and as he himself admits, extremely difficult) issue of the primacy of the parts with respect to the whole, rounding off with some conclusions: in the case of a given being (both an empirical and a conceptual one) its wholeness is posterior (ὕστερος) to its parts, empirical or conceptual, which – as its 'matter' – can exist as substances even when the whole itself has ceased to exist (*Metaph.* VII.17, 1041b; see e.g. [Bäck, 2014, p. 224*ff.*]). The case of the abstract ('immaterial') whole is analogical, as its definition requires earlier (primary) elements–parts, hence the whole 'is posterior to the parts the notion contains…' (*Metaph.* VII.10, 1036a22–23). Further still, the composite empirical (material) whole is always posterior to its form and thus to *a part of this form* (see a detailed analysis in [Wedin, 2000, p. 300*ff.*]). All in all, the thesis that 'parts are *a measure* of the whole' and 'the whole must *be composed of its parts*' every time points to an ontology of externalistic compositionism, which we usually treat as natural or at least reasonable on account of the fact that 'plurality and divisibility are more visible to the senses than indivisibility' (*Metaph.* X.3, 1054a27–28). Finally, Aristotle, as every externalist, tends to accept the option rejected by Plato at *Parm.* 145a that the whole is 'what *encompasses* parts in such a way that they form a unity (ἕν)' (*Metaph.* V. 26, 1023b27–28), thus as a certain encompassing limit renders the entity (as a unity) perfection and finiteness, which makes it immediately possible for us to say in an Aristotelian way – without any metaphysical abuse – that the whole *has form/structure*. For any externalist a whole–substance always requires its prior components–elements (as its *potentia*), without which it would not be able to realize/actualize itself – in fact the 'element' itself is defined as ὕλη, *matter of the whole*, 'into which a thing is divided (...), e.g., "a" and "b" are the elements of the syllable' (*Metaph.* VII.17, 1041b31–33).

All of the Platonic explication in the *Parmenides*, with its climax in the second and third hypothesis, leads much further, though, not only beyond the extensional compositionism given by the PPP and the 'mereological undercurrent' of Eleatic reasoning, but also beyond that externalistic compositionism, moderately realistic, which Aristotle has construed as a foundation of his ontology. It is of key importance in discerning that fact to stress the expression Plato uses for every 'authentic whole', namely μία τις ἰδέα, despite the traditional linking of Platonism with the Socratic doctrine of ideas totally separated from the empirical physical world. It thus becomes clear why at the beginning of the dialogue that is the decisive – and overriding – accusation directed by Parmenides against Socrates. Not only does Plato not suggest a propriety of the extremely convenient and popular expression about the whole 'having a structure/form', but he goes further in saying that the whole 'is a structure', μία τις ἰδέα ἀμέριστος [Harte, 2002a, p. 158*ff.*]. That implies that he rejects externalistic compositionism as being too weak and incomplete or merely partial (and breeding many further problems), and suggests a compositionism which would be its opposite, i.e., intensional and internalistic at the same time. The former one is connected to the theory that wholes composed of the same parts (elements) *are not necessarily* identical to one another, whereas the latter one additionally to the thesis that parts (any elements of any whole) are definable and differentiable *solely* within a whole which – as a structure – defines the relations between them and hence also their identity as distinct elements. It is the ontologically conceived structure which is the whole in internalistic compositionism, rendering the particular elements their meaning which are *nonexistent* outside its scope (and not even definable) [ibid., p. 166*ff.*]. That is an extremely strong supposition, principally different from Aristotelian externalistic compositionism, but one that enables – and the only one at that – a disentanglement from the fallacy of the compositionist 'special pleading' [ibid., p. 163] which demands an actual existence of the primary elements[11]. The primacy of structure as a measure (μέτϱον) in the Platonic understanding of the whole is at the same time a primacy of the category of 'unity over plurality', as it is that very 'unity' that we have in mind when we talk of an 'authentic whole', and not a randomely produced collection/aggregate (fusion) of many elements ontologically independent of it. It is true that the problem of the whole as 'unity over plurality' – and not an isolated problem of the ideas themselves – is a central issue of Plato's mature ontology. Without comprehending that fact it is impossible to explain why Plato has concentrated all of his efforts on the issues of numbers and measures, as is best seen in the *Philebus*. With reference to the Aristotelian definition of the whole, we can present two different, ontological points of view on conceiving that whole and at the same time two completely different *intensional compositionisms* (i.e. nontrivial ones):

**Externalistic (Aristotelian) Compositionism**
The whole = *elements*–matter (:*parts*) and their unity (:external form–structure)

**Internalistic (Platonic) Compositionism**
The whole = *unity* (:*form–structure*) and its parts (:internal elements)

At first glance, both the definitions might seem a bit pedantic, but there nevertheless *is* a definite difference between them, which leads to two fundamentally different ontologies. In both those cases we might be dealing with the whole as 'the one–unity over plurality–many', but only in Plato's case that unity is treated as truly ontically primary, and the internalistic scheme is simpler as well as more

11 A 'special pleading', as Harte has rightly pointed out, consists here of speaking of the whole in such a way as if it were something secondary and weaker with regards to the particular components/elements creating it, to which one always has to first refer to be able to speak about it at all. The essence of the special pleading here is the assumption that before we state that 'a whole exists', we should have some actualized parts that compose it, thus ontic 'base elements' are always treated in a privileged way. It is true that the concepts of the 'whole' and 'parts' are – as Plato frequently states in the *Parmenides* – correlative, they mutually condition themseves, but that does not mean that the parts as elements–individuals (material or abstract) are to be ontically and chronologically primary with regards to the whole (externalistic compositionism). Plato puts that conviction in doubt, speaking about the whole as 'some singular form/idea', 'a perfect form' which points to the primacy of an irreducible, *ontological unity*, and not a material plurality of the prior elements, which must be related/connected to form anything at all.

radical. That leads to several extremely interesting consequences in the realm of construing an adequate ontology of the world, which I shall discuss in Part II of the book. It is to the credit of Harte's reconstruction that the radicalism of the Platonic paradigm has precisely been defined in contrast to the classical (post-Aristotelian) types of compositionism, which Western metaphysics has become accustomed to in the situation when a certain version of 'particularistic atomism' (even in some minimalistic or commonsense forms) is 'everybody's metaphysics'. As we shall see, that intuitive metaphysics fails nowhere as spectacularly as in fundamental physics, thus one might comment that both the metaphysical and physical game with wholes is just beginning.

We could pose the following question: what sort of wholes would certain physical entities be, ones that were not loose collections of elements, such as a particular atom, historical Socrates or the Universe? The Platonic answer would then have to be: they are wholes–'hidden unities', whose concealed structure (μέτϱον = 'measure') has to be uncovered in the course of mathematical cognition or study of the latter. That global structure of all the empirical, observable 'states of affairs', i.e., the Platonic 'authentic whole', is never conceivable directly through the empirical cognition, it has to be 'uncovered from its hiding place', but remains to be, contrary to what we are accustomed to, the *only* guarantee of the consistency and definiteness of phenomena. We never sensually perceive any whole in the *proper sense* of the word, merely certain pluralities characterized by form and material order. A construction of that order in human perception is a continuous manifestation of 'the one over many', 'unity over plurality', or authentic whole, taking place at every level of that cognition. Plato suggests that the speculative postulate of a certain initial plurality of the primordial elements, which are then relationally aggregated into wholes (secondary unities), does not have any sense, as that whole–unity is primordial, both logically and ontologically, and the parts themselves gain an identity only in the realm of the whole they compose. In Part II it shall be demonstrated how adequate that model is when we wish to understand the distinctive nature of quantum holism, which in fact is impossible to reconcile with externalistic compositionism typical of classical mechanics with its 'material points' and the separable physical states.

## 1.2 The division of the whole, Eleatic paradoxes and the Universal Principle of Relativity

An analysis of the logical structure of fundamental physical theories is extremely interesting not only because it reveals their formal properties and possible relations with other theories, but chiefly because it is a way in which physics constructs consistent models of reality. When we take into account the theories termed at present as fundamental, i.e., the General Theory of Relativity (henceforth GR), QM and QFT, then we are in the position to enquire: what logical structures organize their formalisms? An identification of such structures can be conceived to be equal to defining the way in which the cognitive mind orders the knowledge it has of certain sections of reality – in the case of QM that would be, for example, a large group of quantum phenomena. In the situation when physics wishes to construct theories that are very general, unifying ones, the issue of the logic of the models is not secondary any more, but comes to the fore, as it is that very high degree of unification the model offers (the possibility to stretch its scope over greater and greater regimes of physical phenomena) which stirs epistemological and methodological doubts in its wake: why has *such and such* a logical structure been selected as a *universal* language in which to describe nature? It also turns out that one of the most interesting situations arises when we reflect on the meaning of the concept of the 'wholeness of the system' on that fundamental level or on a cosmological scale. In particular, we ask ourselves if the nature of irreducibly holistic natural phenomena requires our cognitive habits to be remodelled and special logical structures to be employed? Quantum systems display totally non-classical, holistic properties, but one of the most astounding aspects of the theory is the fact that it seems to accept contradictory descriptions construed by different observers inside the Universe (see Ch. 5.1). Usually – in line with the 'Copenhagen tradition' – it is analyzed as an epistemological issue, but in fact it is a deeply metaphysical problem, which becomes evident when we reflect upon it in the context of metaphysical holism. Let us first ponder on the old Eleatic paradoxes construed by Zeno of Elea in that very context of the problem of the whole and of the divisibility of a system in physics.

An underlying motivation and at the same time common denominator of all the so-called 'Zeno of Elea's arguments' is an attempt to draw logical conclusions from the thesis that 'plurality exists' (εἰ πολλά ἐστιν – as at *Parm*. 127e and 128d) in such a way that the internal contradictoriness of that thesis is clearly brought out [Olesiński, 2000, p. 154*ff.*]. Traditionally, two groups of arguments have been differentiated there: one dealing with movement and the other referring to the division and real existence of plurality, but their more precise reconstruction allows us to claim that they are all in fact focused on the problem of the relation between the hidden (from the senses), unchanging 'world-whole' (ὅλον), understood as a 'fullness of being' (τὸ πᾶν), and the plurality of beings (πολλά) and their parts to be observed by the senses, and hence it is the very same subject matter that Plato had been concerned with in his late period. In particular, the problem of the many may be interpreted in the context of an infinite versus finite divisibility of time and space – we then gain a set of paradoxes which refer to the movement of bodies. As Olesiński has noticed [ibid., p. 154], in the case of those two types of arguments, two forms of the same problem of plurality can be differentiated there: 1) a *finite* versus an *infinite* division within a given universe (the issue of finiteness in the aspect of quantity), and 2) a *complete* versus an *incomplete* division (the issue of finiteness in the aspect of differentiating the basic elements of the universe). Among the four arguments concerning movement, offered by Aristotle in his *Physics* (*Phys*. VI, 9), one can introduce a further polarity: (1) the Dichotomy and the Achilles; (2) the Arrow and the Stadium [Smolenov, 1984, p. 169]. The first two are grounded in the assumption of an infinite divisibility of space into smaller and smaller parts and the problems that arise from that, whereas the latter two are based on an opposite presumption, namely that 'moments of change' are indivisible (unitary elementariness), which makes them totally stationary (the Arrow), or leads to a conclusion contrary to the assumption which results from the relativity of movement (the Stadium). Let us recall that the content of those paradoxes presents itself thus:

(1) The Dichotomy: Every body which has to cover a certain section of a path must at the same time cover an infinite number of sections which the path can be dichotomically divided into and that is impossible in a finite range of time [Simplicius, *In Arist. Phys.* 1013, 4*ff.*].
(2) The Achilles: The fastest runner can never outrun the slowest one, even if the latter starts off from a point much closer to the end of the race, as the infinite division of the path – as in (1) – makes the distance between them, getting longer all the time, impossible to cover, regardless of the speed of movement [Aristotle, *Phys*. VI.9, 239b].
(3) The Arrow: Any given object in reality does not move, as it remains at rest in every further indivisible element of time ('moment') and hence it also rests in the whole – (sum of) all of the particular moments ([Aristotle, *Phys.* VI.9, 239b], [Simplicius, *In Arist. Phys.* 1015, 19*ff.*]).
(4) The Stadium: Movement is not a real phenomenon, as its relativity results in contradictory conclusions in the reports of particular observers, assessing the intervals of a path covered in indivisible intervals (moments) of time – it is then proved in the course of reasoning that the unitary length of a moment would have to be equal to half of that moment, which is impossible [Aristotle, *Phys*. VI, 9, 240a].

The last argument usually draws the attention of some contemporary commentators as its solution is interesting when it is viewed from the relativistic perspective. Aristotle, when recounting Zeno's ideas, referred to the picture of two rows of objects of the same $n$-dimensions, $A^n$ and $B^n$, which approach one another with the same constant velocity $v$ measured in a third resting system $C^n$, and cover in a given element/interval of time ('moment') $t_0$ a given element of the path (an elementary interval of space) $l_0$. In accordance with the classical law of the addition of velocities, an observer will judge the velocity in every of those moving rows to be $2v$, while the path covered in $t_0$ as two elementary intervals, $2l_0$. The conclusion, as Aristotle points out, proves contrary to the assumption that there exist indivisible (finite) elements, as the time interval $t_0$ turns out not to be so elementary: its half is equal to its whole, which the philosopher judges to be a paralogism. The final intention of Zenonian argumentation is thus to exclude both the physical hypotheses: the continuity of movement and its discontinuity, through a reference to the more general issues of divisibility (in the aspect of finiteness and completeness). From the contemporary relativistic perspective, the Stadium would be a paradox only in the context of pre-classical and classical kinematics, assuming a real absoluteness of time *and* space, as it is easy to find that it is eliminated by transformations based on a constant speed of light and the continuous Lorentz group (suffice it to imagine, instead of rows of such objects as carts, rows of clocks, moving towards each other, and to reflect on what they indicate). Similarly, it is usually assumed that the first two arguments on continuity are solved by a somewhat sophisticated theory of the summing up of infinite series, whose disregard Zeno was accused of for example by Alfred N. Whitehead [Poczobut, 2000, p. 118]. Regardless of whether we agree with the conviction that the theory of the summing of infinite series does indeed *solve* the problem of the measurability of movement, there is something extremely interesting in a wider ontological context in the relativistic solution to any generalized or extended version of the Stadium.

As the quoted paradoxes are in fact based on two opposing hypotheses, it is worth deliberating whether there exists any general way to neutralize them in the form of a 'prescription' to formulate a certain *hypothesis tertia*. If we agree that the first two of them, treated seriously, lead to a 'freezing of change' as a result of a realistic physical interpretation of the continuum, then we can treat them as a suggestion that the basis of the observed changes should be formed by certain physical micro-processes or micro-actions which perform the role of fundamental, atomic spatiotemporal intervals[12]. Hristo Smolenov has

---

12 Also Plato was well-known as a decided opponent of the thesis that the continuum (such as the temporal or spatial one) is 'made up' of points, which he considered to be fiction: in his 'physical geometry' and the ontology of mathematics he defended the theory that all extended objects are composed of 'indivisible lines' (ἀτόμοι γραμμαί,

proposed, for example, that such hypothetical intervals be called 'geometrons'[13]. He writes that: 'The refutation of the assumption about the infinite divisibility of intervals leads to the notion of atomic intervals: in the process of splitting spatiotemporal intervals a limit is reached, beyond which dichotomy is trivialized, if associated with parts, shorter (smaller) than the whole. The atomic intervals obtained are invariant under the transformation of dichotomy. That is the reason why geometrons can be regarded as elements of spacetime (from a logical point of view) or – to put it in another way: the world-points of a spatiotemporal continuum can be considered to be atomic intervals (continua within the external continuum). Invariance under dichotomy can be conceived of as a logical analogue of invariance under Lorentz transformations, roughly speaking' [Smolenov, 1984, p. 170]. Such an assumption adopted on the basis of the first two paradoxes (which should rather be called 'arguments' in that context), naturally gives rise to the following speculative query: what would the nature of those elementary 'intervals' be?

In his famous *Metaphysics* Δ (*Metaph*. V.3, 1014a-b), Aristotle defines his third concept, following that of the 'principle' and the 'cause', namely that of the 'element' (στοιχεῖον = *elementum*) mentioned already in Ch. 1.1, which he links directly to the problem of divisibility. The definition is as follows:

> Στοιχεῖον λέγεται ἐξ οὗ σύγκειται πρώτου ἐνυπάρχοντος ἀδιαιρέτου τῷ εἴδει εἰς ἕτερον εἶδος.[14]

The division of an element can of course frequently continue to go on, but there is no further sense in that due to the *differentia specifica* of the 'species' under analysis – the parts obtained in the course of such a division will not be a further 'sub-species' of the species, as it is elementary *on account* of the division under analysis. (Aristotle gives examples from phonetics, and also logic: 'Elements

---

*lineae indivisibiles*), of which Aristotle informs us (*Metaph*. I.9, 992a22). In the cosmology of the *Timaeus* (*Tim*. 53c–56c), the material elements of the cosmos are created not *in* space, which as a continuum would serve as a stable background, but *from* the space (χώρα), which is subject to global operations of limitation in such a way that 'indivisible lines' are formed ('space quanta'), and next, from them, geometric plane configurations, solids and physical objects, cf. Harte [2002a, p. 235*ff*.]. That is simply a speculative project of a combinatorial geometrization of physics without ascribing any reality to the points. (See also Ch. 2.2 in the context of Leibniz).

13 The term has been coined by Azarja P. Polikarov (1921–2000), a Bulgarian physicist and philosopher of science.

14 'The "element" is a primary, immanent component of something, indivisible into parts of different kinds'.

are also the primary geometric proofs and those which are at the basis of other demonstrations' (*Metaph.* V.3, 1014a37-b2), as they are elementary on account of being 'the basis of conclusions' etc.). What seems interesting here is the possibility of a relational interpretation of an 'element', bred in the operation of dividing into species (we always enquire: on account of what *operation of dividing* has that species/element been specified? [Smolenov, 1984, p. 170*f.*]). The concept of an 'indivisible element', when detached from a *given operation* of dividing, is bereft of sense, therefore the definition hints that an 'element' of time or space is such a component whose further division is pointless on acount of its property 'remaining to exist' or 'stretching' – it might have a mathematical sense, but not a *physical* one, which forms a basis of the property analyzed here. But one should always remember, and that is a decisive point in his massive struggles with Plato, that Aristotle, as a model externalistic compositionist, never takes *genus* as autonomously existing apart from its species, that is, *a whole* as a 'matter' of the division and differences *is in fact never anything really primary as a whole.* So any operation of division *is not internal* to the whole in a strictly ontological sense and cannot be modelled in a Platonic manner, even in principle. But Aristotle also believes that the most general concepts classified as belonging to some *genus* (i.e. undefinable) are often elements on account of their simplicity and lack of differentiation. When we take such a theory of the 'element' (the metaphysical atom) into consideration, we must reconcile ourselves to the fact that it is difficult to speak of an 'absolute divisibility' or an 'absolute indivisibility' – the latter two should rather be replaced with the category of an invariance *with regards to a concrete operation* (division/divisive action). It has to be noticed at this point that a possible hypothesis of spatiotemporal 'atoms' introduced as the result of analyzing two paradoxes of the physical continuum, have to be defended in the above way, as we otherwise fall into two further paradoxes connected with the *absolute* indivisibility of those atoms (the Arrow and the Stadium). It should be stressed that Zeno's intentions went in a totally opposite direction, as both sets of paradoxes were intended to prove the futility of any physical research on the plurality of beings: 'Parmenides's pupil took the same path his teacher warned him of, but he probably took it with the intention of proving how barren it was and that the warning had been justified. In that sense, he supports his master's reflections and can be considered to be his "shield"' [Olesiński, 2000, p. 163]. It turns out that we could just as well treat those paradoxes (particularly the pair: the Arrow – the Stadium) as arguments for the key role of relativity and invariance under transformation (symmetry) with regards to any 'atoms' (elements).

In the instance when we do not apply the Principle of Relativity to our elementary intervals of time and space, we shall obviously obtain contradictory statements, such as (in the case of the Stadium): '$l_0 = 2l_0$' and '$\frac{1}{2}l_0 = l_0$' (Aristotle), or contradictory indications of the clocks from $A''$ and $B''$ in a modern relativistic version, with speed $v$ approaching the physical speed $c$ of light in vacuum. We could ask next what might happen when, in the case of the first of the paradoxes discussed, namely the Dichotomy, we do not accept the hypothesis of elementary intervals of time and space and assume the physical possibility of an infinite division of moments and of every area of space? Due to the logical symmetry between the two pairs of paradoxes we are entitled to expect that we shall thus obtain a certain kind of contradiction – and so do we in fact do. We first have to notice that that contradiction is an interesting consequence of the general principle, *ontologically* understood, which we can simply term, after Leibniz, the Principle of Continuity (*Lex Continuitatis, La Loy de la Continuité*, PC), which puts down that '*la nature ne fait jamais des sauts*'[15]. Leibniz presents it in a longer formula as follows:

> It is to be inferred from there that one always moves from what is small to what is big, and vice versa, through what is in-between both in degrees and in parts (*elle porte qu'on passe tousjours du petit au grand et à rebours par le mediocre, dans les degrés comme dans les parties*). And that no motion is born directly from a state of rest nor does it move into it otherwise than with the help of a weaker movement (*par un mouvement plus petit*), just as one can never run to the end of a line or length before running a smaller path, although up till the present those who have formulated the principles of motion have not noticed that rule as they have believed that a body can be put in motion at any time by a motion that is opposite to the previous one. [Leibniz, 1882, p. 49]

PC and the realistic or even openly physical treatment of infinitesimals leads to many serious difficulties, both mathematical and physical ones, which Leibniz was aware of, and therefore it seems that he treated infinitely small magnitudes as tools helpful in calculations [Poczobut, 2000, p. 126], whereas PC itself, as the more recent research proves, was probably not understood by him in a uniform way and not treated as strictly universal [Crockett, 1999]. A simple and at the same time clear reasoning, pointing out the contradiction arising from employing the PC in physics was formulated and refuted by K. Ajdukiewicz [1977] in his well-known article *Change and Contradiction* (1948). It is based on a physical interpretation of PC in the form of the 'postulate of transition', which is close to Leibniz's above 'law of motion'. In the case of two states, A and *non*-A (*resp*. B), there seems to be an instant when the body is neither in state A nor in *non*-A, which coupled with the rule of transposition employed in predication about

15 'Nature never makes leaps' [Leibniz, 1882, p. 49].

*non*-A = B state, allows us to conclude that the body during that time is both in state A and in *non*-A ([ibid., p. 204]; cf. the commentary in [Poczobut, 2000, p. 167]). We can thus conclude that the application of a *physically* interpreted PC (as in the paradoxes of Zeno of Elea) to make a change of the *transition* type: 'state A' → 'state B' (unequivocally defined), leads to opposing judgments describing the physical state: '*x* is in A and is not in A'. One of the most well-known argumentations in favour of the thesis that the motion of bodies analyzed for any moment contains a real contradiction has been offered by Graham Priest [1985], who is also at the same time one of the well-known defenders of a radically interpreted paraconsistent logic, accepting statements that are true contradictions (*dialetheias*) ([Priest, 1979; 1987], [Priest, Routley, 1984]). In Priest's original formulation, every moment of the body's motion is described by a contradictory judgment treated as a faithful representation of a real, physical *state of change*. The conclusion is a direct result not only of the adopted physical interpretation and universality of the PC (and hence of the thesis of the existence of 'temporal moments' corresponding to points on the real line), but also of the assumption of an existence of an 'internal state of motion' described by the above contradiction (such an 'internal state' is omitted in Zeno of Elea's paradoxes). It therefore seems that the only reasonable solution for anyone who is not ready to accept the *actual* contradiction contained in constant motion, is to treat the PC merely as a mathematical idealization and to assume that on the microscale the real physical process together with its spatiotemporal characteristics is quantized (granular), i.e., there exist 'elements' composing it in the Aristotelian sense. Such a solution probably lies at the root of Leibniz's understanding of continuity, which encompasses an ideal order (a mathematical one) and a phenomenal (physical) one, but not a *real metaphysical* order of individual monads, which are 'atoms of nature' [Crockett, 1999, p. 134*f.*]. But then we still encounter the aforementioned problems with absolute elements.

Priest construes his 'logic of the paradox' by referring to contradictions which according to him are characteristic of fundamental physical phenomena, and also by recalling internally contradictory but non-trivial physical theories, although one can harbour serious doubts as to the necessity of such a solution for the conceptual difficulties of physics. Generally speaking, alternative solutions can be divided into classical and non-classical ones (i.e., ones referring to new concepts and mathematical theories). One of the 'classical' ways of overcoming the paradox of the Arrow is for example G. Vlastos's [1966] well-known diagnosis, according to which speaking of a 'momentary motion' is a categorial error, as the movement as such is meaningful only and solely when we define it over an interval (in the framework of duration), which Ajdukiewicz [1985] also agrees with. But such a solution is unfortunately only seeming and not only does not resolve the paradox, but finally only makes it more acute, as it is based on the ontology of kinematics, in which the time intervals are filled with a 'neutral' or 'innocent' passage of time – of which strategy a detailed and crushing criticism has recently been presented by Constantin Antonopoulos [2004, p. 237*ff.*], who rightly remarks that 'being *at* a point is a concept *constitutive* of motion and, at the same time, a concept destructive to it' [ibid., p. 237]. That means that we are dealing with a sharp alternative there: we can say that: (1) something is *to be found at a point*, but it cannot move then, or that (2) it *moves*, but it cannot be at a point then, as movement can be defined only by their set, i.e., topologically by the neighbourhood of that point. In case (2) we obtain, despite Vlastos's and Ajdukiewicz's reassurances, a contradiction, as the object in its real movement *in* space occupies at a given moment a place where it is not, from which Zeno of Elea draws his famous conclusion on the non-reality of motion [ibid., p. 241]. On the other hand, one of the well-known non-classical solutions in the case of the second paradox, the Stadium, is a reinterpretation of infinitesimals in the framework of a non-standard analysis based on the so-called internal set theory, thanks to which it is possible to represent an infinite set of elements (viewed 'externally') with the help of a finite set of non-standard elements (numbers), seen from an 'internal' perspective [McLaughlin, Miller, 1992]. Also that solution turns out to be deeply problematic, as it gives the impression of being a technical trick which completely sets aside any questions about the nature and micro-structure of motion, whereas the very introduction of non-standard numbers can be qualified as an *ad hoc* mathematical strategy [Antonopoulos, 2004, s. 244*f.*]. It seems that the problem of motion and change, just as that of time and space as such, cannot be solved 'locally', with the help of temporary or *ad hoc* constructions, but it requires a much deeper reflection on the ontological meaning of the concepts of 'spacetime' and 'physical state' as such.

Let us now return to the question posed earlier: what would the nature of elementary units (the 'geometrons'–events) of time and space be, if we were to assume their reality under the influence of the argumentation relying on the Zenonian paradoxes (1) and (2)? Firstly, they would be invariant under the operation of division, i.e., the result of the operation would not change the physical character of the element. That results from the fact, as Smolenov stresses [1984, p. 172], that parts do not differ from wholes when they are analyzed within the same *differentia specifica*: 'two dialectical opposites: whole and parts, appear to coincide with regard to the notion of element' [ibid.]. That suggests that the abstractly understood 'element' (in the Aristotelian sense) can be ascribed with opposing terms connected to its species (in this case: the context of motion and change), which *seem to* lead to the paradox stressed by Aristotle. Secondly, to avoid the rightly criticized by Priest 'cinematographic' theory of movement (e.g. advocated by B. Russell), which is counterintuitive and to a large measure artificial, we should conceive the elements of spacetime as *dynamic units*/action units, and not only geometric or 'idle' (in Antonopoulos's understanding). That means that such an element, *as a part of a series/history*, would not be a mere static shot of the motion, but would contain an inner 'tendency' to get linked with other elements within that *whole series*. In such a perspective the source of motion taking place in time and space would have to be found in the deep, *pre-dynamical* sphere of nature, which has to be conceived *pre-temporally* and *pre-spatially* (temporal and spatial relations would be abstractions of the second order). If we deny that sphere any reality or primordiality, then kinematics based on such static and absolute spatiotemporal 'elements' does indeed undergo a 'freezing', just as in the 'cinematographic' concept, and hence we have to struggle on with the paradoxes of the Arrow and the Stadium. All the metaphysical doubts are now concentrated on the issue: what does it mean that an indivisible element can be ascribed with opposing definitions, which would apparently force us to adopt a paraconsistent logic?

If we analyze the moving object on the lowest possible level of that process, i.e., in micro-spacetime, then the alleged 'quantum of motion', which we shall here also identify solely hypothetically with a 'quantum of spacetime', will exhibit a troublesome property, as both motion and rest are undifferentiable with regards to it if seen as an element *per se*. (We have to remember nevertheless that the quantization of motion does not necessarily imply a quantization of spacetime, hence the above hypothesis is rather radical and is to serve as a speculative option – another, more classical possibility is discussed for example by Antonopoulos [2004]). That is intriguing, as it might point to the relativity of physical motion as having its deep source in its quantum microstructure. Smolenov [1984, p. 172] proposes the following reasoning in that context, concerning a moving object *A*, with relation to a quantum of that movement, *q*:

$$[1] \quad \frac{A \text{ is neither moving, nor at rest,}}{\text{moving and at rest.}} \quad \text{(therefore it is both)}$$

The reasoning is unfortunately highly problematic, as despite the seemingly encouraging symmetry between dialetheism and analetheism (in which certain sentences are bereft of logical value as so-called 'truth-value gaps'), there exists a marked difference between them, discernible not only intuitively, and the choice between them is *not* arbitrary, as for instance Terence Parsons [1990] made out. Antonopoulos [2004, p. 257] justly criticizes that apparent symmetry: 'The argument, in other words, which seeks to turn "neither A, nor – A" into "both A and – A", is simply an argument mistaking an *incomplete* concept for a *complete* one', and rejects the Principle of the Excluded Middle in his analysis of the discontinuity of movement, while retaining the Principle of Non-Contradiction. Let us suppose, though, for the sake of illustration, that the indifferentiability of the movement and rest of *A* regarding a quantum *q*, which the Arrow paradox discloses in a primitive form, would be correctly characterized by the sentence: '*A* is both in motion and at rest', which is a conclusion a potential observer of the physical phenomenon could have made – *eo ipso* we recklessly assume that the observer has the ability to observe any possible opposing characteristics at the same time. It might seem surprising that the contradiction keeps reappearing with regards to an 'element', since we had introduced it lifting the PC exactly so as to get rid of the paradoxical conclusions caused by the assumed physical continuity. Given the 'spacetime quantum', the 'event', we might then say that it both has no characteristics of movement (there is no change existing within it, i.e., analyzed with regards to itself, 'internally', it has no 'state of motion'), and possesses such a property (i.e., analyzed in relation, 'externally', it constitutes an 'element' of change/process). Nevertheless, we can observe a fundamental principle organizing that paradox emerging here, which had already appeared in a relativistic solution of the Stadium paradox (when we had assumed, contrary to Zeno, that each of the moving clocks had its own time and its own kinematical perspective). It is obvious that a contradiction emerges when we link by a conjunction two opposing statements which are made from two different perspectives (and thus two different observers or observations which are not simultaneous) just as if they were being expressed from the very same perspective and absolutely at the same time. We have thus introduced the classical assumption of an *absolute observer* (the ontological *absolute*), which possesses the ability of registering all the aspects of a physical process at the same time, and hence it is not so much aperspectival as physically *pan-perspectival*. We might formulate the hypothesis that a paradox based on a contradiction is directly linked to a universal Principle of Relativity violated by the introduction of a certain empirical absolute into the universe of theory. A working formula of that principle in the context of the ontology of the whole (compositionism) could thus be as follows:

> The Universal Principle of Relativity
> No arbitrarily singled out element of any whole is of an absolute character, which means that it is an element only within the relational structure of that whole and is not in any way ontologically privileged with regards to the other elements of that whole.

In the case when one prefers to formulate the indiscernibility of movement and rest with the help of the sentence: '*A* is neither in motion nor at rest', then we assume the existence of an aperspectival observer who is a certain ontological *anti-absolute*, i.e., in the case of certain phenomena he/she observes, he/she has no observational perspective and is unable to assert any occuring or existing state of some type. It would thus be a violation of the clasically realistic universal Principle of Definiteness, according to which every possible property or state of an entity is always fully definite/truth-definite, matter-of-fact, determined by the reasons of its actuality (in the ontological version) or predication (in the epistemological version). As we have found, both those principles (of Relativity and Definiteness) are to be found in their primitive form in Aristotle in his reflections on the essence of the element and whole, and also on the relation of 'matter' and the 'form' of the phenomenon (the Principle of Relativity), and in treating the real entity as something concrete and fully definite, τόδε τι, in the course of being actualized, ἐνέϱγεια (the Principle of Definiteness). After all, we are dealing here with the following deep relations between general logic and physics modelling the structure of properties (states/information):

> The Principle of Non-Contradiction ⇒ The Universal Principle of Relativity;
> The Principle of the Excluded Middle ⇒ The Universal Principle of Definiteness.

It turns out that we can clear fundamental physical paradoxes of the accusation of their violating the Principle of Non-Contradiction as long as we successively and determinedly remove absolutes from them or, possibly, to some degree, anti-absolutes in the case of those which seem to violate the Principle of the Excluded Middle, which locally could give back the theory its realistic or just semi-realistic character. (As will be seen especially in Ch. 4 and 5, quantum phenomena, and QM in general, are a very subtle, non-classical interplay between the above principles, which forces us to undertake an in-depth redefinition of mechanics itself). In practice, as far as the first case is concerned, that involves a change of predicates of the type: '*x* is in motion', to two-argument predicates of the type: '*x* is in motion relative to *y*', which is of course the essence of the Principle of Relativity working in Galilean kinematics. Every element, when it is treated as a real (particularly physical) absolute, reveals a property of generating a paradox based on contradiction, as it does not fully respect the Principle of Relativity (understood both ontologically and epistemologically): as an absolute it must then contain both an 'internal' and an 'external' perspective at the same time, and therefore a monadic and relational one. That does not imply that that is how the paradox of the Arrow is solved, but we can see what conditions a fundamental physical theory has to meet, so as not to fall into another contradiction nor to become unrealistic (as the 'cinematographic' theory of motion): it should be radically relational from the very beginning and should not be 'aridly' geometrical (that is, it should not be based on points and infinitesimals), as that 'freezes' the dynamics, making *any* talk about physical interactions between systems superfluous (but, as a matter of fact, physics *is* about transfers of information in the Universe). Finally, it is interesting to analyze to what extent applying the Principle of Relativity to a spatiotemporal 'element', although it could eliminate the contradiction, nevertheless leads to a situation that is somewhat analogical to the one Priest [1979] postulates in his 'spread hypothesis', but this time on the playfield of relationism. The hypothesis states that a moving body may be localized at a given moment only with regards to the neighbourhood of the point it is to be found at, whereas the above quantum-relational approach suggests that it is reasonable to speak of every property with regards to an 'element' only in relation to other elements with which it creates a dynamically defined (at least in a proper atemporal generalization of dynamics) 'stream of events'. Therefore Priest might be right when he states that even when one assumes a granular (interval) nature of the emergent space and time that does not fully eliminate the paradox of the Arrow (cf. [Poczobut, 2000, p. 162]), as the very internal dynamics of any thus generalized 'element' as such (creating some dynamical relation with other events/actions meant as transgressing that particular, singled out 'element') would result in it containing something which it is not, and thus a contradiction.

A metaphysician who happens to be an advocate of the relational approach might reply that such an accusation is still rooted in an image of a series of spatiotemporal elements (= 'states') similar to a film reel, in which each of them is a completely realized event-absolute, or, alternatively, *dynamics itself has not been at all relativized*. And more important, all this seemingly simple reasoning is deeply, maybe thus hopelessly *entrenched in the ontology of externalistic, Aristotelian-like compositionism* (its 'elements' have been linked in a bottom–up scheme) which steers all the imagining in physics. However, in accordance with the Universal Principle of Relativity, each dynamical unit is to be conceived solely with relation to all the other units, thus always due to the particular operation of division (singling out some subsystem, as Bohr and Heisenberg has postulated), while the 'possibility' or 'tendency to change' as the foundation of local dynamics is based – expressing this in a purely metaphysical manner – precisely on that lack of self-sufficiency, of *privatio* in every 'element' as a product of division, and never on its monadic completeness. For if there is a tendency in every elementary event to link up with other events, then there is never any instantaneous absolute in the process. In that sense the event as an 'indivisible part' of the process of change should not even be called an 'interval' in the common understanding of the term, to avoid serious misunderstandings and some naïve variabilistic–presentist metaphysics. It looks as if the extreme relativity of the 'elements' and the operation of division itself made necessary a notion of the *real primary whole* as its *ontological ground.* After all, it seems that if the Universal Principle of Relativity is taken seriously, one has to quit commonsense Aristotle and revert to Plato and his internalistic compositionism without any absolute elements (or, more generally, 'base states') as a foundation for cosmology. Relativity is a deadly weapon against commonsense physics and metaphysics, and that battle started well before the simpleton Simplicio and his opponents from Galileo's *Dialogue Concerning the Two Chief World-Systems* (1632).

Let us now reflect on a certain fundamental phenomenon of microphysics, which is directly linked to the above problem of the division of a whole into parts and which seems *prima facie* to provoke an equally troublesome paradox. The phenomenon – *nonseparability* – is of a purely quantum nature and concerns all known micro-processes, both forces and particles of matter (I shall deal with that in greater detail in Ch. 3). Let us consider here, as an example, instead of pairs of photons or electrons, a dimer of mercury isotope atoms $^{199}Hg_2$ dissociated by two laser beams and spatially separated into a pair of atoms being in a singlet state known as a so-called 'quantum entanglement' created by their interaction, where each of them, being in a ground state, has a nuclear spin $I = 1/2$ and total angular momentum $F = 1/2$, while the total nuclear spin of the dimer is $I = 0$ (experiment proposal and details in [Fry, Walther, 1997; 2002]). The photodissociated molecule is still to be found in one entangled fermionic quantum state $|\psi\rangle_F$, and conforms to the Pauli exclusion principle, while what is employed during the measurement is a ionizing laser instead of the classic Stern–Gerlach apparatus (as in a Bohm's version of the experiment, cf. Ch. 3.1). In a system thus prepared, the so-called Bell–Clauser–Horne inequality arising from QM can be tested for a parameter that we wish to measure (it can be a nuclear spin), and the inequality concerns the coincidence rates $R_{\uparrow\uparrow}(\theta_1,\theta_2)$, i.e., a simultaneous registration of a certain value of the parameter (spin component) for a pair of directions $\theta_1$ and $\theta_2$ of two analyzers. It is discovered, in full concordance with QM formalism, that such a type of experiment reveals an extremely intriguing property of atomic systems, exactly identical to that of pairs of photons or electrons: the nuclear state of such an atom is to be considered *solely in relation to the state of the whole system*, $|\psi\rangle_F = \alpha|\uparrow\rangle_x|\downarrow\rangle_x - \beta|\downarrow\rangle_x|\uparrow\rangle_x$ (for complex amplitudes $\alpha$, $\beta \neq 0$, $x$ is a direction in the physical space, while $|\alpha|^2 + |\beta|^2 = 1$), which is expressed in the concept of the 'quantum entanglement' manifesting itself in appropriate quantum statistics. That means that one *might* treat the quantum state $|\psi\rangle_F$ as a local physical 'element' in the Aristotelian sense, and its indivisibility is framed in appropriate physical symmetries, which it is an invariant of – it is pointless for a local observer to analyze separately the quantum state of each atom of mercury before the final measurement interaction, as they constitute a quantum-dynamical whole, although each of the atoms seems to possess an identity of its own connected to its mass. One could, it seems, therefore, state a contradictory 'Zenonian judgment' with relation to element $|\psi\rangle_F$, just as in the case of the former 'spacetime element' *q*:

[2] The system described by state $|\psi\rangle_F$ is both an indivisible whole and is not an indivisible whole.

When we speak here of the 'quantum state' of a pair of atoms after the interaction, we would like to safeguard the principle of the separability of the states of those two atoms, each on its own, so that regardless of their being detected by any external system or not they would always remain totally separate states of affairs in the world (for a general metaphysical discussion on that point see e.g. [Maudlin, 2007, p. 50–77]). Recognizing that expectation allows us to discover that we treat the quantum state of each atom as a certain *absolute*, completely independent of the system which is measuring (detecting) it. It is true that atoms have their physical states registered in the course of a conducted measuring procedure in the form of separate value detections for separate subsystems, but it is also true that those quantum states are totally dependent on the system as a whole, i.e. 'state $|\psi\rangle_F$', as that is an element of the whole physical situation defined by its local physical history *relative to the measuring system*. The division of state $|\psi\rangle_F$ into separate 'component states' existing independently of the detection events is not absurd in the conceptual sense, but is physically nonsensical, if we wish to treat parts as certain micro-absolutes of the physical world, which are real and objective, regardless of the physical interactions (information transfer). Statement [2] does not result in a material contradiction, as the quantum situation described locally by $|\psi\rangle_F$ never occurs together with the physical observation of two different atoms moving in space in the direction of detectors. Atoms $^{199}$Hg are thus independent parts of a system only *with relation to a local measurement action*, when their nuclear states are registered with the help of laser beams, but they are an indivisible whole when no updating of information takes place as a result of physical interaction, that is, they are a whole *in relation to the elementary situation* of the local 'observer' described (coded) in $|\psi\rangle_F$. These are two different physical situations and two distinct relations between systems. Contradiction [2] is apparently an artifact of the cognitive process, which does not respect the metaphysical Principle of Relativity and treats certain entities or states as objectively definite independently of any *physical action/relation*. That means, as also evident from the previous discussion, that the notion of the 'element', and thus the operation of division itself, should be subject to the Principle of Relativity, which has profound physical consequences (see Ch. 5).

That observation reveals another more general rule: paradoxes which refer to ontological absolutes are also generated when the pseudo-external 'observer' attempts to capture all the perspectives on the whole from within, including the 'hidden' ones, which are nevertheless tacitly included in the realm of that representation as 'inaccessible to instantaneous cognition'. It is of course possible to formulate such statements in the framework of a given empirical language, but the introduction of an absolute perspective gives rise to a contradiction of type [1] or [2]. Priest sees in it the great force of the dialetheic approach, which does not have to set any language hierarchy in action to solve paradoxes, and it at the same time defends the expressive universalism (completeness) of a natural language expressing a paradox. As an alternative to that assumption of the absolutism of cognition, we can point to the Universal (metaphysical and epistemological) Principle of Relativity closely connected to that of Non-Contradiction. How far paraconsistent logic can enrich the understanding of the physical sense of QM (see e.g. [Dalla Chiara, Giuntini, 1989; 2000]) remains, of course, a separate matter. I wish to postulate here that the problem of a division of the whole into parts in the context of microphysics and its relation to the principles of Relativity and Definiteness is not simply a non-classical epistemological or logical issue, but is chiefly a deeply metaphysical one, and the paradoxes it generates can be eliminated only in the course of a far-reaching relational redefinition of the concepts of 'realism' and 'physical states'. Quantum physics requires much more than a mere new epistemology or a new predicate logic and that is never as evident as in the context of quantum nonseparability as a physical ur-phenomenon and *physical information resource*, as will be seen further.

## 1.3 The philocryptology of nature, internalistic compositionism and the hidden whole in microphysics

'It is obvious for systemic reasons that we shall be unable to understand Plato's philosophy as long as we ourselves shall not be able to reconstruct *Parmenides*'s content after him. As long as we are unable to do that, can we hope to understand our own philosophy better than his?' – such a question was posed by Carl F. von Weizsäcker [1982, p. 472] in 1970 in his essay devoted to the *Parmenides* and in the context of Plato's dialogue analyzed earlier it seems extremely apt. 'Is it a logical farce, a joke – or is it the highest level of Western theology? (…) Can we expect that we shall learn anything about our own problems from texts as dark as that?' [ibid., p. 471] – he asks as a theoretical physicist and thus one is introduced once again into the Eleatic-Platonic riddle of the whole. Von Weizsäcker has encouraged us to read the *Parmenides* and his hypotheses in a completely novel way, in the light of the 'paradoxical' properties of QM, and hence he has lent it with a cosmological dimension [Blandzi, 2002, p. 162], which we finally find in the *Timaeus*. An appropriate point of departure in the light of what we have already found is Heraclitus' famous statement, passed down by Philo and Proclus: 'φύσις κρύπτεσθαι φιλεῖ' (B 123: 'Nature loves to hide'; cf. [Kirk, 1975, p. 227*ff*.]). Jakub S. Bańka [1998] proposed that this central category of Greek thinking about nature be called 'philocryptology' and encourages us to speak about it, in a Heraclitean spirit, that 'nature (*phusis*) is philocryptic', i.e., 'it tends to concealment' [ibid., p. 21]. It is at this point important to analyze more closely in what sense 'authentic wholes' of nature (as measures–forms–structures), which we have spoken about, are ontologically hidden and how that is connected to the already mentioned problem of empirical non-contradiction.

In *Book N* of *Metaphysics*, Aristotle clearly agrees with the Platonic identification of the whole, unity and form/measure, but at the same time stresses his externalistic compositionism as an alternative to Platonic (non-atomistic) holism:

> The 'one' obviously means the measure (τὸ δ' ἓν ὅτι μέτρον σημαίνει, φανερόν). Every thing discussed contains something else which is a proper subject, for example a quarter-tone in a musical scale, a finger or foot or something alike in a spatial length, a bar or syllabe in rhytms, and similarly some distinct weight unit in weight, and in all other cases alike: some unit quality for a quality and some unit quantity for a quantity. (Such a unit measure for quantitites is indivisible in form, and indivisible with regard to the perceptual recognition for qualities). That means that the one itself does not exist as a self-contained entity. And that is correct: 'one' means the measure of some plurality (σημαίνει γὰρ τὸ ἓν ὅτι μέτρον πλήθους τινός), as 'number' means some measured plurality and the plurality of measures. (*Metaph.* XIV.1, 1087b33–1088a6)

In the case of Aristotle, plurality (e.g. of material parts) or a thing is *composed of* units, and their type determines what sort of whole (substance) they can create (cf. *Metaph.* VII.17, 1041b) – the effect, although it seems close to commonsense thinking (which is to be observed on the examples given by Aristotle), can nevertheless raise doubts, as it is subject to the accusation of a special pleading, as has already been mentioned. The question appears: as parts ('elements') are always primary and independent of the whole as its 'matter', then what is 'form' (a unifying entity, *not* an element or even a 'unifying relation' of any kind), which binds them and orders them into a whole beyond plurality? We are clearly having trouble here with an *externalistic dualism*, as the concrete whole has to be composed of material parts *and* of a 'structure'/'substantial form' that is totally independent of them and being of a completely different kind (it is 'the cause of unity of the parts', *Metaph*. VIII.6, 1045a8–10), but both of equal importance. Plato clearly turns all of his attention onto the wholes–unities not only irreducible and predicated – as the Aristotelian ones – in a sentence (e.g. 'a set of organs is an organism'), but first and foremost those that are primary and not dependent on any spatiotemporal relations or causes operating *in time* and space (such as physical 'contact' or 'viscidity'; ibid., 1045a11–12), and hence, as long as an Aristotelian *might* say that the unified whole (e.g. an organism) *has* structure/form, then a Platonist always has to state it more strongly: the whole *is* indeed a structure, which *is* a true unity, not a 'formal' or secondary (coming-into-being) one. Thus internalistic compositionism demands that in reality the whole/structure has an ontological primacy, and it seems only secondary (e.g. in sensual experience, when things undergo disintegration and are perceived spatiotemporally) that any 'elements' have a prior autonomous and independent existence as some natural 'subjects' or even 'matter' for 'things'. The Platonic distinction of two levels of cognition, *doxa* and *episteme*, finds its concrete illustration in those opposing views: when commonsense and everyday experience (δόξα) are resorted to, each whole/thing having a temporal history can objectively and effectively be split (divided) back into the basic parts/elements (i.e., some differentiable unities, 'ones') it 'contains', whereas according to the Platonic 'cosmological' method of studying (ἐπιστήμη) the atemporal 'authentic wholes' that is effectively impossible to carry out completely, as the real structure (immanent measure *on a cosmological scale*) is *not composed* of anything – it is not dependent on the actual plurality of the recognizable parts (e.g. individual objects, states or events), which in empirical reality and in a given case actualize it spatiotemporally, hence a Platonic 'history of things' is always derivative. To put it shortly, the Platonic view on the whole–unity is a complete reversal of Aristotelian externalistic compositionism, and the Platonic wholes have nothing in common with the Aristotelian wholes *becoming in time* from their existing 'matter'. Platonic, anti-atomistic (in a broad metaphysical meaning of the word) compositionism proceeds 'top–down', or rather – true to its name – 'internally': from the hidden whole to the internally emerging or 'crystallizing' parts, and not in the opposite direction. (In Ch. 2 we shall turn our attention to how Leibnizian holism resurrected Platonic internalistic compositionism, in opposition to Hobbes and Newton, yet at the same time drawing on Aristotle's theory of substance, interpreted as a metaphysical 'element').

We encounter an in-depth development of that subject-matter in the *Philebus*, where Plato moves the problem of the whole onto a higher level and attempts to conduct a metaphysical reconstruction of the process of composition itself. He achieves that by introducing 'four kinds' or general principles, which are treated as ontological components of the process: *unlimitedness/the unlimited* (ἄπειρον), *limit/completion* (πέρας, τέλος), *mixture/composition* (σύμμειξις, κοινωνία) and *the cause of the composition*. The whole, i.e. the composite being a unity, is from the Platonic standpoint, as we have seen, a measure–structure, but it is impossible to assume that without an ontic content which would fill that structure – hence Harte stresses that the Platonic whole is always a 'contentful structure', contrary to the accusation that Platonism is a hopeless abstractionism [Harte, 2002a, p. 177*f*.]. The limit (*peras*) generates the structure of the whole and in itself is a measure (*metron*), whereas what is engaged in the generation of the harmonious whole of the system is the unlimitedness, a factor bereft of any structure, but one which is the foundation of plurality and content. We can therefore resort to Socrates's decided statement at the end of the dialogue:

> Any kind of mixture (σύγκρασις) that does not, in some way or other, possess measure (μέτρου) or the nature of symmetry (τῆς συμμέτρου φύσεως), necessarily dissolves (ἀπόλλυσι) its components and most of all itself. For there is no blend (κρᾶσις) in such cases at all but rather an uncombined disaster (τις ἄκρατος συμπεφορημένη), a real catastrophe (συμφορά) for whatever happens to take hold of it. (*Phil.* 64d9–e3)

We discover here a confirmation of our reading of the *Parmenides*, according to which the lack of a measure–structure of the whole not only makes it impossible to speak of the composite (the whole), but even of parts–elements that would compose it! Before we look with greater scrutiny at that radical conclusion, let us pay attention to the fact that Socrates directly identifies 'unlimitedness' and bare 'plurality', πόλλ' (*Phil.* 24a3), which means that the limit/measure is *per contrariam* not only a principle of defining and determination, but also of individuation. Opposites within plurality are disproportionate, it is possible to specify neither their relations nor value, while the limit, by introducing structure and creating a whole (system) 'introduces a number into them and makes them commensurate and in harmony with one another' (*Phil.* 25e1–2: σύμμετρα δὲ καὶ σύμφωνα ἐνθεῖσα ἀριθμὸν ἀπεργάζεται), producing 'that third one' (τὸ τρίτον). We have to perceive the 'unlimited plurality' (ἄπειρον πλῆθος) as a sheer potentiality of creating wholes–structures, and not a multitude of some ready elements, which are next put together, combined or 'unified' by form, therefore that is why it is so important to properly understand what Plato had in mind, when he writes of 'plurality' in the second hypothesis of the *Parmenides*. That is also the actual reason why Socrates says that composition (σύγκρασις) without structure contains no parts (elements) at all, and hence it is not composition in any real sense.

When the matter is thus presented we tend to oppose it, as accustomed to the Aristotelian externalistic model of composition we assume that we have the right to speak of 'unorganized elements' (the real plurality), which next, as a result of the peculiar physical processes/interactions taking place between them, create ('organize themselves into') singular, real wholes. Plato nevertheless evidently defends the standpoint that any statements about parts have sense *solely* within a prior ontic structure (immanent, global measure), and hence we cannot speak about the actual, not organized plurality 'composed' of some pre-elements. The unity reflected in the structure, or simply *being it*, is also a basic condition of the appearance of any empirical whole. Such a Platonic approach to primacy and the reality of a whole in nature can certainly be called 'holism'; and what is more, each form of ontological internalistic compositionism is a strong holism *per se*. (We shall go back to a more precise definition of holism in the context of physics in Ch. 3.2).

Let us now reflect on Platonic holism in the context of the hidden structure of a physical phenomenon which was, as we know, one of the main areas of Plato's interest in his late period (*Timaeus*). In that context, it is extremely important to keep in mind that Plato clearly differentiated two types of pluralities that remained mutually irreducible: a numerical or arithmetical one (of monads) and geometric-spatial, independent from the former (an excellent presentation of the problem in the context of an intuition of number and the continuum is offered by Z. Król [2005]). Let us now go back to the example of the split dimer $^{199}Hg_2$, whose quantum entangled state is described by the particular product of states of the two atoms composing it (see Ch. 3.2). Of course, true to our habits, we tend to model the situation on a classical scheme of externalistic compositionism: we assume that both atoms within the dimer are independent individuals with their own, intrinsic dynamical properties, but as a result of the experimental arrangement of laser beams we force them to 'reveal' to us the relation that binds them within the structure created by state $^{199}Hg_2$. We can at once agree that the relation ('entanglement') is of a unique character and is typical for microphysical phenomena, and at the same time we can accept that it is irreducible to any other intrinsic properties of the particular atoms. Thus we reject extensional compositionism, as we can see that we shall never be able to split the full information encoded in state $|\psi\rangle_F$, in which the dimer is in relation to the external measuring system, into two disjoint sets of information on the states of atoms, $|\psi\rangle_1$ and $|\psi\rangle_2$. That is a physical fact and we can put it most precisely by saying that some form of nonseparability of state is a fundamental property of QM as non-classical mechanics (see Ch. 3). We thus at the outset tentatively accept classical and minimal (Aristotelian) holism expressed simply in the statement: 'two independent entities (resp. atoms) create a certain whole, as there is a non-trivial, unique physical relation binding them that does not result from any of their intrinsic properties'.

It is obvious that a 'relation' thus understood will not be any physical entity existing independently outside those particulars, but it would nevertheless remain independent of any of their other properties (such as resting mass or charge), being responsible for the symmetric structure of that whole system under study and revealed in the quantum probability distributions of the serial measurement results[16]. When we speak here of *externalistic* compositionism, we have in mind the fact that the relation responsible for physical holism is 'external' here, which means that it is necessary for each atom (or particle) to possess the basic properties of an individual physical entity (or, in fact, any generic intrinsic properties) to be possible for that independent relation to occur. The acceptance of such a holism (we shall call it 'weak') is connected to the thesis that the structure itself, creating a specific arrangement/system of individuals, is – apart from the observable, material entities – an important and *real* component of the whole, which is expressed in the behavior of the system, i.e., in the measurement results registered (we assume that to be a primitive working criterium of 'reality', without resolving yet what kind of 'realism' we are talking about). And hence we would tend to treat each quantum system described by joint state $|\psi\rangle_F$ in a semi-classical manner, in the spirit of externalistic compositionism, as a system of two real objects with their own separate, particular states, remaining only in certain irreducible external relations given geometrically and statistically by $|\psi\rangle_F$. Such a type of compromise realistic holism obviously puts stress on lending the *own states* of the atoms, $|\psi\rangle_1$ and $|\psi\rangle_2$, a firm objectivity (independence of properties from the detecting systems/information transfer) and primordiality (ontic independence of the intrinsic properties from the system they compose). Holistic, irreducible relations might at most modify or constrain here the intrinsic properties possessed by individual objects, but that is the only way the latter are dependent on the wholes they produce. As the above way of thinking about physical phenomena is deeply rooted in classical mechanics built upon the debris of Aristotelian physics, but one which makes use of some of its metaphysical imagery and prerequisites, it will not be an exaggeration if we call it a habitual, and thus by definition a natural one. (As has already been mentioned,

16 In medieval ontology relations as *esse ad aliud* were directly called *ordo*, 'order', or *ordinatio*, external 'structure', to stress that function of theirs. The Aristotelian standpoint on the independence (irreducibility) of relations from the immanent properties of the individuals (*subiecta*) as their foundation was defended also by Albert the Great and Duns Scotus, contrary to Abelard or Ockham.

it found its ardent defender in Einstein when he protested against the so-called Copenhagen interpretation of QM[17]).

Nevertheless, when we look more closely at the situation locally described by quantum state $|\psi\rangle_F$, it becomes clear that the above model of compositionism not only breeds serious theoretical problems (connected, for example, to the status of particles as the postulated base objects of the theory), but, what is more, does not allow us to explain, on the grounds of quantum theory, what exactly happens to systems of those entangled atoms (if we wish to retain other commonsense assumptions, such as the possibility of an independent choice as to the direction of the spin measurement). The more classical the compositionism for quantum systems we seek is, the more hopeless the paradoxes we get. As I am going to deal with that issue later on (in Part II), I am now just signalling the character of the problems. Now, in a purely metaphysical-speculative mode, we can draw closer to that quantum phenomenon of entanglement by using Plato's model from the *Philebus*, which is all the more interesting as the spin possesses a fundamental geometric characteristic, and it was exactly that physical geometry which interested late Plato.

Even in the situation when we ignore the precise mathematical form of $|\psi\rangle_F$ and merely assume that any definite, measured values of the spin of both atoms will always be opposite, there is no empirical sense yet in speaking of those atoms as having their own, independent pure 'states'. As a result, there is even no point in talking about them as creating some *whole* (system), which they are just a *part* of. The fact is to be explained in the following way: there is no sense in the question, in QM, 'what is the spin of atom 1?', as it has been erroneously formulated[18]. The spin can be measured freely in any direction of physical space, chosen by any external system *O*, and hence no atom possesses any spin *independent of the whole system* (possibly comprising even the measuring system

17 I shall demonstrate in greater detail in Ch. 3.1 that Einstein was mainly intent on retaining the Aristotelian condition of a physical separability of elements (such as spatiotemporal events and mechanical states), and hence their primacy with regards to the whole. Perhaps he would have been prone to accept certain global (holistic) properties irreducible to local microproperties (particularly the global properties of fields and spacetime), but that would prove to be a weak form of holism – Einstein always stressed his faith in objectively real, noncontextual 'elementary things' or 'states of affairs'. As I shall argue, that cannot be reduced to the postulate of causal locality.

18 Instead of that question we should rather pose the proper one: 'what is the spin eigenstate $\sigma_1(n)$ of atom 1 in a given orientation *n*, in interaction with an external system *O*?'.

itself), which next, in relation to the spin of the second atom, would co-create the whole structure. That is an inadequate presentation of the situation, as no such thing as a 'spin in general', conceived as an objective characterization of the intrinsic 'state' of atom 1, possesses any physical sense on its own, being solely a potential unlimited *plurality* of all the possible measured values for directions in the physical space. (It is important to stress here that that striking quantum property has nothing in common with the different so-called problem of quantum measurement). If we now say that independent, individual 'states', marked as $|\psi\rangle_1$ and $|\psi\rangle_2$, always physically 'exist', as each of those atoms possesses at every instance unknown mechanical properties such as a 'spin value', but a specific, irreducible 'quantum relation' takes place between those states, which is revealed to us as the 'wholeness of the system', then we shall be responsible for a subtle, yet marked abuse, resulting from a tacit acceptance of externalistic compositionism. All in all, though, what is astonishing about the novelty of the physical situation we are dealing with here arises from the fact that we can say nothing about the definite 'spin states' $|\psi\rangle_1$ and $|\psi\rangle_2$ as separate from 'state' $|\psi\rangle_F$ of the whole system (and its relation to $O$). The whole $|\psi\rangle_F$, cannot be 'put together' or reconstructed from ready, defined 'elements' $|\psi\rangle_1$ and $|\psi\rangle_2$ in such a way that their status as semi-autonomous parts is preserved, as beyond the total 'state' $|\psi\rangle_F$ defined locally for $O$ they themselves as such have no sense, they are merely an undefined potentiality of 'all possible spin values', and nothing else. Putting it shortly: the 'state' of the quantum whole, and that alone, defines the quantum 'states' of the differentiable parts and all the relations between them (physical *correlations* for all values), and hence that is of fundamental importance for the very definition of the former. To put it in Platonic terminology: apart from the whole $|\psi\rangle_F$, spin states $|\psi\rangle_1$ and $|\psi\rangle_2$ are a completely *undefined* (infinite) *plurality*, ἄπειρον πλῆθος; it is *only* the measure–structure that creates the whole they are parts of. And it is only that real structure–unity that enables us to write down the unique relation of both states–parts as:

$$|\psi\rangle_{F,x} = \frac{1}{\sqrt{2}}[|\uparrow\rangle_1 |\downarrow\rangle_2 - |\downarrow\rangle_1 |\uparrow\rangle_2],$$

for direction $x$, or for example:

$$|\psi\rangle_{F,y} = \frac{1}{\sqrt{2}}[|\rightarrow\rangle_1 |\leftarrow\rangle_2 - |\leftarrow\rangle_1 |\rightarrow\rangle_2],$$

for another direction $y$ (orthogonal to $x$), which means that it is only within that structure that there is any sense in defining the 'quantum states' of each atom as 'parts of a whole'. When we mention in that case the 'quantum properties' (such as spin, momentum or position) of both atoms, as if they should be independent of the state of the whole, we are simply employing the special pleading *post hoc*, founded in that case on the commonsense (classical) ontology of a whole composed of the prior parts being something allegedly necessary for metaphysics.

What then is, once more, the decisive difference between externalistic and internalistic compositionism in the interpretation of the above quantum experiment? The former would habitually assume that independently of the quantum whole, described by the form (geometry) of state $|\psi\rangle_F$, both some nuclear spin state $|\psi\rangle_1$ of atom 1 and some spin state $|\psi\rangle_2$ of atom 2 should be seen as independent (separable) and objective – properties of particular, individual atoms (even if 'hidden') – and only as a result of the arranged physical experiment with the dimer prepared and split do they collectively evince as a global constraint the unique property of 'creating a quantum entangled whole'. An advocate of that view might say that the realization and manifestation of the holistic nature of entanglement is of course possible solely when we have at least two atoms, but the fact remains that each of them should already be in the possession of the property of 'having some value of spin' at every instant. In that sense, the structural-relational property of 'being a part of the whole $|\psi\rangle_F$', revealed in the experiment, is to be *external*, in a way secondary to the primordial property of 'possessing some spin value', and we thus obtain an externalistic compositionism scheme (and externalistic holism). An advocate of that view, true to classical or semi-classical mechanics, wishes to avoid at all costs the situation whereby the physical identity of parts (as given by e.g. spin states) would be strongly, or even totally, dependent on the structure of the whole. The structure of QM, though, makes it difficult to formulate a similar ontology to such an extent that it might not seem too encouraging or even utterly disappointing to a metaphysician (which is illustrated by the example of Bohm's early 'toy' mechanics from the 1950's). The standpoint can therefore be formulated in a compromise or minimalistic version: the particular atoms can indeed not have a property of 'possessing a given spin value $m_F$' independent from the entangled quantum whole (whatever it would be), but they have to contain a number of other individual base properties, guaranteeing their *minimal, non-relational identity* as the first-order individuals in spacetime, regardless of the physical interactions in the course of the quantum experiment.

An opposite standpoint – internalistic compositionism (and holism) – stresses that apart from the given structure of a whole represented strictly locally as $|\psi\rangle_F$, one not only cannot differentiate, but even there is no sense in talking about the existence of *any* 'spin states' $|\psi\rangle_1$ and $|\psi\rangle_2$, and, all the more so, there is no sense in speaking of the 'objects' *being* some base *elements* ('pre-parts'), which can next exhibit in this or that physical situation a holistic state $|\psi\rangle_F$. The key accusation formulated there with regards to externalistic compositionism is that of metaphysical special pleading: why should we at all postulate any non-empirical property of 'having some spin direction' independently of the whole system, since it is the structure of the whole and *solely it* that determines all the possible relations between all the possible values of the spin direction? To state that individual but entangled atoms possess the property of 'having some correlated spin value' apart from the structure of the whole, we have to already assume their certain whole – treat them collectively – to be able to state it at all (see [Harte, 2002a, p. 163]). Internalistic compositionism, in turn, places great stress on the fact that the structure (matrix of relations) is important not only for the system as a whole, but equally so for the definition of the parts themselves, which do not exist outside it: 'structure is built into both the parts and the whole' [ibid., p. 164]. That is such a strong holism[19] that it has usually been excluded in the metaphysics of nature as being unrealistic, contrary to the weak holism based on Aristotle's realistic scheme for familiar 'things'. One of the reasons for that reluctance is doubtlessly the fact that the focus is radically shifted from the concept of the 'substances'–objects forming composites with other objects to the concept of the ontologically primordial relational structure on a global scale[20]. Internalistic compositionism – which is to be found allusively expressed in the *Parmenides* and openly in the *Sophist* and the *Philebus* – is born out of the Platonic intuition of the whole as 'unity over plurality', treats that unity seriously as the most original characteristic of that whole and finally expresses the demand for the philosophy of nature to be construed on the basis of the ontology of relations.

The *Philebus* model allows us to state that every composite which is a whole requires two principles to be met: that of *unlimitedness* (ἄπειϱον) and *limit* (πέϱας), which is a *measure* (μέτϱον), and thus creates a structure. The laser-measured nuclear spin states are not any 'structure', but their concrete values exemplify (actualize) it physically, thus to speak of real 'parts of the whole' coded in a state $|\psi\rangle_F$ requires a matter-energy grounding – *apeiron* – of the information structure. Socrates, when analyzing any composite, presupposes its originality and then cognitively distinguishes both principles regardless of what ἄπειϱον is in that special example. As we have seen, though, what makes that situation exceptional is that ἄπειϱον is a condition of the actualization of concrete physical

19 We may call it without any ambiguity 'strong' as opposed to 'weak' (externalistic).

20 Cf. Rickles, French [2006, p. 25*ff.*] in the context of structuralism as a programme for fundamental physics.

wholes–composites, but in itself has no 'parts', 'form', 'order', nor does it create any primitive 'plurality'. Socrates is thus right when he calls ἄπειϱον by the name of 'τις ἄκϱατος συμπεφοϱημένη' ('some uncombined disaster') and συμφοϱά ('catastrophe'), as indeed, the geometric potentiality (not even 'blend') of all the possible spin eigenstates *of all the possible* space directions for each atom is nothing more than internally contradictory ἄπειϱον πλῆθος, the disordered 'unlimited plurality' from the *Parmenides*. Verity Harte [ibid., p. 189*f.*] rightly points out that that wholly undefined nature of *apeiron*–potentiality can be even characterized only negatively, privatively in its name as such: ἄ-πειϱον, thus 'what lacks a limit–structure'. That results, obviously, from the ontology of internalistic compositionism: if something is not defined in the framework of the structure, it is not defined at all! Hence there is no point in talking of any intrinsic quantum spin states of each atom regardless of the state of the whole system and its relation to some external system(s) – and that is in fact the most striking characteristic of the quantum entanglement. Later on, in Ch. 5, we will see that this formula is true in every instance of local quantum interaction/measurement which creates an entanglement of the arbitrary physical systems.

If we wish to better understand that Platonic teaching from the subtle, dialectical exercises of the *Parmenides* and the *Philebus*, we should now return to the problem of plurality and the division of the whole which appeared in Ch. 1.2. We have established that in order to avoid the naïve absolutization of the concept of 'state' we should respect the supposition that in the case of that entangled quantum system, global 'state' $|\psi\rangle_F$ is an 'element' of our description of the *local physical situation*. That means that the division operation generating that element is possible to be envisaged only as an operation of locally defining its structure (and *positions*[21] within the structure) but it has no direct physical sense, as no observation will ever reveal that global structure the way a definite position or an atomic spin in a chosen direction is registered. The structure of the whole $|\psi\rangle_F$ is purely *relational*, i.e., it determines the structural positions of the parts as their mutual relations, without which those very positions (and the parts of wholes) are impossible to be conceived. In his ontology of the whole, Plato postulates that it is structure and only it (and not some 'pure matter' of phenomena, ἄπειϱον) that is the subject of every study/science (σοφία): for instance, dynamics does not study the 'essence' of force and energy as such, but the

21 The term 'position' is used here in a technical sense, in the way it is used by structuralists, particularly in the philosophy of mathematics (cf. Harte [2002a, p. 175], Resnick [1981; 1982] or Shapiro [2000]).

*structure* of dynamical phenomena (in which the physicist refers to the concept of 'energy')[22]. As the differentiation of the parts of the whole takes place only in the framework of those structures, hence – as has already been mentioned in Ch. 1.2 – that is an 'internal' operation setting up proper *measures* (μέτρα), that is, introducing symmetries and correlations 'measured by numbers' (*Phil.* 17d5: δι' ἀριθμῶν μετρηθέντα). The complete structure itself, though, remains hidden as the ground of every local physical action: it is a hidden *logos* of the serial phenomena, which is revealed in the course of the experiment. That very property seems most interesting here, as – as it turns out – it not only deepens our understanding of 'empirical wholes' from Plato's examples, but also leads us in the direction which Plato merely alludes to in the *Parmenides*, although – as I shall claim in Part II – it is of fundamental importance to the understanding of the structure of QM. The problem, as far as I am aware, was for the first time undertaken philosophically in a consistent way, in the context of microphysics and the Platonic tradition, by Carl F. von Weizsäcker [1982].

To capture that Platonic ontological intuition we shall recur once again to the original, Platonic terminology – with reference to our example of quantum entanglement it not only turns out not to be archaic, but interestingly adequate. Although it might at first seem rather speculative, the pedantry and depth of Platonic dialectics nevertheless suggest a fascinating central idea, brought to light by von Weizsäcker. In accordance with the logic of internalistic compositionism, we can employ the same structuralist-metaphysical thinking not only to the parts of an isolated whole, but also to the distinct whole defined locally for some different system – an 'element' of the local physical situation in a quite Bohrian manner. What happens then when that whole–unity described by 'state' $|\psi\rangle_F$ is considered in relation to another whole, represented by another internal 'state'? What ontological dialectics does such a relation set off? In Plato's model, presented in Hypothesis II, the whole – as we have seen in Ch. 1.1 – undergoes a 'fragmentation' once again as soon as it is related to anything 'other'. One becomes 'disaggregated'–'chopped' (κεκερματισμένον) into parts, and thus to a certain degree 'destroyed' (on the condition that it 'exists' in a constant interaction with 'the Other'). That active transition, the disaggregation of the whole as a self-contained unity, is in the context of the quantum theory called a local 'updating of the information' (interactional 'state' actualization), i.e., in fact a local transfer

22 See M. Heller [1986; 1988] or D. Rickles and S. French [2006] concerning the physical structuralism which corresponds to the Platonic definition of internalistic compositionism being analyzed here.

of information, which can be understood to be another relation internal to some whole (see Ch. 4 and 5). Plato's model suggests that one cannot experience unity without having destroyed it as the result of the interaction, which makes 'the One' cease to be 'something one' any more. Hence the wholeness of a phenomenon, which we discover at the fundamental physical level through the correlations by measurements, is also metaphysically the lowest level of the whole, as it is only being uncovered by the experimenter in the series of scrupulously registered phenomena and a local comparison of the available information about the systems. The holistic nature of those quantum processes is fully revealed only when that effect of the fragmentation of the whole, an effect characteristic for the sphere of physical phenomena, is turned round in the course of a cognitive scientific process (ἐπιστήμη). A higher level of 'authentic wholes' is empirically inaccessible, and thus it is always 'hidden', manifested solely in the organization forms of the phenomena themselves (e.g. in series of quantum events). Precisely in that very sense Plato's model is a critical development of Heraclitus' theory of *logos:* 'Nature loves to hide' (B 123) and 'Invisible harmony is stronger than the visible one' (B 54). That central intuition of Plato, the 'philocryptology of nature' and internalistic logic, which I shall return to in a further part of this book, has been exposed in a cosmological context by C.F. von Weizsäcker thus:

> Let us go back to the quantum theory once more. The way in which an object first conceived as fully isolated can nevertheless be an object, and thus be properly – is its interaction with other objects. Yet, it for that very reason ceases to be exactly that and even to be *a single* object. One can say paradoxically: any property of some object becomes observable only by nature of the fact that the object comes to lose that very property. Classical physics or the classical ontology on which physics is based constitute an approximation with which we can abstract from that loss. But it is only with a classical approximation that we can make and express observations. In that sense, all physics is based in some important way on an approximation. In an individual case, that very approximation can every time be physically described and in that way corrected – only when, though, we shall make use of that in another place.
>
> Bohr described the situation with the help of the concept of complementarity. This has often been seen as his resignation on account of the incomprehensible empirical difficulties with measurement and, as a consequence of Bohr's application of the concept to other disciplines, was treated as an unjustified hypothetical generalization of one problem in physics. Yet, already in Plato's *Parmenides* we can discover a clue to the ground of complementarity. It is classical ontology that does not in fact come up to the level of Parmenides's reflection (both the old Eleate and Plato's Parmenides), it does not comprehend the fact that its own application assumes its own falsehood. The Universe can itself *be* only to the extent that it is not one, but many. All that plurality, though, does not exist for itself, as logic and classical ontology describe. It exists only in an unconceivable

> One. (…) The One in itself does not constitute a principium: when it *is*, it is *many*, but at the price of contradiction. [von Weizsäcker, 1982, p. 490*f.*]

Von Weizsäcker took up the Platonic trail in a fascinating manner when he reflected on the dialectics of the 'hidden whole' and its destruction (καθαίρεσις), disaggregation–split–division (διαφορά, στάσις) in empirical experience. In one of his later works devoted to the phenomenon of time, he also proposed a modification of Goethe's famous statement [1979, p. 45] from his *Farbenlehre* – calling colours 'the deeds and sufferings of light' ('*Taten und Leiden des Lichtes*') – as he poetically wrote that isolated parts, which come to light as 'objects' and 'states of affairs', can be called 'deeds and sufferings of the whole' ('*Taten und Leiden des Ganzes*') [von Weizsäcker, 1992, p. 985]. Just as to Goethe colours are not components which white light is 'composed of', so according to Weizsäcker single quantum states or objects are no 'parts' from which any quantum whole could in any summational way be 'composed of'. Weizsäcker writes about the 'loss of the property by an object' and the intrinsic contradiction impossible to eliminate from quantum cosmology, but – as often happens in such cases – he seems not to take into regard the radical consequences of internalistic ontology and limits himself to stating the paradox. The most basic problem which appears in the context of QM is nevertheless, as evident from Ch. 1.2, the issue of the physical *relativity* of the 'whole' treated here, since the structure–element represented locally by $|\psi\rangle_F$ is simply an abstract piece of information one physical system has about another system inside the Universe. Despite that, a purely subjectivistic approach to that piece of information is inadequate: the quantum correlations between systems and the quantum entropy have a strictly physical sense, which is most aptly illustrated by quantum thermodynamics. Internalistic compositionism applied consistently in fundamental physics requires one authentic single whole to exist in the end – and that is simply the Universe, **Ω**. That means that – as opposed to externalistic compositionism, which can always work beginning with independent individuals, such as cats, stones and cells as the basis of metaphysics – the former always requires a *cosmological context* to be primary. The question about the 'authentic whole', the 'Platonic whole', is by necessity a cosmological question and, I think, the problem of quantum entanglement and quantum nonseparability is insoluble in the framework of quasi-classical approaches based on externalistic compositionisms by nature of their programme abstracting from the most general cosmological assumptions. That is perhaps a situation where physics itself would require to be remodelled and the metaphysical imagination to be radicalized.

In line with Strabo's testimony, Plato called philosophy–science 'music'/'art', while Socrates, as we know from the *Phaedo*, at a call from a dream, 'create music and practice it!' (*Ph.* 60e6–7: μουσικὴν ποίει καὶ ἐργάζου), treats philosophy itself as the 'ultimate music/art' (*Ph.* 61a3–4: μεγίστη μουσική). The term *megiste mousike*, 'ultimate music', may denote not only a cognitive process (*episteme*), which reveals those hidden structures, harmonic order–measure 'through' and 'behind' the natural phenomena, but first of all – in a truly Pythagorean sense – that very order itself. In the *Timaeus* (35a–36d), Plato presented the vision of a harmonics lying at the root of the universe and, though that outruns any narrow understanding of 'music', it nevertheless intrigues even today with its consistency and beauty (see M. Wesoły's reconstruction [1993] and V. Harte's comments [2002a, p. 212*ff.*] as well as M. Heller [1995]). In view of what has already been said about Platonic internalistic compositionism and the ontology of the hidden structure, speaking of the *megiste mousike* of nature unfolds some nontrivial sense: it is an issue of an internal, global harmony (ἁρμονία) of all processes, which requires us to stir up a physical imagination different from the one we have become accustomed to. Such attempts have been extremely rare in the philosophy of nature, hence it might be worthwhile to look more closely at one of the historical examples of such a mature project.

# Chapter 2
# 'The Adaptation of All Things': Leibniz's Series and the Foundations of Microphysics

## 2.1 *Mathesis quaedam Divina*: relational states, the Connection and a mathematizable process in the Leibnizian system

David Finkelstein, one of the most original theoretical physicists in the second half of the 20th century, described his quest to understand the nature of physical laws and the sense of QM during a conference devoted to Whiteheadian inspirations in physics [Finkelstein, 1993]. What seems interesting in that account is the idea that the popular distinction between dynamics and kinematics, and even the attempts at hierarchizing them are nothing other than an artefact, a byproduct of a more basic, erroneous habitual way of thinking about the spatiotemporal split as a separate concept of time and space. It would be quite justified to call a theory in which kinematics would be tantamount to dynamics wholly 'processual', but the question remains what its mathematical formalism would have to look like for it to meet that criterion. Finkelstein stressed that the beginning of that processual revolution in physics was QM itself, which becomes all the more apparent when we appreciate the consistency with which Bohr, Pauli or, first of all, Heisenberg formulated their understanding of the microphysical process. The Heisenberg algebra of observables is just such a processual language, whereas, as Finkelstein notes, the very controversial term 'non-objective physics' used by Heisenberg was probably modelled on the phrase 'non-objective art' (*nicht-objektive Kunst*) coined by Wasilij Kandinsky (1866–1944) ([Finkelstein, 1993, p. 2; 1999, p. 450])[23]. That remark is not just a loose analogy, as it was also in art where Kandinsky's abstractionist revolution of the 1920's bore fruit in antirealistic and antirepresentationist contemporary art, starting with cubism, through Pollock's abstractionist expressionism right up to the process art of the 1970's and to performance art. The changes which brought about such

23 Perhaps 'non-objectival' might indeed be better suited, but the sense would be the same: Heisenberg's approach goes very far, right up to the conclusion that 'there are no external, independent objects'. The approach was at the time realized in art also by the great Dutch artist Piet Mondrian (1872–1944) in his so-called compositions.

varied phenomena in Western culture had different sources, but they all led to singular transformations in postmodern consciousness with its various, fluctuative characteristics. Whitehead, and Finkelstein in his wake, can be perceived to have been consistent supporters of similarly radical changes in understanding nature: in their view, the Newtonian vision of the world of objects has been undermined much more deeply than it is pointed out to in the popular comments on quantum physics. And still that programme has not been fully realized yet and it is already certain that standard formalisms and a mere interpretation of the models are not sufficient for physics to forego 'objects'. We could ask therefore, in the manner of Finkelstein: is it possible for our understanding of nature to reach beyond the phenomenological distinction between space and time so that the dynamics itself become fully relational?

If we wish to draw closer to the problem, we should look into Leibniz's 17th century metaphysics to find an outline of one of the most fascinating modern conceptions of nature. The ontology (see e.g.: [Woolhouse, 1994], [Rescher, 1991], [Gloy, 1996, pp. 37–70], [Futch, 2008]) is much more radical than a cursory survey suggests and has at least two different levels, with the best-known one, that of the 'monad theory', being most external and even secondary, also in the manner of its formulation. There obviously has long existed an awareness that a relational physical theory should be construed on the Leibnizian model, but its association with 'monadology', 'pre-established harmony' and philosophical theology has discouraged some theoreticians, or at least it has raised doubts with regards to its extremely speculative nature. Leibniz can nevertheless be perceived to be one of the greatest metaphysicians of the process in modern philosophy, as it is the category of the immanent 'action' (together with the role of relation and representation) which constitutes a central characteristic of his ontology [Rescher, 1996, p. 12*ff.*]. In his short discourse on the fundamental metaphysical principles, *On the Final Source of All Things* (1697), Leibniz analyzes the mutual reference of the metaphysical necessity and the physical necessity governing the cosmos. The latter is presented as operating at a lower level than the former: Leibniz's cosmos acts according to the law that the greatest possible physical effect (and the greatest diversity) is attained with the smallest possible input (work), 'just as if dice used in a game were to be placed in such a way that their greatest possible number would fit in a given field' [Leibniz, 1890a, p. 304]. The law is a physical manifestation of the great metaphysical principle, with the aid of which the Absolute designs and selects such and such a universe as actual, and the operation in that context is called by Leibniz, incidentally, 'some Divine Mathematics and thus a Metaphysical Mechanism' (*Mathesis qaedam Divina seu Mechanismus Metaphysicus*) [ibid.]. The reader does not learn from the text what exactly that

*Mathesis* is, but we can find the answer in another field of his intensive studies: mathematics (cf. e.g. [Busche, 1997]). That hidden layer of Leibniz's thought was uncovered several decades ago by his commentator, Leroy E. Loemker [Loemker, 1969]. The main categories Leibniz's thinking revolves around are the concepts of 'function'/'representation', for – as Loemker has aptly pointed out – they most fully express the radically novel character of his metaphysics.

Although the concept of 'function' is fundamental to the whole of mathematics, it was known in antiquity only in the context of concrete applications as 'tables', mainly in arithmetic, trigonometry and astronomy. An awareness of the abstract nature of a function came only in the late medieval period as the result of the work of theoreticians of the philosophy of nature from the great metaphysical and natural science schools in Paris and Oxford (the so-called *calculatores*), which were the first to reflect on the 'logic of the laws of nature' (the fact was stressed among others by the Russian historian, Adolf P. Youschkevitch [1976/77]). It was noticed that the origin of the geometrical concept of function could already be discovered, for example, in the works of Nicole Oresme (c. 1323–1382), translator and commentator of Aristotle, who had debated the correspondences of quantities taking into account the division into variables today termed as 'dependent' and 'independent'. In its fully developed, formal shape, the concept appeared only in the mid-18th century together with the publication of the famous, revolutionary *Introductio in analysin infinitorum*[24] by Leonhard Euler, the basis of contemporary mathematical analysis, in which it plays a central role – its first chapter already bears the title *De Functionibus in genere* (see e.g. [Lützen, 1983], [Thiele, 2007]). It has rightly been observed that that moment marks a revolution in mathematics not only because the function in its new notation $f(x)$ becomes a basal object of analysis, but also because its geometrical understanding is set aside into the background – a hundred years after Descartes – to give way to a purely algebraic formulation (cf. [Fraser, 1989])[25]. Euler perpetuated that algebraic approach applied to the concept of a series in his monumental textbook *Institutiones calculi differentialis cum eius usu in analysi finitorum ac doctrina serierum* (*The Foundations of Differential Calculus with its Application in the Analysis of Finite Quantities, and the Theory of Series*, Saint Petersburg 1755), in which he gave vent

24 *Introductio in analysin infinitorum auctore Leonhardo Eulero*, published by Marcus-Michael Bousquet in 2 volumes, Lausanne 1748; also in Teubner's edition: L. Euler, *Opera Omnia*, ser. I, vol. 8–9, Leipzig 1922 and Zürich 1945.

25 For example, in the *Introductio*... Euler completely replaced the old, geometric definitions of conic curves (by sections) with a purely algebraic analysis, thanks to which – by generalization – he also obtained a classification of quadrics.

to the ideas he had adopted from his great teacher, Johann Bernoulli, and made them flourish. The most visible manifestation and at the same time an issue that interests us most here is his remark in the *Institutiones...* that what is essentially the object of the analysis are not the 'vanishing increments' of functions, and hence not some mysterious infinite quantities, which had worried commentating philosophers so much (among whom bishop George Berkeley was also to be found), but the geometric ratios of 'vanishing increments', i.e., *correspondence* or mutual *dependence* (relations) between quantities, which is finite. In the end, though, as we know, in that very same work, he stands by Newton's side and allots infinetely small differentials of functions $dx = p/\infty = 0$, according to the 'law of continuity', a full reality as an *actual* vanishing (although Leibniz's opinion, founded on the very same law, was exactly reverse, treating them as fictions, *fictiones mentis* [Leibniz, 1879b, p. 305]). Nevertheless, the impetus of Euler's work is derived not from Newton's geometrical curve calculus, but from Leibniz's algebraic approach, although Euler, as a fervent, orthodox Christian and participant of theological debates, remained an opponent of 'freethinkers': not only materialists and spiritualists, but also 'monadists' – Leibniz and Wolff. Putting it more precisely, Leibniz used the term 'function' also in a geometric way, treating it still at the beginning of the 1690's as construct associated with curve points treated as primary ones [Thiele, 2007, p. 69], which nevertheless is not as pronounced as Newton's geometrico-physical picture, Newton being a perceptive reader of his teacher, Isaac Barrow's *Lectiones geometricae* (London 1670). Newton himself wrote of *fluents* which spatially moved ('flowed') in *fluxions* along the curves, and not of functions. (According to the original definition from *De quadratura curvarum* of 1704, he understood fluxions–derivatives a bit vaguely to be 'augments of their fluents begotten in the very smallest equal particles of time' [Newton, 1981, p. 123*f.*], and hence speeds of those fluent movements generating lines, planes and cubes – that is a kinematic perspective, '*per motum continuum*', for as Newton himself stresses, mathematical values are created in 'continuous motion'). As the abstract function seemed to be irrelevent in Newton's picture, it seems justified to seek the proper source of that mathematical idea in the work of Leibniz and Bernoulli, and in their vast correspondence (1694–1698). The general concept not only of time-dependent 'function' as such, but also that of 'parameter', 'variable' and 'constant', are to be derived from Leibniz: he had used the first of the concepts for the first time in writing in August, 1673, in his work *Methodus tangentium inversa, seu de functionibus*, treating function as a certain definite relation, *relatio*, and that laid down the foundation for his later *Nova methodus pro maximis et minimis*, 1684. That was the key step to defining 'function' as any 'analytical expression' (*expressio analytica*), which Euler had introduced in his *Institutiones...*

Leibniz in a natural way arrived at a new, extra-geometrical understanding of the meaning of function on the basis of his metaphysics of the universal Union of beings (*Verhängnis*), as the mathematical function, which he had already analyzed in the field of geometry, was to him a *singular instance* of a deep natural order (*ordo*) responsible for the regularities studied by physics (for an excellent historical analysis see especially [Busche, 1997]). In the monad theory, there is a need to ascertain an eternal, dynamic compatibility between the particular original substances, a harmony in their inner transformations ('states'), and that compatibility is guaranteed by the Absolute, which harmonizes those states – not like a clockmaker, but like a creator of the particular monads ('ideas'), whose content is eternally established (cf. [Mercer, 2004, p. 231*ff.*]). Leibniz expresses that in the most open formula in §XXXII of his *Metaphysical Treatise*:

> Similarly only God creates a connection or mutual communication between substances (*la liaison ou la communication des substances*), and only thanks to Him do phenomena proper of some substances converge and conform with the phenomena proper to other ones, and hence our perceptions contain something real (*et par consequent qi'il y a de la realité dans nos perceptions*). [Leibniz, 1880a, p. 458]

The relation of any given number of quantities, which we put down as function $f(x_1, x_2, \ldots, x_n) = y$, was to Leibniz the proper sense of the concepts of 'order' and 'law', as the function is an orderly relation defined on certain sets[26]. And hence a consequence of his metaphysical relationism is that even when he used the term 'substance', he always concentrated his attention on the relations between the 'states' described symbolically by the variables, with the symbol of the function as such referring to the *law* (*Gesetz*) related to those states – which Loemker had placed at the centre of his reconstruction. Hence, when Leibniz writes of the 'law' ascribed to each of the infinite number of monads (defining it), or rather of the law *being* it, what he has in mind is the set of all the relations to other 'laws' (:monads), with completeness implying that those relations exhaust all of that unique law, and thus the *essence* of substance. It is clear that there is in fact no full definition of 'substance in general' in Leibniz (e.g., as a certain type of being that is unchanging in time, forming the basis of properties etc.), and when we ask: 'what is the definition of a given original substance $x$?', the answer cannot be of the type: 'it belongs to substances of the $T$-type' or that e.g. 'its most important properties are: $a_1$, $a_2$,..., $a_n$', as the definition (nature) of a given substance, defined on set $R^\infty$ of all the functional relations, is wholly *unique*. Having

26 Of course, the symbol $f(x)$ was first used only by Euler in *De infinitis curvis* (1740); Bernoulli had earlier used the notation $fx$ [Thiele, 2007, p. 75].

underlined that radical Leibnizian relationism, we are able to understand why, despite his not knowing the purely analytical definition of mathematical function, it was 'function' as such, in the contemporary understanding of the term, which as 'the law of representation' constituted the main category of his thinking about reality. The metaphysically interpreted function defines the order of elementary states – it is also, as he writes in a letter to Burchard de Volder on the 21st of January, 1704, 'the law of constant succesion', '*lex ipsa quae involvit continuatam successionem*' [Leibniz, 1879a, p. 263].

It is obvious that when we assume that, we undertake a certain reconstruction and impose a certain interpretation of that Leibnizian concept which historically was available much later, and hence we cannot be certain if a more general mathematical object would not have turned out to be more suited to the needs of that metaphysical project, that is distribution or *category*, where both objects and morphisms can be any abstract entities, which only meet certain conditions, such as that of the associativity of composition. The condition for the categories to be apropriately 'regular' can be met by limiting ourselves to the choice of appropriate toposes as quasi-classical categories for physics, retaining at the same time Leibniz's relational paradigm and its built-in postulated local or partial truth logic (cf. [Butterfield, 2002]). The chief theoretical motivation for the introduction of toposes into quantum cosmology may in fact be the programme of a radical relativization of the quantum theory, that is an attempt at a full realization of Leibniz's relationism, initiated by that very crystallization of the metaphysical meaning of 'function'. Of course all that greatly exceeds what probably might have been accepted by Leibniz, as there exist toposes in which the axiom of choice is not met, or others where the Law of the Excluded Middle does not hold, or there exist no objects inside them defined as natural numbers. In the second case, the abandonment of the principle '$p \vee \sim p$', and thus the use of a non-Boolean topos makes it possible to deal with quantities that are infinitely small in such a way as if they were 'normal' (but infinitesimal) objects, without contradiction (Newton and Euler would definitely be satisfied with that)[27], which is the basis for the construction of synthetic differential geometry ([Kock, 1981], [Lavendhomme, 1996]). All that seems to enable the use of certain nonstandard toposes in the modelling of spacetime, as they allow us to abandon points and sets in such a way that the concept of regions not composed of spacetime points, satisfying the axioms of non-Boolean algebra, becomes primary. We can also

27 Other mathematicians followed in Euler's wake with regards to that, such as H. Weyl and Ch.S. Peirce.

choose the category of von Neumann algebras as a fundamental structure in a natural manner and attempt to formulate all of QM in the categorical language (cf. the proposals by C. Isham and J. Butterfield [1998]). Although one could raise some objections as to whether all of physics could be built on such general structures (Penrose [2004, p. 960] writes of the 'deceptive power' of the category theory), nevertheless it is interesting to historically trace how the originally almost geometrical concept of the function (connected to study of the curves) undergoes a radical algebraization in the realm of the analysis, in the end to attain a maximally general form in the field where topology, algebra and logic overlap. Leibniz, when seeking the most general, fundamental *mathesis universalis* for his theory of nature, dreamed of such types of languages, characterized by a maximum of logical simplicity and constructive power.

When we consider how it was possible for Leibniz to treat a mathematical function as a singular case of a precisely defined metaphysical law, we finally get down to the problem of his way of conceiving substances which after all, if we were to analyze them in a classical, Aristotelian way as the substrate of properties, would be archaic or eclectic. Nevertheless, as Loemker or Rescher have noticed, it is rather the case of a clash of two great traditions of perceiving substances, of which the Aristotelian one was merely conventional, superficial, assumed by him as part of his invocation to the philosophical tradition, while the second, radical and relational-processual, proved the depth of his metaphysics and was a completely unique attempt at construing a model of the hidden whole of nature. We can put forward the hypothesis that the latter one (being logically primary) inspired Leibniz's studies on infinitesimal values, differentials and integrals. What brings us closer to that relational-processual model of the world is Leibniz's later concept of *series*. (We are using the metaphysical concept of 'series' here, and not the mathematical notions of series or more generally sequence, to underline its more primitive, ontological sense). The Law of Series defines substance '... *cujus natura in perpetua quadam ejusdem seriei mutationum lege consistit, quam inoffenso pede decurrit*'[28] [Leibniz, 1879a, p. 171]. We find a short presentation of that model in *Monadology*, intertwined with reflections on the metaphysics of substance:

> §10. I also consider it to be an ascertained fact that any created entity undergoes changes, and hence that also includes a created monad, and what is more, in each of them that change is continuous (*changement est continuel*).

28 '... Whose nature is a certain universal law of the series of changes through which it runs without any disruptions'.

§11. From what we have said the conclusion is to be drawn that natural changes in monads are derived from some inner principle (*d'un principe interne*), since no external cause can exert an influence on their interior.

§12. It is nevertheless indispensable that there be, apart from the principle of change, some particular singularity of that which changes (*un detail de ce qui change* = a unique *series*), which – so to speak – would cause the specification (*la specification*) and variety (*la varieté*) of simple substances.

§13. That singularity ('detail' = *series*) has to encompass a multitude in unity (*ce detail doit envelopper une multitude dans l'unité*), and hence in something simple. For since every natural change takes place gradually (*par degrés*), something changes and something remains, hence a simple substance has to contain a multitude of modifications and relations (*pluralité d'affections et de rapports*), despite not containing any parts (*compl. by M.W.*). [Leibniz, 1885a, p. 608]

The fundamental process takes place in Leibniz's model on the lowest level possible (it is a principle of the inner change, *principium mutationis intrinseca* [Leibniz, 1879a, p. 258]), totally independently of spacetime relations and hence it is in no way merely a macroscopic phenomenon. Every observable physical change is the global result of microchanges in base objects: '*omnes res singulares sunt successivae seu successioni obnoxiae*' [ibid., p. 263]. But when we enquire what those base 'microobjects' are, Leibniz surprisingly says that they are indeed a mere processual unity of a great number of relations with other objects and we cannot speak of any properties of those 'objects' other than relational ones, hence the medieval concept of the individuality of substances, *haecceitas*, here gains an exceptionally different meaning as a coherent stream of elementary actions–information. When Leibniz speaks of 'a plurality of modifications' (*pluralité d'affections*) and of 'a temporary state, containing and representing multitude in unity' (*l'etat passager qui enveloppe et represente une multitude dans l'unite*), he is directly using the language of r e l a t i o n a l  s t a t e s and their modifications (*modificationes transitoriae*), which create *series* as some global, pre-empirical singularity (= *haecceitas*). Every unique 'law' (:a global series) is a unique substance itself: '*Legem quandam esse persistentem, id quae involvat futuros ejus quod ut idem concipimus status, id ipsum est quod substantiam eandem constituere dico*' [Leibniz, 1879a, p. 264]. Thus when he writes of 'monads', the fundamental entities of the world, he always bears in mind ordered series of states, which are metaphysically integrated with each other so much that to speak of a separate 'own state' of such a substance would only be a certain linguistic convention leaving aside the global harmonization of the state relations. The monad, as a 'centre of power', is in fact a stictly ordered 'stream of activity' or 'pure process' [Rescher, 1996, p. 12]. For Leibniz, the fabric of the world is not a net of things but pure, structured forces. One would have to answer the question about what a given substance is by giving a full sequence of its relational states or its 'force' (a 'series' = 'form' as a 'particularity in multitude' = a function with infinitely many arguments), that is, mathematically speaking, to write down all of its corresponding infinite series of numbers (for those series are to Leibniz mathematical abstracts of truly existing series of states in nature).

When we speak of such a Leibnizian 'state' (*etat*), though, we have to stress two of its particular properties. The first refers to the philosopher's theories that every state is at the same time a perception (*perception*), as Leibniz's Universe is a universe of a hierarchy of degrees of consciousness – microstates of the world are at the same time microstates of consciousness (so-called *petites*), of which each is a consequence of the preceding one, *ex praecedentibus derivantur* [Leibniz, 1879b, p. 372]. On its most fundamental level, nature is a great procession of perception, only those microperceptions are, as Leibniz writes, unconscious [1885a, p. 609, §14], and hence human perceptions are simply their singular form, a synthesis of a great number of unconscious perceptions – two centuries later the idea was resurrected in a similar form by A.N. Whitehead. We can comment on this Leibnizian metaphysical scheme and agree that such a type of idea is an interesting radicalization of the idea of relationism understood in a strictly ontological (and not only formal or logical) sense: each information belonging to a 'series of actions' – and thus being an element of a process of 'real production' – is exactly, as that elementary 'perception', purely relational, which evokes a seeming contradiction to the commonsense image of substance, which would be supposed to contain some non-relational 'core', independent of the rest of the universe. The second property, already mentioned, is linked to the thesis that each state is a unique expression (*expressio/exprimare*) of all the other states, which gave rise to Leibniz's famous comparison of the monad to a looking-glass mirroring the Universe, '*miroir de l'univers à sa mode*' [ibid., p. 618, §63]. There is nothing to find, in fact, in perception apart from relations (cf. e.g. [Barbour, 2003, p. 49]), which the philosopher vehemently expresses:

> Further, every substance is in a way a whole world and is a sort of God's mirror or that of the whole Universe, which it expresses in its own way (*qu'elle exprime chacune à sa façon*), just as the same city is seen in different ways depending on the place it is being viewed from. In the same manner, the universe is multiplied as many times as there are existing substances (*l'univers est en quelque façon multiplié autant de fois qu'il y a de substances*), and God's glory is magnified by that number of different representations (*representations*) of God's work. [Leibniz, 1880a, p. 434]

Leibniz's universe is in that analogy *a mirror universe*, in which all processes are their own mutual reflections, while through the infinitesimal transformations of each function (definition of substance) we could obtain a current description of the state of each different monad. 'Internal operations' or the streams of internal monadic activities, are in truth correlates of the operations of all the others and hence the essence of each monad is relational: '... *consequens est omnis substantiae hanc esse naturam, ut vi sua agendi aut patiendi, hoc est serie suarum operationum immanentium exprimat totum universum*'[29] [Leibniz, 1890c, p. 316*f.*]. Leibniz's Principle of the Series of States puts down that each base substance in the universe is exhaustively defined by a certain infinite series (function) of its states, strictly determined by the infinite number of series of the states of other 'substances'. The Law of Universal (Pre-established) Harmony in nature in turn narrows down the previous thesis to the theory that all such series are precisely tuned, which formally implies that the functions defining all (past and future) microstates in nature are mathematically harmonized with each other – it is the 'Connection' (*Liaison*), 'an adaptation of all things to each and of each to all' ('*accommodement de toutes les choses creées à chacune et de chacune à toutes les autres*' [Leibniz, 1885a, p. 616]), which constitutes the *deepest layer of ontology*, as the *haaeceitas* of a given series is completely dependent on it (cf. e.g. [Rescher, 1981, p. 70*ff.*]). The only way of thinking that would be adequate in such a situation is the radical holistic approach, in which each process, regardless of how cognitively separate it is from the other processes on account of idealization or cognitive limitations, is immanently global and pre-temporal. Leibniz seems to suggest that to perceive the whole, hidden harmony of the world, one need not envisage a broader and broader panorama – it is sufficient to look inside, into the activity of the smallest elements construing nature, as every *state* as 'perception' contains an infinite depth ('perspective') of reflecting the whole:

> §61. ... For since everything is filled – and that also binds all of matter – and since every movement fully exerts an effect on distant entities depending on the distance, so that every body is affected not only by those bodies which touch it and it feels in some way everything not only that they themselves experience but also through their mediation it experiences what those that touch the former and which then touch the latter directly, which all means that a mutual communication extends over every distance. And thus every body feels everything that takes place in the universe (*tout corps se ressent de tout ce qui se fait dans l'univers*), and to such a degree that the one who sees all could read in each (state) everything that takes place everywhere, and even what had taken place or

29 '... What results from that is that the nature of each substance is such that *through the force of its actions or affections*, that is *through a series of its internal operations* it expresses the whole universe' (*emphasis – M.W.*). The dynamical law and the relational law of the series (of the succession of states) are here strictly bound in a way typical of Leibniz.

would in the future, seeing in the present that what is distant both in time and in space – σὐμπνοια πάντα (*everything is intimately connected*), Hippocrates used to say. Yet the consciousness can only perceive what is clearly represented in it, as it cannot uncoil everything it contains at once (*tout d'un coup ses replis*), as those coils tend to be infinite (*car ils vont à l'infini*). [Leibniz, 1885a, p. 617]

The above statement is extremely interesting as Leibniz treats the process of cognition as an 'unfolding of coils' of the spirit, a revealing of successive layers of the order of things, using that metaphor in a direct analogy to the 'uncoiling' (*explicatio*) of the whole cosmos in Neoplatonic philosophy (in that of Nicholas of Cusa's in particular). There can be no 'gaps' in the order of the world or any ontological elements ontically independent in the strong sense. Similarly, no given observer in such a cosmos is privileged, as every substance, even the most elementary one (a micro-process as a whole) is in fact another observer who, as Leibniz beautifully writes, posseses its own, unique *scenography*, that is it is at every moment an ontological system of reference, a *point de rapport*. Let us quote at this point a fascinating rough copy of a letter to des Bosses (1712), in its Latin original, in which – as far as I am aware – he outlines for the first time in history the Universal Principle of Relativity (in the *cosmological* sense) with a relativization of the finite, physical observer:

Si corpora sunt phaenomena et ex nostris apparentiis aestimantur, non erunt realia, quia aliter aliis appareant. Itaque realitas corporum, spatii, motus, temporis videtur consistere in eo ut sint phaenomena Dei, seu objectum scientiae visionis. Et inter corporum apparitionem erga nos et apparitionem erga Deum discrimen est quodammodo, quod inter scenographiam et ichnographiam. Sunt enim scenographiae diversae pro spectatoris situ, ichnographia seu geometrica repraesentatio unica est; nempe Deus exacte res videt quales sunt secundum Geometricam veritatem, quanquam idem etiam scit, quomodo quaeque res cuique alteri appareat, et ita omnes alias apparentias in se continet eminenter[30]. [Leibniz, 1879b, p. 438]

30 'If bodies are phenomena and they are conceived from the perspective of our perceptions, then they are not real, as they present themselves differently to different observers. Therefore the reality of bodies, space, motion and time consists in the fact that they are phenomena to God, or objects of the science of perception (*scientia visionis*). And the difference between our way of seeing bodies and the way God sees them is the same as that between scenography (*scenographia*) and ichnography (*ichnographia*). From the point of view of the observer there exist different scenographies, nevertheless a certain ichnography, or geometric representation, is only one and unique; for God perceives things exactly as they are in their geometric truth, although He also knows in what way each entity appears to each other entity, and hence how all the other phenomena are eminently contained in it'.

It becomes clear in the light of the above remarks that Leibniz's central programme was to bind, in the framework of a consistent metaphysics, a processual approach to nature with a rather radical version of relationism, constituting the motivation to seek an appropriately general mathematical formalism on the border between logic, algebra and geometry. Hence, it is not surprising that Leibniz, when embarking to penetrate his understanding of 'nature's harmony' with his characteristic persistence, immediately passes on to the most fundamental concepts in mathematics, which he interprets in the metaphysical context (including those which will much later be developed into a basis of whole areas of mathematics). '*Sequiturque in universum, Mundum esse κόσμον, plenum ornatus, seu ita factum ut maxime satisfaciat intelligenti*'[31] – he wrote in a cryptic, Platonic summing up of his metaphysics [Leibniz, 1890b, p. 290] and that is in a sense an elegant tribute paid to Plato himself. In that presentation, the relation between the world of physical phenomena and the cognitive mind points to one of the deepest properties of nature itself, as that relation is a singular instance of a wider phenomenon of representation (expression) of one process by another, which is made manifest in physics itself as a relational nature of all the measured quantities and the mathematizability of the whole process. '*Une chose exprime une autre (...) lorsqu'il y a un rapport constant et reglé entre ce qui se peut dire de l'une et de l'autre. C'est ainsi qu'une projection de perspective exprime son geometral. L'expression est commune à toutes les formes, et c'est un genre dont la perception naturelle, le sentiment animal, et la connoissance intellectuelle sont des especes*'[32] [Leibniz, 1879c, p. 112]. The relation of 'expression' has in the psycho-physical sphere its realization in the 'mutual feeling' (συμπάθεια) of the body by the mind, but in a more general sense it refers to all of the Universe:

> For such expression takes place everywhere, as all substances sympathize with all the other ones (*toutes les substances sympathisent avec toutes les autres*) and undergo a proportional change responding even to the smallest change (*répondant au moindre changement*) taking place in all of the universe, though that change may be more or less noticeable depending on whether those other bodies and their actions will have a greater

31 'And hence that leads to a general conclusion that the world is a cosmos (the ordered world), fully decorated, and thus created in such a way so as to bring the greatest satisfaction to a rational being'.

32 'For (...) one thing expresses another when there exists a constant and regular reference between what can be said about the former and what can be said about the latter. In that way a perspectival transformation expresses a given geometric figure. Expression is common to all forms and is a kind whose species are natural perception, animal feeling and intellectual cognition'.

> or lesser reference to our own. (…) For all the movements of our bodies have counterparts in more or less clear perceptions or thoughts (*certaines perceptions ou pensées*) in our soul, and thus the soul will also have a certain thought about all the movements in the universe (*l'ame aussi aura quelque pensée de tous les mouvemens de l'univers*) and, according to me, every other soul or substance will have their certain perception or expression. [Leibniz, 1879c, p. 112*f.*]

Perhaps, from the perspective of the historian of metaphysics, Leibniz's greatest merit is that he did not step back from the radical consequences of his holism and hence his ideas are so intriguing as being complementary to Plato's intutions. The 'Connection' as a transcendental condition of ontological consistency of the whole network of perception relations (cf. e.g. [Leibniz, 1880a, p. 440]) offers here a peculiar version of the blockworld ([Rescher, 1981, p. 71]) based on the B-theory of time and with built-in atemporal pre-dynamics (cf. [Futch, 2008, Ch. 6; 2002] and the anti-Eleatic criticism by N. Rescher in [Woolhouse, 1994, p. 179*ff.*]), whose apt illustration (also in its semantic-logical formulation) is to be found in the famous fragment of the *Discourse on Metaphysics* (1686), where Leibniz skilfully illustrates his radicalism using 'Alexander the Great' as an exemplary complete 'law'/'idea'–series–history, or, as he says, '*haecceitas*' (cf. [Futch, 2008, p. 137*f.*, 141; 2002, p. 134*f.*]):

> Hence when the connection between things (*la connexion des choses*) is properly conceived, we can claim that Alexander's soul contains eternal remains of everything that he has experienced and the signs of everything that he would encounter, even the outlines of everything that has been taking place in the universe… [Leibniz, 1880a, p. 433]

The ontological model Leibniz thus gained evinces very interesting features, but at the same time classical weaknesses: although it is grounded in distinct, individual substances-monads understood as processes or streams of activity, then nevertheless what is logically primary is their hidden whole or the completely global context, i.e., the 'pre-established harmony' with its 'ichnography', beyond which the states are unconceivable or bereft of any reality: '… *cum unaquaeque res influat in aliam quamcunque ita ut si ipsa sublata aut diversa esse fingeretur, omnia in mundo ab iis quae nunc sunt diversa sint futura*' [Leibniz, 1879a, p. 226]. Although the perceptions are relational, yet the Leibnizian whole series–histories are separable, that is they do not enter into any causal relations between themselves, while the terminology of 'expressing' (like 'mirroring') can mistakenly suggest non-relational intuitions concerning that what is 'expressed'. Nevertheless, Leibniz's Absolute, when choosing 'the best of the possible worlds', does *not* separately reflect on the sets of the data of individual monads, but on exactly all those logically possible, complete and timeless structures or global 'ichnographies'. An authentic whole (= 'a true substance') is never created *par aggregation*, 'by aggregation', and by external links between pre-existent spatio-temporal things (as their 'composites'): such wholes as a pile of stones or an 'office' ('all the officers of the Dutch Indian Company') are merely composites compiled by the mind, 'phenomena', 'abstracts' or 'connections' (*rapports*), and not *proper entities* [Leibniz, 1879c, p. 101]. Leibniz invokes Plato and his view 'with reference to everything that is purely material', and thus he rejects extensional compositionism, distancing himself at the same time from externalistic compositionism, which is based upon a certain supposition (Platonic *doxa*) leading to misunderstandings: 'it is the lot of those who limit themselves to semblances and of those who treat all the abstractions of the mind (*les abstractions de l'esprit*) as real and conceive number, time, place, motion and shape to be separate entities' [ibid.]. An authentic whole, i.e. unity, is metaphysically guaranteed by the 'principle of unity' (*principium unitatis*), or – formally – by the principle of series (= singular process): each history as an 'authentic whole' is thus a direct realization of the principle of internalistic compositionism. In that sense, Leibnizian 'monadology' constitutes an intermediary or uniquely conciliating model, oscillating between externalistic Aristotelian compositionism (holism) and internalistic Platonic compositionism (holism) (cf. [Blandzi, 2002, p. 93]). Leibniz attempted to construe a truly radical holism of the Platonic type with a concept of the global structure of relations seen as logically primary (cf. [Barbour, 2003, p. 53*ff.*]), while retaining *pro traditione* a rather stiff, Aristotelian-Scholastic intuition of concrete substances, which he nevertheless processually construed as complete streams of atemporal forces, and hence as purely (also atemporally) pre-dynamic objects without any absolute 'now' (thus as the B-series of actions). In the end, Leibnizian wholes (unique relational laws = complete global series = individual 'ideas'/'forms' = '*haeceitates*') are simply Platonic wholes.

That philosophical cosmology is at the same time as much attractive to us as it introduces the relativity principle at the very basis of an action or information transfer in nature (strictly ontological 'perspectivism'). What is yet a classical weakness here, apparently theologically mediated, is the Absolute itself as an element of the theory: Leibniz wished to retain both the Universal Principle of Relativity and at the same time the pan-perspectival observer, transcendent with regards to the universe, thus the *classical logic* for his 'system of perspectives'. Regardless of the usefulness of such a model in any fundamental physics, I believe we can learn from those Leibnizian speculative exercises – just as from Parmenides's paradoxes – something really useful, when we encounter generically quantum 'paradoxes', which one habitually, on account of the long atomistic tradition speaking about 'particles' and their 'states', tries to face in the context of externalistic compositionism and resorts to the counterfactual reasoning and the time asymmetry. However, one always has an option to take those elusive atemporal Platonic wholes seriously.

## 2.2 Action, the universal algebra of the process and the foundations of quantum physics

The main problem we have to tackle when constructing a relational cosmological model with *internal, relativized dynamics* is the definition of the basic concepts of that model so as to make it possible to construe a fittingly general algebra, and hence we should be able to answer the question: how are we to understand 'state' and 'process'? QM itself is different in that respect to such a great degree from classical mechanics that we can even seriously start doubting if the term 'mechanics' is at all appropriate. Firstly, the very use of the concept of 'state' is in itself dubious, as then it seems natural to enquire: the 'state' of what? (For example, Heisenberg [1969, p. 221; 1959] used to speak of '*stationäre Zustände der Materie*'). David Finkelstein [1996; 1999a; 2003a] has recently proposed a complete change in the way fundamental elements of the quantum theory are understood, based on a consistent foundational reflection on the meaning of the applied formalism. As is well known, a 'quantum state' of the physical system is represented by a vector in the appropriately chosen complex vector space, so that we can assume that the proper, fundamental symbolic formula of the theory formulated in Hilbert space, i.e., an algorithm for calculating expectation values or outcome probabilities, has the form of:

$$\omega = \langle \varphi | \Omega | \psi \rangle \tag{1}$$

where $\Omega$ is a linear Hermitian (in particular idempotent) operator acting on that Hilbert space. The equation does not in fact mean anything other than an elementary *series of actions* which we shall at present write down in a generalized form (with the complex amplitude $\omega$) as:

$$\omega = \langle \boldsymbol{Out} | \boldsymbol{In} \rangle \tag{2}$$

The above formula is simply a symbolic record (a 'flow chart', as Finkelstein puts it) of a local physical experiment conducted on a certain 'object', which could be an atom or a particle, while $|\boldsymbol{In}\rangle$ is an *input action*, an 'injection' preparing some 'state' of the system, $\langle \boldsymbol{Out}|$ is an *output action*, 'extraction', which enables us to obtain information about the process, represented by an appropriate covector of dual space. 'These injective and extractive actions are theoretical elements of a kind that is alien to classical physics. They play the same fundamental theoretical role for quantum physics as the force concept for pre-quantum physics, and they meet resistance from Newtonian physicists much as forces did from Cartesian physicists' – Finkelstein writes [1999a, p. 448]. Taking (1) as the point of departure, we can say that each process of 'observation' (measurement, in the general sense) depends on: **[1]** preparing or *importing* (*injection*) of the quantum system into the measurement process; **[2]** a certain medial, selective action (which can also be a whole series of operations), taking the system from the initial state to the final one where the physical information is obtained; and **[3]** an output action, i.e., *extracting* or *exporting* the information by the 'observer' (external system) on the running of the process under study, which in formalism is represented by a complex covector in an abstract space modelling the whole experiment (see [Finkelstein, 1996, p. 42*ff.*]). We can otherwise say in the Leibnizian manner that **[1]** and **[3]** correspond to the directed actions ('*appetitions*') being at the same time the fully relational 'perceptions', while a choice of operator Ω means the choice of a specific frame of reference (f r a m e) for the measurement, consisting of defining a *complete system of (commuting) operators* satisfying the axioms of Boolean lattice. Elements **[1]** and **[2]** thus refer to the fact that the information obtained by the 'observer' about the physical process is at this very phase limited in a fundamental way by the choice of a particular frame [Finkelstein, 1996, p. 20; 1999a, p. 452][33], which consists of a type of local, classical 'window', through which an interacting system gains knowledge of only a fragment of all of the relations (Leibniz's *Liaison*) between the physical processes in the Universe which it is an *integral* part of. We might expect that the statistical Born rule and the probabilities of the outcomes of measurements should be deeply embedded in the non-classical logic of quantum action frames and the general noncommutative algebra of quantum operators: as an observer in some spacetime region cannot gain any information about states beyond his/her causal horizon (a relativistic limitation), so a quantum observer cannot know the results of measurements in a different, incompatible quantum frame (a quantum limitation). Noncommutative operators do not in general possess common eigenvectors, which can be interpreted physically in such a way that the frame which contains one of them does not constitute a possible *physical context* for the actions given by the other frame, containing the operators noncommuting with the former (see Ch. 4.2).

The projective operators themselves, inserted into (2), can be seen as a record of certain selective operations entering a final product – a kind of formal

33 A *quantum frame* can be defined in Finkelstein's terminology as a 'maximal orthogonal collection of injective actions' [Finkelstein, 1999a, p. 452]. In classical mechanics there exists only one frame of all possible actions on any system, hence it can be said that that introduces the Postulate of the Absolute Observer.

representation of the physical actions, where $|\omega_i\rangle$ are eigensubspaces of the complex state space, corresponding to certain eigenvalues of the observable, and outcome probability $p_\omega$ is real-valued:

$$p_\omega = \Sigma_i \langle\psi|\omega_i\rangle\langle\omega_i|\psi\rangle \tag{3}$$

If we interpret the vectors of the state space in such a way, we obtain a rather clear picture of the situation: in reality, wholly separate, objective states of the 'external' nature do not exist, as the formalism we use refers more to interactional operations represented, like classical forces, by vectors [Finkelstein, 1996, p. 55*ff.*; 1999a, p. 448]. Hence, we should rather be speaking of 'vectors of quantum action', and not of 'state vectors'[34], and thus turn from a classically ontic thinking to an informational-operational one in quantum theory, from naïve reism to ontological praxism [Finkelstein, 1999a, p. 451]. The first hint leading us in that direction is the fact that most of the interpretive problems and confusion in QM arises from the constant attempts at introducing the noncontextual 'states' and the laws governing them irrespectively of the actions of the systems (a varying frame of action). The language of the 'states' falsely suggests that the universe is composed of transitorily stable, separable 'objects', which possess temporarily stable *intrinsic properties*, but QM with its radical contextuality seems to keep suggesting that information is fully relativized to actions and their contexts. For any putative 'objects' – particularly quantum ones! – are not discernible by their 'intrinsic states' (a part of the Aristotelian heritage), but rather through their *actions* and *reactions* [ibid.]. We can even conclude that to assume the existence of the 'states' of some stable 'objects' is in a way excessively non-operational, as the wave function of the pre-selected system never refers to an objective, essential 'state of the object', independent of the acting external system, but rather to von Neumann's '*reiner Fall*', i.e., to many possible quantum systems, which the experimenter may pre-select [Finkelstein, 1993, p. 12]. In general, function $\psi$ and formula (2) are the 'bookkeeping devices' or 'information transfer model', enabling a complete representation of a physical process and the assessment of probabilities from the perspective of an acting local, and only local, 'observer' pre-selecting the systems. Formula (2) – hence also (1) and (3) – is *not* constrained by any 'time flow'/time asymmetry or pre-selective action only: one can use it to generalize the notion of the action–'state' itself, which is pre-temporal, and the whole relativized process can be described fully time-symmetrically as

34 'This is a good reason not to call them states; for in pre-quantum physics one imagined that the system under study truly carried a state, a complete determination of its responses to past and future actions, within itself' [Finkelstein, 1999a, p. 449].

an array of the entangled multiple-time actions and amplitudes (cf. [Aharonov *et al.*, 2009]). *The relational quantum 'process' is atemporal.*

One of the most important characteristics of Leibniz's philosophy of nature, which is important here, is its pandynamism, as a 'cinematographic' image of nature in the form of merely ordered series of separate states was to Leibniz simply nature's static caricature. He clearly preferred to speak of a *conatus* present in every process, that is 'impetus' (*impetus*) or 'tendency' (*tendentia*), without which nature would be a dead mechanism. These two aspects of the process, the static and the dynamic one, are in turn opposed within human consciousness as 'perceptual' and 'volitional', which are commonly conceived to be completely autonomous and having nothing in common – hence the cybernetic image of the cognition process being merely that of a passive gathering of information from the surroundings (input), while any action, such as a physical experiment, is an effect of the exertion of a certain direct influence on purely external conditions (output). The mind is then perceived as a 'black box' acting as a go-between between the two poles, which results in serious paradoxes and a strong dualism in conceiving its essence (in her in-depth study, Susan L. Hurley [1998] presented a precise and convincing analysis of the perplexing issue). The common conception that perception of the physical state of an environment is merely a passive registering of the objective information about it can be called a 'myth of pure input/data' or 'myth of the given', whereas the conviction that the choice of the conditions of perception, such as the decision about the frame of reference or the measured quantity, is a totally non-complex, one-track activity of our consciousness and independent of input processes can in short be termed a 'myth of pure output/choice' or 'myth of the giving' [ibid., p. 73*ff.*, 240*ff.*]. According to Leibniz, though, states–perceptions cannot exist without the incessant action of internal, atemporal forces (*conatus*), and likewise those forces atemporally define global histories. One of the most astonishing properties of QM in its operator-algebraic formulation is the fact that the mutual autonomy of data and actions, evident in classical physics (the noncontextuality of information), *is relativized* [Finkelstein, 2003a, b; 1999a, p. 450], as we have the right to speak of an 'action' and its specific 'outcome' only in a certain chosen frame, when we make an arbitrary cut which sets the boundary between the observed and the observing system. What is more, as we have noticed, the mathematical objects we use – vectors – represent conjunct operations rather than some passive 'objective states of things': the pre-selective action of the measuring system 'excluding itself', in a way, from the rest of the universe as its injective subsystem (:a contravariant element) and the coupled action of the 'excluded subsystem' represented in the model of the experiment by an extraction vector (:a covariant element). The same applies to the fully time-symmetric account of (2): those operations are just the coupled pre- and post-selective actions within the global quantum history.

One can suppose that a great deal of the paradoxes in microphysics is rooted in the habitual transposition of naïve metaphysics of the 'data'–'action' separatedness onto the microworld and nature as such, as well as in the concept of the separation of the cognizing mind (the 'black box') from the external world, contact with which is made solely by the gathering of information (a passive input) and by exerting a one-track influence (an active output). Finkelstein's praxistic approach to the quantum theory nevertheless offers the possibility to consistently get rid of those constant habits creating certain Baconian 'idols' in the sphere of cognition, which then serve as intangible orientation points, or else as absolutes of the observer's mind ([Finkelstein, 2003a, b]; on the example of the relativization of time and space: [Wigner, Inönü, 1952]). When we look at quantum formalism in that way it becomes clearer that it represents rather the logic of actions inside the Universe than complete information on the intrinsic 'states'/properties of its singled-out objects. Those basic conclusions are surprisingly anticipated in Leibniz's approach, in which each 'monad' – as a consistent history or series – acts internally *as if* it were independent from the rest of universe, which it 'perceives', and those 'perceptions' of finite informational content are ordered in infinite series by a global 'force–action' itself immanent to it (*conatus*) or, which is exactly the same, by the *dynamical law* (*lex/idea*) *of each global series*. When construing quantum cosmology, we thus have to take into consideration what keeps being attempted to be removed from the foundations of quantum physics by supplementing it with certain absolutes from pre-quantum mechanics – that *all* dynamics is purely internal and relativized, and one has to learn how to incorporate such a dynamics into the atemporal blockworld physics. That property makes it pointless – as Finkelstein (and many others) have been right to observe – to describe the whole Universe as a mere 'quantum object', as there is no physical system which would be in a condition to gain information about (and thus 'act upon'/'extract from') the Universe sharply set apart from it. That is connected to the fact that we cannot, strictly speaking, even assume an 'entropy of the Universe' [Finkelstein, 1993, p. 14]. On the list of quantum observables there is also no such thing as a 'state' (there is only 'position', 'angular momentum', 'energy' etc.), one cannot in fact enquire what the state of that 'object' is, and hence there is also no sense in asking what the general, objective 'state of the Universe' is from the perspective of its particular internal subsystem (the question would be incorrectly posed) ([ibid., p. 12], cf. e.g. [Smolin, 1997]). It is not possible to conduct a quantum experiment on the whole Universe, nor does it have a prepared 'quantum state' one could physically act upon from within, and thus extract any information about it as if it were an electron. Thus, the internalistic physics of the subsystems brings into play the problem of metaphysical compositionism as if QM were strictly inseparable from quantum cosmology.

Before, in the spirit of Leibniz, we tackle the issue of the most basic mathematical structures which could describe global relations between those internal processes in the framework of a quantum cosmology, we should reflect on the foundations of the way change as such is perceived (we are not at the moment concerned whether change or flow of time is an 'illusion' or not – what only interests us is whether an internal observer in the Universe can abstractly describe the directed process or ordered actions). If we wish to formalize the problem of the internal process/action, we need a very general algebra (Leibnizian 'ichnography'), which would enable us to analyze the 'process' and 'transformation' in the possibly greatest number of contexts and cases (that is, mathematically, for example in an infinite number of dimensions and for any physical system). What is more, it should be general enough to be possible to apply, as a certain 'quantum topo-chronology' [Bohm, 1965], to model processes in the smallest scales without referring to differentiable manifolds, which at the Planck scale could be topologically an inappropriate model [Hiley, Fernandes, 1997]. What is looked for is thus a certain type of 'universal algebra of the process', in which that indivisible process with its inner symmetries is a primary concept, and the singled-out 'states' (its 'phases' or 'elements') are secondary constructs, relativized to the perspective of any arbitrarily differentiated subsystem. Thus an 'action' and a 'state' should be always treated on equal footing. To obtain such an algebra, let us begin with the simplest algebraico-geometrical constructs (as Finkelstein does), without worrying yet about a fully axiomatic approach.

The most primary (and abstract) concept of the process is mathematically given in algebraic 'operations', while geometrically it can be represented as certain operations on objects, which can be then differentiated as their 'phases' ([Hiley, Fernandes, 1997], [Hiley, 2005]) – it shall be our simple geometric (and, firstly, classical) toy model. Let us imagine a certain ordered array of elementary processes (Finkelstein's 'actions'), which we shall graphically present as a series:

… ● ● ● …
**A B C**

The singling out of individual 'objects' ('elements of the process') is arbitrary, hence we can compare it to the choice of three points on a certain projective plane $\mathbf{\Pi}$ (we at first do not require it to be a metric space – we are interested in a more primitive, non-metric approach to the process, cf. [Bohm, Hiley, 1993, p. 378]). Let us therefore assume, for the sake of simplicity, that firstly – and only firstly – it occurs in a real projective space $\mathbf{\Pi}(\mathbb{R})$, in which we select such three arbitrary, different projective points. We shall also represent those projective points in the algebra of the process $\mathcal{A}(V)$ as three *vector-processes* of a certain real vector space $V$: $|\mathbf{A}\rangle$, $|\mathbf{B}\rangle$ and $|\mathbf{C}\rangle$ (cf. [Hestenes, 1988], [Hestenes, Ziegler, 1991]). Since the points of plane $\mathbf{\Pi}$ constitute here merely an abstract model of the 'moments' of the transformation process – and hence they themselves can be treated as processes – then we shall further on mark them in the same way as their 'developments' into lines on our plane $\mathbf{\Pi}$, i.e., with the help of vector brackets (we thus represent those lines as $|\mathbf{AB}\rangle$, $|\mathbf{BC}\rangle$ and $|\mathbf{AC}\rangle$ in space $V$). We can therefore now geometrically envisage that composition of the point-processes, such as $|\mathbf{AB}\rangle \doteqdot |\mathbf{A}\rangle \circlearrowleft |\mathbf{B}\rangle = |\mathbf{A}\rangle\, |\mathbf{B}\rangle$, as a process of a higher order, as the 'unfolding' ('stretching out') of all of plane $\mathbf{\Pi}$ from its simplest elements (points) into its line-processes. To avoid the irreducible dualism of the 'state' and 'action' at the very initial stage, let us stress at the beginning that just as we shall consistently consider 'states' $|\mathbf{A}\rangle$ and $|\mathbf{B}\rangle$ to be just singular actions (given as vectors in $V$)[35], we shall similarly treat actions $|\mathbf{AB}\rangle$ or $|\mathbf{BA}\rangle$ as certain 'states' modelled geometrically by the projective lines in $\mathbf{\Pi}(\mathbb{R})$, upon which further actions can be defined: an exclusive differentiation of 'states' and 'actions' could be termed a constructional (cognitive) artifact, though sometimes useful[36]. We treat them on equal footing from the very beginning, which is so important in the operational structure of QM.

Let us therefore assume that the three actions singled out by us in $\mathcal{A}(V)$: $|\mathbf{A}\rangle$, $|\mathbf{B}\rangle$ and $|\mathbf{C}\rangle$, are to be *fully independent* of one another, that is the vectors they are represented by are orthogonal in space $V$ (thus the respective points in $\mathbf{\Pi}(\mathbb{R})$ are not collinear). Let us therefore set down two basic rules: the multiplication of vectors by scalars in $\mathcal{A}(V)$ refers to the relative *strength of the processes* (dilation

---

35 'Actions' are here understood in the sense indicated above, a praxist-ontological one. The 'state' as a process of a lower order, 'on' which other actions are conducted, is always by definition also linked with the action itself ('how' a system interacts) – in that sense scalars in the framework of algebra operationally represent actions similarly to vectors, only they are processes of a different type.

36 The distinction between vectors and scalars is of that character – nevertheless, in order to unify those concepts, a much wider framework and a much more general concept is needed, such as $k$-blade [Hestenes, Sobczyk, 1987]: scalars can then also be geometrically interpreted as 0-blades.

of vectors), while their addition always gives another vector and is both associative and commutative, i.e., $|\mathbf{A}\rangle + |\mathbf{B}\rangle = |\mathbf{B}\rangle + |\mathbf{A}\rangle$, which is a proper translation of plane **Π**. $|\mathbf{AB}\rangle$, $|\mathbf{BC}\rangle$ and $|\mathbf{AC}\rangle$, in turn, represent movements–developments of the process in three basic directions, something of the type of an oriented expansion or spread in the direction (*directed areas* swept out in $\boldsymbol{V}$ by $|\mathbf{A}\rangle$, $|\mathbf{B}\rangle$ and $|\mathbf{C}\rangle$, thus the projective lines on **Π**). The basis is orthonormal, as we can always choose some orthogonal unit vectors spanning $\boldsymbol{V}^3$, and we also introduce the rule of tensor contraction for vectors: $|\mathbf{A}\rangle^2 = |\mathbf{A}\rangle \circ |\mathbf{A}\rangle = ||\mathbf{A}\rangle|^2 \in \mathbb{R}$, where: $||\mathbf{A}\rangle|^2 = ||\mathbf{B}\rangle|^2 = 1$, and $|\mathbf{A}\rangle \circ |\mathbf{B}\rangle = 0$. Thus, by generalizing the associative inner product $\circ$ over all the actions–'states' of our space $\mathcal{A}(\boldsymbol{V})$, we obtain a metric that enables us to talk later on of perpendicular lines in $\mathbf{\Pi}(\mathbb{R})$ or compare the intensity of processes. The choice of such a triple of projective points (elementary 'moments' of a process) is thus equivalent to defining a vector basis spanning $\boldsymbol{V}$, and when we assume the distributivity of scalar multiplication over vector addition we have an algebra $\mathcal{A}(\boldsymbol{V}^3)$ over field $\mathbb{R}$. We would now like to understand (and that is a key issue, if we are interested in a *universal* algebra of the relational process) what exactly is implied by that special composition or 'unfolding' of some processes in our two-dimensional space **Π**, that is, is there any and if so what sense in the following operation: $|\mathbf{A}\rangle|\mathbf{BC}\rangle \doteqdot |\mathbf{A}\rangle \circlearrowleft |\mathbf{BC}\rangle = |\mathbf{ABC}\rangle$, or another one, $|\mathbf{B}\rangle|\mathbf{AC}\rangle \doteqdot |\mathbf{B}\rangle \circlearrowleft |\mathbf{AC}\rangle = |\mathbf{BAC}\rangle$[37]? The notation for such a process of composition of two processes in space $\boldsymbol{V}$ may at first make us slightly uneasy, as we undoubtedly cannot apply any additive contraction inside the vector bracket, whereas the resulting process of the third grade, although it represents a subplane in **Π**, is surely not a simple action vector in algebra $\mathcal{A}(\boldsymbol{V})$, just as neither is action–'state' $|\mathbf{AB}\rangle$ nor $|\mathbf{BC}\rangle$. An analysis of the properties of those products should shed light on the general nature of the 'process', as it is algebraically given in $\mathcal{A}(\boldsymbol{V})$ and geometrically in $\mathbf{\Pi}(\mathbb{R})$, and thus, to put it shortly, explain what the enigmatic, unclear at first operation '$\circlearrowleft$' at all is.

Let us take as a two-dimensional geometric toy model any pair of orthogonal unit vectors on a real plane $\mathbb{R}^2$ or, in other words, two basic processes, independent of each other, that we are going to study: $\boldsymbol{\sigma}_1 = |\mathbf{A}\rangle$ and $\boldsymbol{\sigma}_2 = |\mathbf{B}\rangle$, where $|\mathbf{A}\rangle$ and $|\mathbf{B}\rangle$ are 'elementary actions' of that general process in $\mathcal{A}(\boldsymbol{V}^2)$, just as they had corresponded to the points in $\mathbf{\Pi}(\mathbb{R})$. By taking the possibility to add vectors (to

37 We have to remember, in order to avoid any misunderstandings, that vectors $|\mathbf{A}\rangle$, $|\mathbf{B}\rangle$, $|\mathbf{C}\rangle$ belong in $\mathcal{A}(\boldsymbol{V}^3)$ to a completely different category than $|\mathbf{AB}\rangle$, $|\mathbf{BC}\rangle$, $|\mathbf{AC}\rangle$, and they in turn must be different from $|\mathbf{ABC}\rangle$! We can express that by calling them three different *steps* of the processes that can take place in our geometric space **Π**, that is processes of the order $k = \{0, 1, 2, 3\}$, which also generate appropriate subspaces in $\boldsymbol{V}$.

conduct the translation of plane $\mathbb{R}^2$), we shall apply the Pythagoras' theorem to see what internal order is contained in that process:

$$|\mathbf{A}\rangle + |\mathbf{B}\rangle = |\mathbf{C}\rangle,$$
$$\boldsymbol{\sigma}_1 + \boldsymbol{\sigma}_2 = \boldsymbol{\sigma}_3, \text{ thus:} \tag{4}$$

$$(\boldsymbol{\sigma}_3)^2 = (\boldsymbol{\sigma}_1)^2 + (\boldsymbol{\sigma}_2)^2,$$
$$(\boldsymbol{\sigma}_1 + \boldsymbol{\sigma}_2)^2 = (\boldsymbol{\sigma}_1)^2 + (\boldsymbol{\sigma}_2)^2. \tag{5}$$

And hence after having applied the tensor contraction to the squares of the vectors:

$$\sigma_1^{\ 2} + \sigma_2^{\ 2} + \boldsymbol{\sigma}_1\boldsymbol{\sigma}_2 + \boldsymbol{\sigma}_2\boldsymbol{\sigma}_1 = \sigma_1^{\ 2} + \sigma_2^{\ 2}, \tag{6}$$

from where we obtain:

$$\boldsymbol{\sigma}_1\boldsymbol{\sigma}_2 = -\boldsymbol{\sigma}_2\boldsymbol{\sigma}_1. \tag{7}$$

That means that the orthogonal vectors on a real vector plane $\mathbb{R}^2$ anticommute, and hence the 'elements' of the process which we have described algebraically, are noncommutative. The process or action described as $|\mathbf{AB}\rangle \doteqdot |\mathbf{A}\rangle \circlearrowleft |\mathbf{B}\rangle$ is not the same as $|\mathbf{BA}\rangle \doteqdot |\mathbf{B}\rangle \circlearrowleft |\mathbf{A}\rangle$, as $|\mathbf{AB}\rangle = -|\mathbf{BA}\rangle$, and for the process of a higher grade we also have: $|\mathbf{A}\rangle|\mathbf{BC}\rangle = -|\mathbf{A}\rangle|\mathbf{CB}\rangle$, thanks to what we get a sort of very primitive algebraic notion of internal, relational 'chronology' ('proto-chronology', see: [Hiley, Fernandes, 1997], [Hiley, 2001; 2005]). The constructs thus obtained have a very interesting property, which we might not notice at first glance, and we shall therefore write it explicitly:

$$\boldsymbol{\sigma}_1\boldsymbol{\sigma}_2 = -\boldsymbol{\sigma}_2\boldsymbol{\sigma}_1 = |\mathbf{A}\rangle|\mathbf{B}\rangle = |\mathbf{AB}\rangle = -|\mathbf{BA}\rangle, \text{ and hence} \tag{8}$$

$$||\mathbf{AB}\rangle|^2 = |\mathbf{AB}\rangle|\mathbf{AB}\rangle = -|\mathbf{AABB}\rangle = -||\mathbf{A}\rangle|^2 \circ ||\mathbf{B}\rangle|^2,$$

$$||\mathbf{AB}\rangle|^2 = (\boldsymbol{\sigma}_1\boldsymbol{\sigma}_2)^2 = -||\mathbf{A}\rangle|^2||\mathbf{B}\rangle|^2 = -1 \tag{9}$$

It is therefore obvious that 'action' $|\mathbf{AB}\rangle$, whose model is constructed by composing the action vectors in $\mathcal{A}(V^2)$, is not a vector as such any more, but a totally new object, which we shall call a 'bivector'. It possesses very interesting properties on plane $\mathbb{R}^2$ and represents in that space a relational 'process' of the highest grade, just as a scalar is a process of the lowest (zero) grade. Hence, we can call the bivector here – also for even more principal reasons – a 'pseudoscalar' in $\mathbb{R}^2$, which has the image in $\mathbf{\Pi}$ as a projective line. And thus the space of algebra $\mathcal{A}(V^2)$ under scrutiny is four-dimensional, spanned by the following four elements: $\{\boldsymbol{\sigma}_0 := 1, \boldsymbol{\sigma}_1, \boldsymbol{\sigma}_2, \boldsymbol{\sigma}_1\boldsymbol{\sigma}_2\}$, i.e., a unit scalar element, two independent unit vectors and a unit bivector, referring to the three grades of the process of 'unfolding' plane–space $\mathbb{R}^2$. The whole process in that simple grade-2 algebra comprises two polar sub-processes: that of contracting $\mathbb{R}^2$, which is represented by internal product $\circ$ referring to the 'rolling up', enfolding action, or *implicatio*, and that of spanning $\mathbb{R}^2$, represented by outer (Grassmann or 'wedge') product $\circlearrowleft$, referring to the 'spreading out', unfolding action, or *explicatio*. What is most interesting in that general language is that we can study those actions–'states', given as directed quantities, without referring to any coordinates, and the algebra, namely that full Clifford algebra $\boldsymbol{Cl}(2) \cong \mathcal{A}(V^2)$ of the two-dimensional Euclidean space, can model an order much deeper than the usually explored Cartesian one. For we have analyzed here the case of a real vector space $\mathbb{R}^2$, but we can notice already at this point that the 'action' of the highest grade in $\mathbb{R}^2$, constructed as $|\mathbf{AB}\rangle \doteqdot |\mathbf{A}\rangle \circlearrowleft |\mathbf{B}\rangle$ (the greatest unfolding of an order in that state space), is a purely dynamic element, expressing the process of rotation of the whole plane, and hence it acts upon all the processes of the lower grade (it can also have different 'intensities': $1x|\mathbf{AB}\rangle$). The bivector in $\mathbb{R}^2$ is simply a rotor, an operator of rotation. The dimensionless scalar object $\boldsymbol{\sigma}_0$, in turn, represents the greatest (zero) enfoldment of the whole space, 0-blade, which is given only as a pure 'intensity' or 'strength' without any extension. There is nothing to deter us from unifying the notation of objects in our algebra, each time giving only grade $x$ of the process (its 'unfolding'): $\langle\mathbf{A}\rangle_x$ (see e.g. [Hestenes, Sobczyk, 1987]).

With the help of that Clifford universal (graded) algebra we can encode all of the projective geometry in $\boldsymbol{\Pi}(\mathbb{R})$, but what is most important, it turns out that property (9) of the pseudoscalar makes it the geometric model of an imaginary unit $i$ in $\mathbb{R}^2$, which is no longer any mysterious arithmetical entity, but a 'process' of a higher order, defined in real space on other representations of the process called vectors. (Let us immediately notice that the universal nature of that algebra allows us to analyze not solely the positive quadratic space $\mathbb{R}^2 = \mathbb{R}^{2,0}$ – there is no reason not to analyze the order of unfolding in the negative space $\mathbb{R}^{0,2}$!) We can now add processes of completely different grades in $\boldsymbol{Cl}(2)$ without worrying about doing something that is not allowed or dubious (apparently clumping together entities of totally different categories), for instance we can note such a projective operation on different elementary 'actions':

$$\langle\mathbf{A}\rangle_0 + \langle\mathbf{B}\rangle_0 \circ \langle\mathbf{CD}\rangle_2 \doteqdot x\sigma_0 + y\sigma_0\sigma_1\sigma_2 \doteqdot x + iy, \tag{10}$$

which is nothing more than a representation of complex numbers in $\boldsymbol{Cl}(2) \cong \mathcal{A}(V^2)$, giving a certain new 'state', which can be manipulated in that algebra. Such geometric treatment in a Leibnizian spirit makes each complex number in fact a notation of an operation, which can be expressed by the instruction: 'translate and rotate the plane', and not merely a certain static algebraic entity, a rigid construct used for calculations.

As Roger Penrose [2006, p. 73] aptly put it, 'the magic of complex numbers' is the foundation of QM, and we can see that – only on the example of the algebra (order–*logos*) of the process for a simple space $\Pi(\mathbb{R})$ – that the quantum theory itself, as far as formalism goes, contains ideas which can be expressed in a purely processual way, in the language of operations, if we redefine (generalize) it ontologically. If we treat the algebra of complex numbers as part of a wider processual algebra of a real plane, we can also easily discover the variety of possible operations in such a simple space as $\Pi(\mathbb{R})$ seems to be. For example: it contains easily identifiable rotation symmetries given by the SO(2) Lie group, which as an abstract group is isomorphic with spin group *Spin*(2), as our algebra contains the sub-space of two-dimensional spinors – it is in fact a symmetry of the processes defined in that $\Pi(\mathbb{R})$ space. What is more, when analyzing those simplest examples, we immediately invoke that projective space as the geometric model of our toy algebra of process $\mathcal{A}(V^2)$ and it turns out that, though that might seem an unnecessary complication, the relation is important when attempting to comprehend the purely geometric sense of QM formalism. Suffice it to note here that the space of quantum physical states is not so much Hilbert space $\mathcal{H}$ as the space of rays in $\mathcal{H}$, that is the *complex projective space* $P\mathcal{H}$ (where if $\mathcal{H}$ is infinite-dimensional, then so is $P\mathcal{H}$). Quantum states can be represented by points of symplectic manifold, and physical observables as real functions on that space which, through an analogy with the symplectic formulation of classical mechanics, enables us to term $P\mathcal{H}$ *a (generically) quantum phase space*. (In point of fact any space $\mathcal{H}$ can be seen as a symplectic manifold, or a phase space, if only we represent it as the real vector Hilbert space equipped with a complex structure – see A. Ashtekar and T.A. Schilling's [1998] detailed lecture on the subject). It is easy to guess that the most interesting, deeper mathematical and physical levels of the order are revealed when we turn to spaces and processes of higher grades and dimensions defined by Clifford algebras $\mathcal{A}(V^n)$ and higher Clifford numbers. What we thus gain is the possibility to study the relations between those processes without the need to introduce a 'background', as what we are interested in are solely differentiated elements of the whole process, algebraically expressed by objects of different grades[38]. Perhaps such a calculus would fully correspond to what, alongside combinatorial topology, Leibniz had dreamed of in his *Characteristica geometrica* and in his letter to Huygens in 1679

38 A systematic study on the above ideas concerning the algebra of the process (represented by appropriate orthogonal and symplectic Clifford algebras), in the context of the foundations of QM as a 'theory of actions' is also to be found in the work of Basil Hiley [2001; 2003; 2005], who develops the 'ontology of the process' of late Bohm (e.g. [Bohm, Hiley, 1981], [Bohm, Davies, Hiley, 2006/1982]).

as an '*analysis (geometria) situs*', that is a method of operating with geometrical objects in the same way as with ordinary numbers.

Finkelstein's remark thus seems more understandable as the quantum wave function, whose values are of form (10), represents not so much an intrinsic property of some separable 'object' which is independent of the observer at moment $t + \delta t$, but rather the relative action of the external system, which is itself a part of the Universe being a hidden whole (let us denote it as $\mathbf{\Omega}$). That internal process is modelled directionally, therefore algebra, which can represent any 'pre-topo-chronology', should have the form of a universal algebra of directed quantities. Thus, by using vectors or bivectors (for example in the electromagnetic theory or in special relativity), we can imagine *modo Leibnitiano* that we are using simple mathematical models of directed actions in nature itself, of which the 'observer' is a part of. That is also one of Leibniz's most profound ideas which always stresses the primeval *conatus* ('endeavour') of every elementary event in nature (also in human consciousness). That fundamental element of action creates a kind of pre-chronology and we should expect quantum cosmology to find a fundamental law, i.e., the quantum statistics of nature, which rules the directed processes in the vicinity of Planck's regime (one of them could be the so-called Clifford statistics built on elements of algebra $\boldsymbol{Cl}(n)$ called *cliffordons* and studied by Finkelstein's group, see e.g.: [Baugh *et al.*, 2001], [Finkelstein, Galiautdinov, 2001]). On the fundamental level such a model would thus contain ordered arrays/atemporal flows of elementary actions, which from the perspective of processes of a higher order would be a sort of relational 'perceptions' which can be accounted for as the *relative* 'states' of the subsystems. One can imagine that exactly at that point fundamental physics and metaphysics of nature could fruitfully coincide.

If we now treat $\langle \mathbf{A} \rangle_1$ or $\langle \mathbf{CD} \rangle_2$ represented by (bi)vectors as certain kinematical, static (devoid of a causal power) physical 'states', an external factor '$\mathbf{Q}$' is necessary then – a postulated external 'force' (*energeia*) – explaining their 'topo-chronology': it is a certain imaginary scheme, which (as we usually hope) is to enable a reconstruction of all of physics and it was such a solution that Leibniz defended still in the 1670's. We then gain a strange mechanicism transposed onto the level of a fundamental description of nature ('quantum mechanicism'), in which series of states are ordered by a certain external principle (dynamics *imposed* on base elements), which is in full harmony with externalistic compositionism:

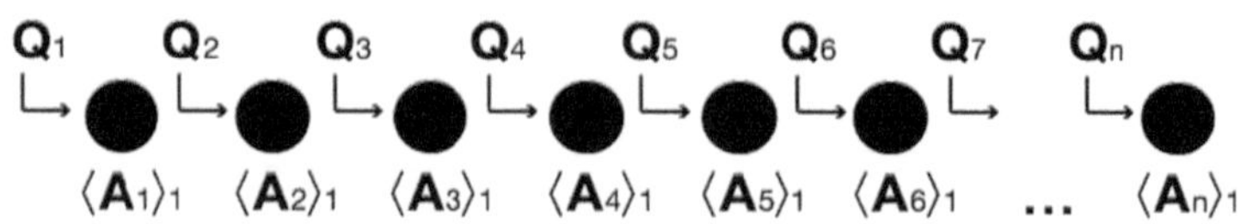

Mature Leibniz's key thesis is the conviction that at a fundamental level there cannot exist any dualism of 'states' and 'non-states' (= actions), as that would be equal to a picture of the universe in which any real (and hence also physical) connections between the processes are purely external, they 'interlock' physical events or objects just as certain mysterious, nonreducible 'qualities', *vis occultae* (that is the deeply metaphysical foundation of Leibniz's protest against Newton's mechanics and his approach to the force of gravity). Classical mechanics has been construed in such a conceptual frame which has been very successful and has shaped a great part of modern physical imagination, but that does not imply that it is proper when we wish to draw nearer to the modelling of fundamental processes in the framework of quantum cosmology. In Leibniz's metaphysical scheme, the relations between the phases of the process are solely internal, directly mirroring the immanent link between 'perception' and the 'endeavour'–'force' in monads/*series*. Thus we are dealing here with elementary 'states' (Leibniz's 'small perceptions') represented as $\langle \mathbf{A}_i \rangle_n$, which should be seen at the same time as 'tendencies'–'forces' generating further 'states', both of the same grade and those belonging to higher grades: $\langle \mathbf{B}_i \rangle_{n+1}$, $\langle \mathbf{C}_i \rangle_{n+2}$ etc. They are in fact actions rather than the caleidoscopic properties of certain 'things' (their 'force', **Q**, is immanently contained in them, and hence we are facing here an internalistic 'quantum processualism'), which renders us Leibnizian series of actions described by a singular global 'law'/form (*lex seriei*):

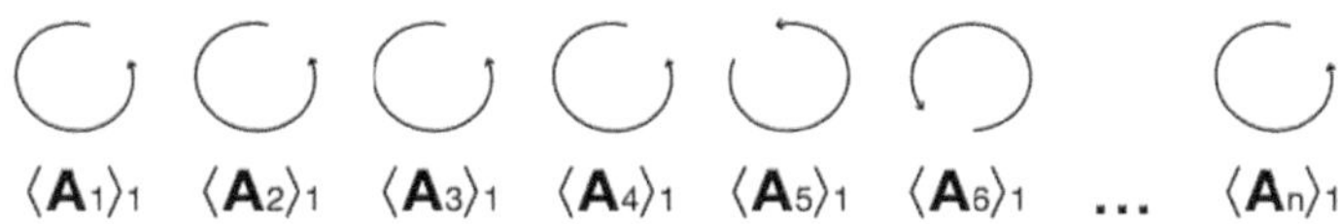


(The fact that a relational series is presented here by the circular arrows of the actions with the help of bracket symbols of 1-blades, does not mean of course that the physical actions themselves are 'vectors', 'bivectors', 'trivectors' or anything of that kind, but only that that is an abstract way in which those processes of various grades are relationally represented in the Clifford algebra in faithful coincidence with the original intentions of Leibniz himself).

There is a deeper ontological argument, though, which inclined Leibniz to reject the passive kinematical (i.e. Cartesian) states and assume point-like actions–'endeavours' to be fundamental. When analyzing the problem of a continuum and of elements, Leibniz stumbled across the same problem on the borderline between mathematics and physics as the one that had earlier caught the attention of Galileo and Hobbes in the area of physics itself, namely the issue of the finite division of matter and the character of its indivisible building blocks. Galileo postulated the existence of finite space volumes between finite material elements in such a way so as to make the cohesion of physical objects and their perceptible forms in space and time possible to explain. Hobbes, in turn, postulated in his 'first philosophy' the existence of infinitely small units/elements of motion, operations 'instigating motion', which he called *endeavours* (= *conatus*), forming the basis of both physical and mental processes (cf. [Arthur, 1998, p. 113*ff.*], [Wahsner, 1991, p. 53*ff.*]). With relation to those speculative proposals (particularly that of Hobbes, on whose theses he could model his own solutions), Leibniz noticed that the continuum cannot be an entity actually existing *in nature*, but it can merely be an abstract, as – by analogy to the argumentation about the relativity of time and space – its parts are indiscernible from one another, contrary to the matter taken concretely, physically, and not abstractly or metaphysically (like the *materia prima* of Aristotle and scholastics, which is *ex notione* homogenous) [Arthur, 1998, p. 111]. Still, according to Leibniz, it is also impossible to imagine any moments of the process when 'nothing takes place' (he refers to Gassendi's idea of an absolute interval in nature, adopted by him when he was young), therefore also talking about an enduring 'state' as something real is impossible to uphold – a 'state' can be understood only in an abstract sense, just as a continuum: '*Motus est continuus seu nullis quietulis interruptus*'[39] [Leibniz, 1880b, p. 229] (see also [Arthur, 1998, p. 130*ff.*]).

In his *Theoria motus abstracti* in 1671, Leibniz still defends his view which is close to Hobbes and assumes that a point is to be seen as an entity '*cujus extensio nulla est, seu cujus partes sunt indistantes, cujus magnitudo est inconsiderabilis, inassignabilis, minor quam quae ratione nisi infinita ad aliam sensibilem exponi possit, minor quam quae dari potest*'[40] [Leibniz, 1880b, p. 229]. According to that thesis, just as a point posseses a real (but 'imposible to define', *inassignabilis*) quantity, so elementary 'endeavours'/'actions' are real, though infinitesimal component elements of motion in nature (though they themselves are not motions – Leibniz distances himself from Hobbes here): '*Conatus est ad motum, ut punctum ad spatium, seu ut unum ad infinitum, est enim initium finisque motus*'[41] [ibid.]. It is only later, after 1675, that Leibniz sees endeavours/actions as *real* motions (nature's

---

39 'Motion is continuous, that it is not interrupted by any moments of rest'.

40 '... which has no extension or whose parts are not separated by distances, whose size is slight, not possible to define, smaller than what could be expressed by its relation to another sensually perceptible quantity, smaller than any relation that could be given'. Cf. [Arthur, 1998, p. 114].

41 'Endeavour is to motion just as a point is to space, or as a unit to infinity, for it is the beginning and end of motion'.

elementary 'micro-motions'), and not infinitely small quantities, and in the years 1676–1677, on the basis of his reflections on cohesion and the collision of bodies, he even accepted the fact that movement could not be continuous, but had to take place by means of leaps [Arthur, 1998, p. 117]. Taking aside all of that course of evolution of the philosopher's ideas, let us simply stress that finally, under the influence of his mathematical studies, he came – as is well known – to a redefinition of the concept of a continuum, returned to the thesis on the continuity of movement and put down the law of the conservation of energy [Leibniz, 1860] and the relativity of motion (*motus respectivus*), which jointly led to the construction of monadology. The studies Leibniz thus conducted ended in his conviction that the only way of explaining the external and internal differentiation of physical objects was not merely through a reference to certain empirical properties connected to the mass of the bodies and their motion, but to the processes constituting their ontic foundation, the incessant 'actions' which can be defined only in the framework of the *global internalistic dynamics* of the Universe. And therefore, according to the philosopher's argumentation, there can be no *minima* in the world of nature, points in the mathematical sense composing it (of a zero extension) as it is 'real pseudo-points' that have to compose it, understood as processes, i.e., discernible 'point-actions' ordered into series that are possible to study algebraically and combinatorially. Enduring, non-relational states (like geometric *minima*) in the physical world are thus to Leibniz always merely an abstraction from the global relational process: in his universe, every phenomenon is metaphysically embedded in a pre-temporal flow (series/idea/form) and there is no absolute physical 'existence at a moment'. All the phenomenal individuals without exception are from that perspective the bundles of the atemporal 'serial actions', 'a continuous expressions of universal laws' [Loemker, 1969, p. 30], with Leibniz later calling those 'endeavours' the 'appetitions' as the essence of the atemporal series/forms. If therefore we want to model the most basic structures of the world with the help of geometric algebra, we should drop our habit of treating those points, lines, vectors or numbers as models of static entities – they should also (as sensual phenomena) be treated as pure actions (cf. [Arthur, 1998, p. 130*ff.*]). Noting in our geometric algebra a projective point in the form of a vector $|\mathbf{A}\rangle = \langle \mathbf{A} \rangle_1$, and a scalar as $\langle \mathbf{A} \rangle_0$, is thus perhaps no mathematical extravaganza, as what it says is that even a mental object, which we ascribe a 'zero extension', 'pure intensity', in an abstract way mirrors only a certain relational action, e.g. as a ratio of amplitudes. Resignation from abstract points and their sets can thus lead us much further, beyond the standard algebraic topology: we need in such a case a certain strong generalization of the concept of a 'point', so that we do not need to use it at all in fundamental physics. That is a quite radical postulate, as all of heretofore physics has been built upon the topos of sets (**Set**) and *n*-dimensional differential manifolds, hence that would imply that to construe a quantum cosmology we would have to construct all of physics on the basis of different, non-classical categories without the concept of point and its neighbourhood. That is certainly an intriguing possibility on account of the hard to overcome difficulties in fundamental physics connected to the reappearing infinities and mathematical singularities.

The statement that *energeia* is always contained in the 'states' themselves corresponds to Leibniz's frequently repeated thesis that all arrays of those 'states' at the lowest level have to be 'spontaneous' (*spontanea*), and therefore cannot be generated by any external influences or causal factors (as in Newton), but only by their internal dynamics, the purely global 'law' (*lex*) of the series (= forms) themselves. '*Si nihil sua natura activum est, nihil omnino activum erit; quae enim tandem ratio actionis si non in natura rei?*'[42] [Leibniz, 1879a, p. 263]. In 1705 Leibniz wrote in a letter to de Volder that '(…) *vires primitivas manifestum esse censeo nil aliud esse posse quam tendentias internas substantiarum simplicium, quibus certa suae naturae lege de perceptione in perceptionem transeunt, atque inter se simul conspirant*'[43] [Leibniz, 1879a, p. 275]. (We have to remember that Leibniz's monad is in no way a standard physical object and is neither spatial nor temporal, hence its conception in the categories of reistic ontology is a grave misunderstanding). Those 'primitive forces', which are the basis of what Leibniz usually calls 'appetition' [Rescher, 1991], should be conceived as 'tendencies' inherent *to the whole process* (each *global series*): '*ses tendences d'une perception à l'autre*' = '*les principes du changement*' [Leibniz, 1885b, p. 598] are the internal structure/form of the universe, which remain inaccessible to physical experiments, they can only be studied theoretically. Even the physical (derivative, spatiotemporal) force as a special case, in standard mechanics represented by such objects as vectors or tensors, could also be seen as nothing more than '*ipse status praesens dum tendit ad sequentem seu sequentem praeinvolvit, uti omne praesens gravidum est futuro*'[44] [Leibniz, 1879a, p. 262]. What

---

42 'If nothing were to be active by its very nature, nothing at all would be active. For what reason of action would there be, if not the nature of things?'.

43 'I consider it to be obvious that primitive forces can be nothing other than the internal tendencies of simple substances, through which they pass, on the basis of a certain law of their own nature, from one perception to a successive one and, further, they are found to be in harmony with each other'. The specific nature of the link Leibniz sees between 'law' and 'force' requires us to reject any commonsense conceptions about the former.

44 'The present state itself, when it tends to a successive one, or contains a successive one, as the one formerly coiled in it, just as all of the present is brimful of the future'.

startles Leibniz's reader accustomed to the schematic, almost reistic understanding of his metaphysics when reading that statement formulated in his key letter to de Volder in 1704, is the equating of 'forces' (appetitions, endeavours) and the 'state', which seemingly seems to be a misconception arising from an erroneous blurring of the difference between what is by definition passive ('state' or 'perception'), and that which is active ('force' or 'appetition'). But that is the reader's evident mistake, as Leibniz devotes all of his letter to a clear (and imbued with a disappointment at de Volder's stereotypical thinking) distancing himself from the concept of kinematical 'actual states' in nature, which take place one after another as a result of the external actions upon them. Perception/state cannot be deduced from anything non-dynamical, since it '*in actione consistat*' [Leibniz, 1890d, p. 328], and power (*puissance*) is action in one context and perception or 'sensation' in another (cf. [Leibniz, 1882, p. 155*f.*]; also [Leibniz, 1885a, p. 615, §52]: '... *ce qui est actif à certains égards, est passif suivant un autre point de consideration*'[45]). I claim that in Leibnizian-type ontology the 'state' and 'action' are relative and should be considered as the same internal process (*les actions internes* [ibid., p. 609, §17]) treated from different perspectives, thus different only *in abstracto* (cf. [McRae, 1976, p. 40, 60]; for a complex problem of the relation between dynamics, perceptual action and substantiality in Leibniz see e.g. [Rey, 2011]).

All of that subject-matter, as we have seen, is of fundamental importance to the understanding of the foundations of QM, as a lack of comprehension what exactly those 'state vectors' or operators as outer products of those vectors represent, draws a mix-up of quantum categories with classical and mechanical-reistic schemes of thinking about reality and objectivity. To Leibniz, the appetition–*conatus* is in a certain way a relational state, whereas the particular perception (state) as a representation of complexity in a relational whole is always an action, directed at the following states: their differentiation is accidental, depending on the specific perspective of the observer, hence it is also relativized with regards to other actions, 'accordingly to one's own reference or point of view' ('*suivant son rapport ou point de veue*' [Leibniz, 1880c, p. 475]). What is more, Leibniz stresses that one set of actions–states contains ('is brimful with') successive states, hence all of nature's life is a harmonious process of 'enfolding'/'envelopment' and 'unfolding'/'development' (Latin: *involvere–evolvere*; French: *enveloppements–développements*). The foundations of our description of the cosmos should contain not a physical calculus based on stable entities moved by external forces but

45 '... Whatever is active in one respect is passive considered from another point of view'.

pure combinatorics or an algebra of actions. Applying those anti-dualistic, internalistic ideas of Leibniz, we are able to obtain a tentative at least view of what contemporary fundamental theory could look like[46]. Leibniz was the first philosopher in the modern period to attempt to construct a model of the world in which information ('perspective') is radically relativized, and which at the same time imposes limitations on the possible extent of that local knowledge. The fundamental theory – as can be expected – cannot contain an absolute perspective, describing the Universe as if it were a 'thing', on which one looks 'objectively' (and 'object-wise') and which has non-relational states that are completely noncontextual. Instead, the theory should offer information on what internal 'states' (phases of the internal process or 'flows') are physically possible, that is: *what elementary operations can be realized by a subsystem* and *what informational relations take place between those actions of the local subsystems*.

We therefore have reasons to suspect that in order to give an answer to such general questions on the elementary operations in nature and their relations, an equally fundamental language is required, one from the borderline of pure algebra and topology, a certain type of *mathesis universalis*, whose quest Leibniz had postulated (cf. David Finkelstein's comments [1977]). For such a *mathesis* should use a strongly generalized concept of function, so as not to place prior restrictions on what can be the basis of the relations under study (their *relata*): that results from the fact that at the outset we are always dealing with different physical actions/information transfers given in distinct local frames, but we assume that there have to exist strict categorical relations between them which are guaranteed by the consistency criteria. That is exactly the Leibnizian condition of general consistency, *correspondance/liaison universelle* [Leibniz, 1880a, p. 439*f*., §14]. One of the strategies of dealing with that situation in microphysics, undertaken on the basis of the prerequisites described above, is the project to apply the topos theory to quantum gravity ([Isham, 1997], [Isham, Butterfield, 1998; 2000], [Butterfield, 2002]), so that it could serve as a convenient formalism to formulate a generally covariant physics of any local system, i.e., its 'local physical world'. Constructivist strategies in particular could for example be assumed in modelling a causal (discrete) microstructure of spacetime, whose evolution would be modelled by Heyting algebra (cf. e.g. [Markopoulou, 2000]). Leibniz himself, as a classical metaphysician endorsing Boolean physics, would probably have

46 Some interesting remarks on that issue, particularly those concerning the fundamental relationship between dynamics and causality, have been made by M. Heller [2001, p. 77*ff*.].

protested against any constructivist attempts at solving the problem of infinitesimal quantities by suspending the Principle of the Excluded Middle, but perhaps he would have supported equally strongly any attempts at reformulating fundamental theories into the physics of actions modelled combinatorially ([Finkelstein, 1999a], [Baugh *et al.*, 2001]). One of the reasons of such a state of affairs is clearly the fact that Leibniz – much more so than Newton – was concentrated in his metaphysical search on an attempt to mathematically represent the global order of the processes in nature in such a way so as to avoid any impassable dualisms (for example that of the corpuscular matter and space or of the 'state of motion' and force). Despite his purely dynamical account he constructed a form of relational blockworld ontology [Woszczek, 2011], in which the nonseparable whole is simply the global structure of relativized actions, the atemporal 'system of correlations' (*le systeme de la correspondance*) [Leibniz, 1880c, p. 476]. Thus, after all, Leibniz transferred Platonic wholes, under the guise of atemporal histories–'forms', onto the playfield of modern cosmology *ante litem motam*, before the blockworld controversy over the theory of relativity was even launched.

## 2.3 Quantum history as a hidden whole: the delayed choice quantum eraser experiment

A closer perusal of Leibniz's intution in the context of microphysics leads directly to its most intriguing and paradoxical phenomena, which are revealed in experiments. The ontology of 'monads', which are global, ordered series of actions, might to a contemporary philosopher of physics seem irritatingly speculative (as it, at least in some respect, is), but it would be worth considering how useful it could be to understanding those surprising quantum phenomena, analogically to Plato's earlier intuitions. For in Leibniz's relational and holistic universe there could occur microphysical phenomena of exactly the type predicted by QM (in fact we should even directly expect them). To fully become aware of that specific nature of Leibniz's ontology of the 'hidden whole' with relation to microphysics, one should place the stress not only on its dynamical character (since that pre-temporal, relativized dynamics is definitely *not* what commonsense metaphysics means when speaking about a 'process'), but first and foremost on its apparently 'teleological' one, which we have up to now ignored. That is exactly where the Leibnizian model is at odds with the externalistic compositionism married to the time-asymmetric presentism.

Leibniz's universe is not a mechanical one, similar to some model 'mechanism', as for example Babbage's Difference Engine, as the most profound causes which operate in it have – as in Aristotle, but in a sharp contrast to Cartesianism – the character of final causes and remain in complete harmony with efficient causes (see e.g. [Leibniz, 1860b, p. 243]). That assumption has also led to his adoption of the terminology of 'appetitions' (*conatus* = *nisus*, *appetites*), which by defintion are directed at aims, even if humans are not exclusively concerned here. Basic substances are here, as we have seen, series of aimful actions:

> I found then that their nature depends on force (*la force*), that that leads to something similar to feeling and endeavour (*au sentiment et à l'appetit*) and that they should therefore be treated by analogy with the conception we have of souls (*que nous avons des ames*). [Leibniz, 1880c, p. 479]

The adoption of Leibniz's conception of 'aimful endeavours' as an apt analogy for fundamental physical processes renders that ontological model of microphysics a special character: *for the particular phases of the local micro-process are nonseparable here from the hidden whole of quantum history of which they are a part.* Leibniz is clearly intent on incessantly stressing that every relational state is meaningful (and has to be considered) always in a global context (*lex* = *series/history*) not only of his causal past, *but also* of his causal future. Such a philosophical assumption is nothing other than an application of internalistic (Platonic) compositionism in the context of the ontology of the process. That of course raises a number of doubts from the perspective of classical mechanics based decidedly on externalistic compositionism and modelling with the help of time-asymmetric causes: the transfer of the properties of human 'endeavour' onto micro-processes in nature seems to be either a metaphysical extravaganza, or simply an anthropomorphic error. Instead of concentrating on all of Leibniz's ontology and on its assessment, we can merely check how far that analogy turns to be useful in clarifying the peculiarities of QM. One can at least see how Platonic wholes might work in physics.

One of the most astounding of those peculiarities is revealed by the quantum experiments lately conducted, which make use of two experimental schemes: that of 'delayed choice' and of the 'quantum eraser'. The first of those protocols was suggested by John A. Wheeler over thirty years ago ([Wheeler, 1978; 1979], [Miller, Wheeler, 1983]) as *Gedankenexperiment* – a modification of the experiment with slits – and then carried out in a laboratory (e.g. [Hellmuth *et al.*, 1987], [Lawson-Daku *et al.*, 1996], [Jacques *et al.*, 2007], [Peruzzo *et al.*, 2012], [Kaiser *et al.*, 2012]), bringing spectacular results consistent with the QM predictions and widely discussed in literature. The second one, that of the 'quantum eraser', was in turn proposed by M.O. Scully and K. Drühl [1982], and also by Scully, B.-G. Englert and H. Walther [1991], and then carried out with much success in many experiments of that type (e.g. [Zajonc *et al.*, 1991], [Kwiat *et al.*, 1992], [Herzog *et al.*, 1995], [Pittman *et al.*, 1996], [Walborn *et al.*, 2002], [Trifonov *et al.*, 2002], [Ma *et al.*, 2013], [Weisz *et al.*, 2014]). In the year 2000, a team of physicists from the University of Maryland together with Marlan O. Scully from the Max-Planck Institut für Quantenoptik in Munich [Kim *et al.*, 2000] proclaimed to have carried out an experiment in which both those protocols had been applied, thus carrying out to the full the original idea of the experiment from the article by Scully and Drühl [1982] and at the same time an extremely refined version of the experiment with slits. The obtained results are, true to their expectations, correctly predicted by QM, but at the same time are wholly counterintuitive, which is usually stressed when their philosophical implications are referred to. As no experiment can serve as a verification of any ontology, one should carefully approach the formulation of such 'implications'. In that context we are solely interested in the way one could try to understand the obtained results in the framwork of a certain consistent ontology, so as not to hastily agree to the necessity of accepting insoluble 'quantum paradoxes'. Nevertheless, classically they *are* paradoxes indeed.

The experiment with the quantum eraser (Fig. 1) is based on the quantum phenomenon of entangled pairs of orthogonally polarized photons, which are randomly separated and then separately undergo measurement procedures: one of them, measured earlier, is called the *signal photon* (*sp*), whereas the second one, measured later and transmitted through a system performing the role of an optical interferometer with equal-path arms, is called the *idler photon* (*ip*). (Of course that sorting of photons is completely random, and the terms refer to the photon's history in the experimental setting and not to any physical properties they might contain). The parameters of the arms of the setting are designed in such a way to enable the measurement of the *ip* state with a delay of at least 8 ns with regards to that of *sp*. Pump laser beams of photons fall upon two strictly defined nonlinear optic crystal regions (*NOC*), A and B, performing the role of slits from the classic quantum experiment, and next the entangled, orthogonally polarized photon pairs from either A or B are separated into *sp* and *ip*. Signal photons pass through the lens (*L*), thanks to which the distance between *NOC* and the mobile detector *D*0 can be regulated, with the latter measuring the position of *sp* (the measurement itself does not of course inform us about the source of the *sp*, i.e., the region of the crystal where the emission began, and neither does it supply information about its precise spatial path to the detector). Idler photons in turn fall upon prism (*P*), where their trajectories are modified depending on whether they come from region A or B. Each *ip* then has a 50% chance of passing through the 50–50 beamsplitter straight into the detector: in the case of a photon from region A that is beamsplitter *BSA* and detector *D*3, while in the case of a photon from region B it is beamsplitter *BSB* and detector *D*4. In the instance when the *ip* gets through one of the beamsplitters, though, it reaches the mirror (*MA* and *MB* respectively) and reflected further on passes through joint 50–50 beamsplitter *BS* (the probability of letting through or reflecting the photon is again 50% there). If a photon coming from region A is let through *BS* further on, then it will be detected by *D*2, while if it becomes reflected, then it will be detected by *D*1. And on the contrary, a photon from region B let through *BS* will reach detector *D*1, while reflected by *BS* it will reach *D*2. The recordings from all the detectors are finally gathered in the coincidence circuits *CC* registering detection rates for the pairs of joint signals (from entangled pairs of *sp* and *ip*): *D*0–*D*1, *D*0–*D*2, *D*0–*D*3 and *D*0–*D*4.

*Fig. 1: The scheme of the delayed choice quantum eraser experiment. The whole arrangement is construed from two subsets: one of them consists of a lens L and a detector D0, while the second one is a complex subset comprising the 'quantum eraser'. NOC: nonlinear optic crystal with two separate regions of photon beam emission, A and B (circa 0.7 mm apart), and P: Glen–Thompson prism differentiating orthogonally polarized, entangled pairs of photons from each region; BSA, BSB: 50–50 beamsplitters marked 'A' and 'B' for photons from respective regions of NOC; BS: 50–50 beamsplitter performing the role of a quantum eraser for the idler photons; MA, MB: mirrors for the photons generated at A and B; D0, D1, D2, D3, D4: photon detectors, including mobile detector 0; CC: registering device for signals from the detectors (coincidence registrator). As in [Kim et al., 2000, p. 2].*

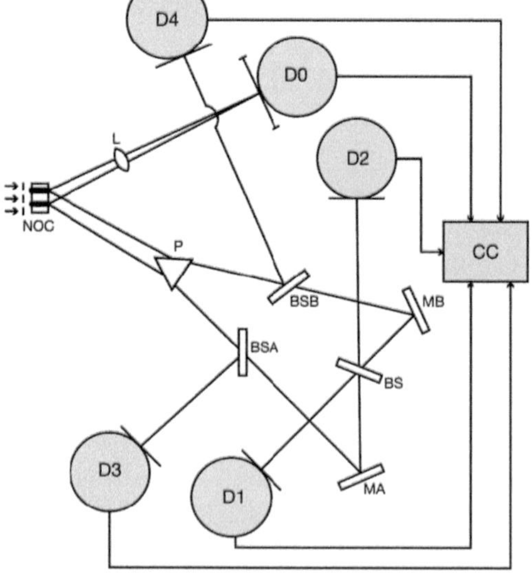

The novelty of the experiment consists of an at first glance seeming possibility to bypass, with the use of the entangled particles, the fundamental limitation arising from QM principles: it would be apparently possible to precisely define the position of the *sp* by *D*0 and at the same time of the trajectory it had travelled solely thanks to registering *ip* at detector *D*3 or *D*4, while retaining its interference pattern at *D*0 intact. If QM is an adequate theory, then such a situation of course cannot take place, as gaining knowledge – thanks to *ip* in statistically half of the cases – which crystal region also *sp* comes from (and hence finding out also its trajectory from *NOC* to *D*0), should affect the phase of the quantum wave and completely destroy the final interference pattern created at *D*0 after the emission series. Beamsplitter *BS* in the interferometer subsystem prepared for *ip*-s is here a 'quantum eraser', as in the case of half of the idler photons it 'erases' the information about which crystal region the particle registered at *D*1 or *D*2 comes from. Another important property of the experiment worth noting here is the fact that the physicist–observer does not take any decisions at any of its stages, and it is nature itself which randomly 'decides' about a given trajectory and about the result (particularly at *NOC*, *BSA*, *BSB* and *BS*). Scully and Drühl suggested exactly that type of *Gedankenexperiment* as a good test of the paradoxical nature of QM, arising in fact from its holism manifested in the phenomenon of entanglement (cf. Ch. 3). For one can expect that if the entanglement takes place without exception as a result of quantum interaction, then the arrangement like the one presented in Fig. 1, would enable us to obtain in a certain number of cases knowledge of the trajectory of the particles and *at the same time* retain the interference pattern at *D*0, which must have been generated *already earlier*, still before taking any measurements on *ip*. Yet if nature, in accordance with QM expectations, does not enable the realization of such a possibility, then the obtained result gives that acute 'quantum paradox' connected to the process of 'erasing the past' of part of that *sp* ⊕ *ip* system (in that case of the earlier *sp* state), and hence clearly a certain serious *temporal-causal paradox*. That is a common feature of all experiments of this type.

The delayed choice quantum eraser experiment always brings results fully accordant with QM predictions, and hence *sp*'s position on the screen *D*0 *depends* (is *correlated*) on which trajectory the *ip*, entangled with *sp*, will *later* realize: when *ip*-s are registered by detectors *D*3 or *D*4 (which-way information is accessible), then the registered positions of *sp*-s are to be found along the *x* axis forming an ordinary distribution curve for two summed probability waves (without interference), whereas when *ip*-s are detected by *D*1 or *D*2 (information about the region of emission is inaccessible), then the serially registered positions of *sp-s* create a clear interferential curve along the *x* axis, typical for interfering quantum waves from the experiment with slits. The problem is that in line with the way of thinking taken over from classical mechanics the history of *ip* in the quantum eraser subsystem cannot bear any influence on the state of *sp*, as at the moment the information about the position of *sp* is scanned and transmitted through *D0* to the circuits *CC*, the history of *ip* is not realized yet (*ip*'s presence is not registered by any of the remaining four detectors). Clearly, in all of that reasoning which assumes a classical treatment of the history of the $sp \oplus ip$ system there lies an error or simplification leading to a completely mistaken assessment of the whole experimental situation. Artur Zajonc quite rightly remarked, when writing about the conservatism of the physical imagination in the context of Goethe's nature-philosophical studies:

> Understanding appears to require an image, and the inevitable trend is toward the machine. The imagination of a mechanical universe has provided science with a powerful means of understanding a large but finite range of effects. There exist, however, even within the domain of physics, phenomena such as those described above that simply cannot be thought of in mechanical terms without spectacular violations of logic or simple common sense. Here enters the arrogance of the tradition. What cannot be imagined mechanically cannot be imagined at all. One can compute and predict on the basis of computation, but one must forego the old pleasures of understanding, or at least modify our traditional sense of what it means to understand [Zajonc, 1998, p. 308*f.*]

The first, key element of the classical mechanical interpretation of the situation in Fig. 1 encompasses the 'evident' reality solely of past-time, time-asymmetric ('efficient') causes and the causality based upon them, accordant with the 'objective' thermodynamic arrow of time. The pair of photons $sp \oplus ip$ cannot behave in any correlation with its hidden *future* states, as neither physical future is determined in some of its aspects nor does it bear any influence on the earlier physical states. (Of course, the future photon states are limited to a certain set of possibilities, but that is a sort of Aristotelian 'potentiality', a physical limitation imposed by the laws of nature, and not the physical future history of those photons, which co-determines their actually registered local states). The classical image is a replica of the explanation scheme using the simple ontology of externalistic compositionism: the situation *elements* here should be the temporary, momentary physical 'states' of each photon, which then successively 'compose' wholes – their *real histories in real, flowing time*. In the end (after the experiment is over) two separate, real histories are produced, that of the signal and the idler photon, which shall together create some non-classical whole $sp \oplus ip$ by the formal external relations between their elements. The scheme seems to be of no use in the case of the quantum eraser experiment, though, as it leads to a paradox: for later action seems to 'erase' the occurence of certain already registered physical states and change the whole past history, which is causally prohibited. Even more, the delayed choice quantum eraser scheme can be used to prepare the time-reversed EPR experiment entangling the systems *from their future* [Elitzur *et al.*, 2002]. There additionally appears the paradox of violating the principle of the conservation of information: the information would have to be – literally – *physically* annihilated. There must be something fundamentally wrong with the whole analysis of that sort, which presupposes the base elements for compositionism.

From the perspective of the earlier analysis (Ch. 1.2 and 1.3) it clearly seems that the classical imagining introduces specific *absolutes* into the physical interpretation in the form of temporary, intrinsic photon states (independent of the state of the other photon and that of the registering external system), and also the absolute of the 'flowing now', which demarks two ontological spheres: that which is already realized (non-relational past 'facts'), and that which is still undetermined/unrealized (non-relational future). A radical relativization of those absolutes seems to be promisingly fruitful, but at the same time forces us to completely discard the ontology of externalistic compositionism in favour of an internalistic one. What is more, the admission of a kind of 'teleological' causation, in which future states are physically (nontrivially) correlated with past ones, makes the old Leibnizian vocabulary with 'appetitions' or 'endeavours' seem to be to a certain extent apt. Both in the Platonic and in the Leibnizian-type ontologies, 'momentary states' as seemingly autonomous, noncontextual 'facts' cannot be elements composing an accreting history, as they are justified solely in the regime of a global, atemporal hidden whole (*structure of events with the timelike correlations* determining only *purely relational facts*). The experiment with the quantum eraser becomes comprehensible when we assume that the *atemporal* quantum history has to be defined globally (although inaccessible as a whole to any internal subsystem) and the past need not be 'erased' in reality by some backward-directed physical process of deleting what has already occured as a seemingly non-relational (classical) 'fact'. The system with a quantum eraser behaves simply *as if* the signal photon 'knew' the whole history, hence also the future of the idler photon, and *vice versa* (so they are *entangled in time* as a whole quantum history). A similar 'teleological' character is revealed by the even more astonishing so-called Hardy's quantum paradox, which has recently been experimentally realized ([Yokota *et al.*, 2009], [Lundeen, Steinberg, 2009], [Fedrizzi *et al.*, 2011], [Karimi *et al.*, 2014]), despite an earlier conviction that it was purely conceptual. In actual fact, that strongly non-classical behaviour of quantum systems is certainly no exception, but is a *direct* result of fundamental QM principles. Thus an apparent 'paradox' is a rule.

There is another important component of the mechanical image, though, which plays the part of an absolute in the classical analysis of the whole situation: the 'observer' himself. The illusion is upheld by the very way in which the experiment is presented, as it suggests that the measuring external system itself does not play any part in the physical process taking place between the pump laser, beamsplitters, detectors and registering device. Using the earlier terminology, therefore, one can notice that the 'observer' as a physical subsystem is also treated here, alongside the measured 'factual states', as an absolute, irreducible element of the whole experiment, separated from it as being independent of the registered quantum history, which the mechanical framework requires. Since, which has already been stressed, the 'observer' does not take any decisions during the experiment, then that impression is magnified even more, and we are thus prone to assume that its role is limited solely to the preparation of the system and to checking the obtained coincidence rates. When we analyze the situation in the light of internalistic compositionism, then such a rigid objectivization of the internal and the external system ('observer') does not have any sense: both the former and the latter are part of the hidden whole (a global, purely relational structure), and any local events can after all be described only from the *action perspective* of one of its subsystems, arbitrarily selected. If we approach the problem in a more speculative way (*in modo Leibnitiano*), we are entitled to assume that from the 'perspective' of some pair of photons or electrons in atoms which compose mirrors the experiment looks different. It is reasonable to assume that the whole system in Fig. 1 together with the fixed 'observer' are just arbitrary sections of the quantum history, which is described here only from a certain special perspective treated as canonical. The singled-out 'perspective' is one that is always relative, as we assume that there is no such thing as an internal physical 'observer' (a part of the whole), who has a 'view from everywhere', registers the whole experiment as it 'really' happens. Therefore the full, globally defined quantum history is always a hidden whole, never empirically accessible to any particular subsystem.

The above conclusions, motivated by the wish to eliminate absolutes from the physical theory (D.R. Finkelstein), can appear to be convincing, but what still turns out to be acute is the issue of an adequate imagining of the whole physical situation, of which Zajonc writes. If the correlations between all the physical 'states' (= actions) are globally, formally defined within the structure of a 'hidden whole', how are we to imagine that whole, if we consciously discard the metaphor of the universe composed from some elementary, non-relational components? The question naturally assumes that some imagining in the metaphysics of nature is at all necessary, but we can critically reply here that in the case of comprehending certain general theories about nature there always appear some guiding metaphorical images, even if that is involuntary. It is also clear that such an image or metaphor of the nonseparable 'hidden whole' of the universe shall radically diverge from the common working metaphors employed by classical physics and its status has to be to a large extent problematic, which was strongly stressed by Plato ('pictorial discourse', *eikōs muthos* of a lower grade, as a cognitive limit in cosmology, *Tim.* 29c–d), as well as Leibniz (the presentation of nature's causal structures by convenient analogies, cf. [Leibniz, LPP, p. 284]). One thing is a purely mathematical model, the other its possible visualization if that exists at all. Nevertheless, such an alternative imagining has appeared in the history of metaphysics in different contexts (not only in the West), though as a rule it was associated with the 'idealistic' tradition in philosophy. In modern Western metaphysics such metaphors were gradually marginalized by an image of nature as a mechanism, which must have been directly connected to the discovery of the verge escapement and the growing popularity of the mechanical clock which 'divides time uniformly into discrete segments' [Whitrow, 1989, p. 99*f.*] in Europe from the 14th century onwards. Nonetheless, the image of nature as a machine is of course older than that: already Lucretius [1917] wrote in his *De rerum natura* (V.96) of '*moles et machina mundi*' ('the great construction and machinery of the world', see e.g. [Balme, 1941]), and next the metaphor was used in the West by such various thinkers as Nicholas of Cusa, Bruno, Descartes, Vico, Boyle or d'Alembert ('*fabrica mundi*', '*orbis machina*', '*machina caelestis*', Newton's God as *mechanikos* etc., see e.g. [Mayr, 1986], [Mittelstrass, 1988], [Pulaczewska, 1999, p. 163*ff.*]). The atomistic genealogy of the image, later supported in the mass imagination by the common presence of clocks and other complex mechanisms, unanimously points to certain types of thinking about the construction of nature which we have earlier on called extensional and externalistic compositionisms. Internalistic compositionism, in turn, which introduces the unconditional primacy of the whole over the parts, requires a completely different type of imagining and that sort of innovation has historically been quite rare in modern metaphysics.

For example, one of such non-mechanistic images had been tried on an idealist ground by Arthur Schopenhauer in the 19th century, who in his *Parerga und Paralipomena* first published in Berlin in 1851, reflected on the interconnections between events in nature, as well as randomness and necessity. What is interesting is the fact that Schopenhauer (just as Friedrich W.J. Schelling earlier on) had constructed a dynamistic ontology of nature, in which the key issue remained to be not what the unchanging essence of every entity was, but how it acted (for Schopenhauer, the fundamental category in ontology was blind 'ur-will' or 'ur-force', *Urwillen–Urkraft*, which, as in Leibniz, has the aspect of 'desire', *Begehren*). In the work above mentioned he analyzes the nature of chains of events in nature and invokes the picture of a sphere (let us call it

*Schopenhauerian classical sphere*), whose meridians are a reflection of 'chains of causes (*Kausalketten*) running in the same direction as time', entwined in different ways into the 'pervasive, interconnected net' ('*ein großes, gemeinsames, vielfach verschlungenes Netz*'), thus creating a global 'history of the world' (*Weltlauf*) [Schopenhauer, 1851, p. 204]. Through this sort of imagining Schopenhauer introduces a certain type of global superdeterminism, as every event is always 'a part of a certain chain of causes and actions' ('*das einzelne Glied einer Kette von Ursachen und Wirkungen*'), and hence 'nothing is absolutely accidental' [ibid., p. 203]: randomness and unforeseeability are the result of the local position of the acting subject. What mostly exhorts that metaphysical image is 'the interconnectedness of all things' ('*Gesammtheit der Dinge*' [ibid., p. 204]) and the contextuality of the parts (events–actions), whose context are the relations determined by the whole. Any event as a process in such a Schopenhauerian cosmos would not, strictly speaking, have an absolutely open future or the characteristic of an 'atom of history' or an 'absolute accidentality', but would be globally determined already within the hidden whole of nature. 'Thus everything is mirrored in everything (*Alles in Allem*), every entity rings out in every other (*Jedes in Jedem*)' [ibid.] – he writes of this all-encompassing interconnection and quotes the same motto by Hippocrates which Leibniz had quoted in his *Monadology* (§61). To Schopenhauer, inspired by Kant and oriental thought, even the passage of time (and the arrangement of processes in space) as a certain mental presentation is only a form of a local observer's apprehension of the connections between series of 'actions' (*Wirkungen*). In that sense, any 'algebra of the process', just as the above picture of the 'Schopenhauer (classical) sphere' spread by series of actions represented by points, would be a mental representation (*Vorstellung*) of an atemporal whole of nature. A much more complex and interesting theory was created by Schelling in his middle-period *Naturphilosophie* as a speculative 'Spinozism of physics' [Schelling, 1927, p. 273], as he himself called it. He was striving for a model of the universe as the pre-temporal, global 'Organization' with an internally relational structure and combinatorial, local physics of actions (*Aktionen*) in the wake of the Leibnizian tradition. Here one has a primitive, post-Leibnizian example of the blockworld with the built-in elementary dynamics, or even the concept of self-organization.

Further, the Leibnizian image of actions mirroring each other ('*miroir de l'univers à sa mode*' [Leibniz, 1885a, p. 618]), i.e., a holistic metaphor of the 'mirrored universe', is much more complex than the Schopenhauerian sphere. It is surprising that such analogies have not appeared solely in the metaphysics of the West, but also have their direct counterparts in the philosophy of the Far East: the most well-known of those is probably Fazang's 'Hall of Mirrors' (the Buddhist master of the *huayan* tradition, China, 7th/8th c.), whose links with Leibniz's and Whitehead's metaphysics have already been noticed and critically studied (e.g. [King, 1979], [Liu, 1982], and most fully in [Odin, 1982]). It turns out that the idea of a mutual mirroring and 'Adaptation of All Things' (Leibniz) has its direct structural counterpart in the Buddhist doctrine *li-shih wu-ai* ('the interpenetration of parts and wholes') and *shih-shih wu-ai* ('the interpenetration of parts and parts'), illustrated by Fazang's famous performance. Odin's [1982] analysis additionally highlights the central importance of the relation of the part to the whole, which plays a fundamental part in the ontologies of both Plato and Leibniz: the latter even uses formulations astonishingly similar to the technical terms in the *huayan/kegon* tradition, defining the events–perceptions, or the 'transitory states' of nature as 'multitude in unity' and 'unity in multitude' (e.g. '*l'ětat passager qui enveloppe et represente une multitude dans l'unité*' [Leibniz, 1885a, p. 608]). While the greater part of Indian Buddhism adopted a reductionist standpoint (the primacy of the part and the reducibility of every whole), in China the problem of the relation of the whole (*li*) and the part (*shih*) underwent a deep reformulation precisely in the framework created by the *huayan* school, starting with its first patriarch, Dushun (6th/7th c.) and his treatise titled *Mirror of the Mysteries of the Dharmadhātu of the Huayan*. The hidden whole was differentiated here from the sum of visible parts (phenomena) and as irreducible, but present in all phenomena, identified with *śūnyatā*, 'emptiness', i.e., the lack of a non-relational, essential nature (a detailed analysis in [Jiang, 2001]). What is more, in the analyses of *huayan* and in particular in Fazang's treatise *Sanbao zhang*, time is a radically relational concept, dependent on the relation of a concrete state (*dharma*) to other ones, so that every one of them is at the same time past, present and future in the net of those relations, and future states are coupled to past ones and they even influence them [Vorenkamp, 2005], which will also appear as a possibility in the quantum relational blockworld in Ch. 4. From the perspective of the 'hidden whole' (*li* = *śūnyatā*), radically different from the empirical-phenomenal perspective, there is no real flow of time in such a universe. As Fazang writes in his extensive commentary *Dasheng qixin lun yiji*, all 'passage' – though that is difficult to encapture by an imagination shaped by sensual perception – takes place 'in one moment', a 'before and after' does not exist in it (quoted from [ibid., p. 208]). Such intuitions are by no means proper solely to oriental philosophy nor are they essentially burdened with a suspicious 'mysticism': in the fragment of the poem Περὶ φύσεως (*On Nature*) by Parmenides of Elea, which has survived thanks to Simplicius (*In Arist. Phys.* 145, 5–6 = B 8 DK), we can read of the Being (World): οὐδέ ποτ᾽ ἦν οὐδ᾽ ἔσται, ἐπεὶ νῦν ἔστιν ὁμοῦ πᾶν, ἕν, συνεχές ('Nor was it ever nor will it be, since it is all now, the Whole, One, holding together in itself'). We should understand Parmenidean physics not so much as the one in which (as Karl R. Popper once phrased it) 'nothing takes place', 'no time does flow', but as one in which the ἐόν as the supra-empirical whole (πᾶν) simply *is* in an unperturbable 'now' (νῦν), so *reality as such is nontemporal* (see [Mourelatos, 2008, p. 105*ff.*]. The subject-matter is just a primitive, purely metaphysical version of the issue of the quantization of the closed universe as a physical system – as we shall see, what is needed here is a formal reconciliation of the local, totally relational action frames of physical subsystems with global, timeless physics. One can guess that without a proper model of intensional compositionism for physics that task is doomed to fail.

The exceptional character of such imagining of the structure of nature is based on the fact that none of the particular empirical states or any class of them is singled out, and hence there is some joint relativization of the event, observer and information built into the model itself. Also the non-classical 'empirical whole' (such as the quantum 'state' of two entangled particles, modelling their physical relations) is always relative: it would refer to information relativized to the internal 'observer' which registers some strong correlations between events that cannot be explained by assuming externalistic compositionism. The whole of Universe, **Ω**, in turn, or the 'quantum history of the Universe', cannot be an empirical concept in the common understanding of the word (i.e., having anything to do with physical interactions between the systems), as that whole is never accessible to any empirical subsystem: that precisely seems to be the sense of speaking of a cosmological 'hidden whole' in the metaphysics of nature. It will be interesting now to analyze if there exists any real link between the relativization of the parts and the local breaking of the symmetry of evolution in the quantum frame of reference of a subsystem that is updating the information about a different subsystem. We shall deal with that problem in the context of QM in Ch. 4.2 and 5.1, but first what seems a key issue here is a precise reflection as to what extent microphysics should be treated as radically holistic, thus a subtle controversy between the *signal/spacelike nonlocality* in a quasi-classical world with the time asymmetry and some 'elements of reality', and the *global nonseparability* in a generically quantum blockworld **Ω** with the time symmetry and without any noncontextual 'states of affairs'. That in fact is a problem every interpretation of quantum theory has to face, in one way or the other, and behind that there is always the more fundamental, metaphysical problem of intensional compositionism.

# PART TWO

# Chapter 3
# Nonseparability as the Ur-Phenomenon of Mechanics and Holism in Microphysics

## 3.1 Is quantum mechanics magic? Einstein's Local Realism, the Bell–Kochen–Specker Theorem and quantum contextuality

When we trace the development of QM from its beginning right up to the 1970's and 1980's, with its culmination in the recent theoretical and experimental research on quantum computation, teleportation and cryptography, it seems that little by little those of its elements have been being uncovered and treated seriously which had earlier been seen as 'purely metaphysical'. The most astounding phenomenon in microphysics is surely the phenomenon of entanglement and the apparent quantum 'nonlocality' implied by the former, or so it seems, which is interpreted in many, often contradictory ways – including such which openly violate relativistic physics with its local causality. (In that context one can quote the words of Daniel M. Greenberger, somewhat provocatively: 'Quantum theory is magic'[47]). How that 'nonlocality' should be understood physically is of direct importance for defining the type of holism we are dealing with in QM and even after eighty years one still has to position oneself against the background of the old Einstein–Bohr debate. What is of key significance here is analyzing at the very outset the logical content and ontological assumptions of the so-called 'EPR argument' in favour of the incompleteness of QM. Without such a positioning a clear tracing of the metaphysical core of the debate is sometimes impossible, since, I shall argue, it is exactly metaphysics itself which is at play here, and not just some physical epistemology.

The EPR argument was put forward by Albert Einstein, Boris Podolsky and Nathan Rosen in their well-known article in the *Physical Review* in 1935, titled *Can quantum-mechanical description of physical reality be considered complete?* The construction of the argumentation might make an impression even today, ending in a decisive philosophical conclusion: 'No reasonable definition of reality

47 D.M. Greenberger at the conference *Fundamental Questions in Quantum Mechanics*, State University of New York, Albany, 12–14 April 1984; quoted from [Mermin, 1985, p. 38].

could be expected to permit this' [Einstein *et al.*, 1935, p. 780]. What 'reasonable definition of reality' were the authors writing about and why should we confront that challenge even today? The essence of the EPR argument lies in the experimental proposal, even if that was at first only a Gedankenexperiment – from the beginning of the 1980's frequently realized with a systematically growing refinement and the precision of laboratory technology (e.g. [Tittel *et al.*, 1998], [Pan *et al.*, 2000], [Julsgaard *et al.*, 2001], [Zhao *et al.*, 2004], [Hasegawa *et al.*, 2004], [Go, 2004], [Mohring *et al.*, 2004], [Hafner *et al.*, 2005]) – which imposes a conceptual choice between the assumption about the completeness of QM and the assumption about the locality of physical operations demanded by the theory of relativity. Einstein's most serious accusation, being directed at QM already since the beginning of the 1930's, even before he wrote the article together with his assistants from the Institute for Advanced Study in Princeton, grew from his conviction that, to put it simply, the 'Copenhagen' formulation of QM annihilated all of the realistic and commonsense programme of the sciences that had developed since the 17th century. The programme, according to Einstein, was to uphold – as an axiom – the theorem about the complete independence of the studied physical events from the observer and his/her experimental operations, hence he considered the interpretation of QM formalism Bohr had introduced, particularly that of 'quantum complementarity', as being shockingly antirealistic (see e.g. [Held, 1999]). The accusation, in Einstein's opinion, was connected to the statistical nature of QM predictions and to his faith in determinism, but strictly speaking, it constituted an independent and deeper foundation of his reluctance with regards to the quantum theory. All of his efforts were directed at a possibly clear and irrefutable demonstration of how QM forced a choice between an *a priori* statement of its theoretical incompleteness and the 'magical' physical effects destroying the axiomatic foundation of the theory of special relativity (henceforth SR), to which the conclusion quoted above referred to. When studying the classical form of the EPR argument, we have to keep in mind that it concerns the nonrelativistic QM (henceforth NRQM), and its extension, for example, to the algebraic quantum field theory (henceforth AQFT) requires a suitable adaptation of the formalism, in this case defining the correlations of the projections in local commutative von Neumann algebras $\mathcal{A}(R_n)$ for appropriate regions $R_n$ of Minkowski spacetime $\mathbf{M}^4$. Nevertheless, it turns out that – as I shall point out in Ch. 4.1 – even NRQM with its noncommutative Heisenberg relations seems to possess a deep connection with $\mathbf{M}^4$ symmetries. The EPR argument, in turn, became thirty years later a direct inspiration for formulating Bell inequalities (and their generalizations) and the (Bell)–Kochen–Specker theorem, imposing fundamental restrictions on every consistent interpretation of the quantum theory, not to mention the whole field of experimental ingenuity. Thus the old EPR story is still central to quantum ontology.

It should be ascertained what the logical structure of the EPR argument is, as that could help to at least preliminarily assess what ontological concessions the quantum-mechanical description of the microworld requires from us. Broadly speaking, the original argumentation contained in the article was formulated on the basis of two simple theses and two assumptions, and also commonsense 'conditions', which jointly, together with the use of the entanglement effect predicted by QM (being a direct consequence of formalism), were to lead to the conclusion that theory had to be incomplete. The argumentation begins with distinguishing two questions a researcher may pose when probing the 'success of a physical theory': (a) 'Is the theory correct?', and (b) 'Is the description given by the theory complete?' [Einstein *et al.*, 1935, p. 777]. After a preliminary description of the specific nature of the measurement of a state in QM, thesis (1) is introduced, claiming that: 'the quantum-mechanical description of reality given by the wave function is not complete', and thesis (2) stating that 'when the operators corresponding to two physical quantities do not commute, the two quantities cannot have simultaneous reality' [ibid., p. 778]. To understand those statements in the way the authors intended them, the concepts of the 'completeness of the theory' and of 'having reality' need to be clarified first, as they both condition the whole argument and at the same time support it. The first appears at the very beginning of the article:

> Whatever the meaning assigned to the term *complete*, the following requirement for a complete theory seems to be a necessary one: *every element of the physical reality must have a counterpart in the physical theory*. We shall call this the condition of completeness. [Ibid., p. 777]

The term 'elements of reality', which appears in that text, is not defined for, as the authors write, perhaps accurately, it 'cannot be determined by *a priori* philosophical considerations', nevertheless an intriguing criterion is formulated, offering the basis for formulating the 'sufficient condition of reality':

> We shall be satisfied with the following criterion, which we regard as reasonable. If, without in any way disturbing a system, we can predict with certainty (i.e., with probability equal to unity) the value of a physical quantity, then there exists an element of physical reality corresponding to this physical quantity. It seems to us that this criterion, while far from exhausting all possible ways of recognizing a physical reality, at least provides us with one such way, whenever the conditions set down in it occur. Regarded not as a necessary, but merely as a sufficient, condition of reality, this criterion is in agreement with classical as well as quantum-mechanical ideas of reality. [Ibid.]

The first assumption in that argument, **(*1)**, is that the two above propositions exclude each other, i.e., there is a true alternative: '(1) ∨ (2)'. It is as a matter of fact a definition of a commonsense expectation, which we have the right to put forward even with regards to as non-intuitive a theory as QM: if physical quantities always had their actual definite values, despite their noncommutativity, and hence 'had reality' then, consistently with the above interpretation of the *completeness of the theory*, they would constitute, jointly and objectively, a complete description of quantum reality, but QM as such, with its noncommutativity of the operators, would unfortunately be incomplete, that is: '~(2) → (1)'. On the other hand, if QM is objectively a complete theory, then – in accordance with its plain formalism – noncommutative quantities cannot at the same time be real and definite: '~(1) → (2)'. And it is at this point that the second, key assumption of the argument appears, **(*2)**, which claims that if QM were to contain that complete description of reality, then it would simply allow us to find those real values of all the physical quantities of interest to us (and all the possible mechanical quantities, as a matter of fact), and thus: '~(1) → ~(2)'. There is nothing extraordinary in (*2), as it also results from the 'criterion of reality', while the truth of thesis '~(1)' is, as the author writes, 'usually assumed' (what is kept in mind there is the unofficial 'Copenhagen school', among others Bohr, Pauli, Jordan and Heisenberg, defending the completeness of QM). Hence, if we agree that (*2) is a true implication with a concurrent assumption that '~(1)' is a true statement (the completeness of QM), then we obtain a contradiction, as the alternative in assumption (*1), in accordance with de Morgan's logical law, turns out to be false. Simply speaking, when we assume the thesis that QM is complete, that is '~(1), the result are two opposing assertions, (2) and ~(2), given in the assumptions, and hence by *modus tollendo tollens* we obtain (1), which ends the proof (cf. [Chibeni, 2005]). Einstein, Podolsky and Rosen do not, nevertheless, merely choose such a simple 'conceptual' argumentation, but in their famous second part of the article wish to directly demonstrate that in certain situations, QM nevertheless allows to find the values of noncommutative quantities, and thus '~(2)' takes place, which is made possible by the phenomenon of the entanglement of quantum systems. In order to demonstrate the physical peculiarities in the case of systems that will later be called 'entangled', the authors propose considering a measurement of two different quantities (represented by noncommutative Hermitian operators) on spatially separated systems and putting them to the test with the use of the above 'reality criterion'. That very operation turns out to be not only controversial, but also important from the ontological point of view, as alongside the reflections on the entanglement situation, the authors introduce in that part *two other* extremely important assumptions underlying their argumentation (one that is concealed and the second openly expressed) although they are not particularly distinguished in the whole text nor are they later recalled. At the same time, that point of the argumentation was, as we know, the target of Bohr's criticism, the latter having never accepted Einstein's strong 'reality criterion', introducing instead a *radical contextuality* of every measurement (information transfer). And, indeed, that point is of extreme importance for ontology, opening at the same time a Pandora's box of metaphysical ambiguity.

The complication of the whole argumentation that appears in the text, written in fact by Podolsky himself after many discussions together with Einstein, and also the stress put on its technical aspect as a result of the presentation of the entanglement problem, were something Einstein did not like[48]. He later on simplifed the 'EPR argument' and construed it in the cript form it is summarized in today (the history and excellent presentation of the whole argument, also on the basis of archive materials: [Fine, 1996], [Howard, 1985]; also [Deltete, Guy, 1991], [Chibeni, 2005], and, first of all, [Held, 1999]). The second part of the article introduces the situation where measurement of the incompatible observables $P$ and $Q$ (represented by noncommutative operators) is conducted on two systems I and II, spatially separated and prepared by interaction in a 'combined' (quantum entangled) state I⊕II, hence each system can be ascribed one of the eigenfunctions of those operators, $\psi_k$ and $\varphi_r$, respectively (it need not be so, but the authors wish to point to situations where the operators cannot be constructed upon the same basis of the state space). Thus a state of any subsystem can be after that represented as e.g. $\psi_k$(I) or $\varphi_r$(II). The key aspect of the situation is connected to that very use of the 'reality criterion' presented earlier: the measurement on one of the subsystems of any of the two observables, e.g. $P$ and the value $p_k$ obtained, allows the experimenter to determine (with a probability equal to one) a corresponding value of that chosen quantity for the second subsystem, and vice versa (exactly as in Ch. 1.3). In line with the 'reality criterion', therefore, both physical values–states, a measured $p_k$(I) and deduced $p_l$(II), constitute 'elements of reality', are 'real'. That very same argumentation can be applied with regards to the functions $\varphi_r$ and the corresponding eigenvalues $q_r$ of the operator $Q$. Hence, in the authors' own words, '(…) in the first case we must consider the quantity $P$ as being an element of reality, in the second case the quantity $Q$ is an element of reality. But, as we have seen, both wave functions $\psi_k$ and $\varphi_r$ belong to the same

48 After the material was published in the *Physical Review*, Podolsky publicized the issue and an article appeared in the *New York Times* on 4 May 1935 proclaiming that the authors had proved the 'falsehood of QM' – Einstein was so furious that he broke all contacts with Podolsky [Peres, 2004].

reality' [Einstein *et al.*, 1935, p. 780]. We are dealing here, though, with a certain logically not self-evident physical conclusion from counterfactuality: the very possibility to perform measurement of *P* or *Q* on system I or II would enable them to be 'elements of reality'. We therefore ask: why? A discerning reader of the article will at once note that such a counterfactual conclusion about two totally independent 'elements of reality' requires further conditions to be met there for the phenomenon of entanglement of the system (I⊕II) to be possible to be applied in a similar reasoning. The conditions are: that of locality and that of separability, which in the end played a much more important role in the discussion on the EPR argument than the definition of the 'reality condition' itself. That was an opinion of Einstein himself.

The (causal) locality condition finds its direct formulation in the text already at the very beginning of the argumentation, when the authors write that 'we suppose that there is no longer any interaction between the two parts' [ibid., p. 779]. (Einstein believed that we should 'absolutely hold fast' by that condition [Einstein, 1949a, p. 85], as it is the foundation of mechanics and all of modern physics as such). And further, clearly repeated once again as the 'locality of interaction':

> On the other hand, since at the time of measurement the two systems no longer interact, no real change can take place in the second system in consequence of anything that may be done to the first system. This is, of course, merely a statement of what is meant by the absence of an interaction between the two systems. [Einstein *et al.*, 1935, p. 779]

On the other hand, in the course of the argumentation we discover a further, hidden condition, s e p a r a b i l i t y – so commonsense that the authors probably did not consider it particularly important at that stage to highlight it. The condition points out that two separate systems possess, even in the situation when a quantum measurement is being performed on one of them, their own *ontic identity* and *independence*, i.e., the 'reality of each' (to use Einstein's wording) does not depend in any significant way on the reality of the other. In that way we guarantee that despite the possible interdependence of the states, revealed in the correlations coded in the wave function, two systems are *ontologically autonomous* and cannot in a physical way influence the conditions of their 'reality' (we shall return to a more precise definition of separability in Ch. 3.2). It should be noted here that it is only those two conditions together with the 'reality condition' that make the EPR argument work correctly, but at the same time it should be stressed that while the above quoted commonsense criterion of reality is a *sufficient condition* to state the reality of a certain 'state of affairs', the conditions of locality and separability are jointly a *necessary condition* for Einstein (if something is to be physically 'real', it cannot be subject to the causal influence of spacelike distant events and, especially, its particular identity and actuality cannot depend upon any other event, even a local one). Such a set of conditions thus guarantees that separate segments of reality (of matter in spacetime) gain independence and measurements conducted in one of them do not disturb or create a reality in another, and every one of them is fully governed locally by deterministic laws. Naturally, an additional condition, present in fact from the very beginning, is the correctness (adequacy) of QM predictions, which is indeed confirmed in all the conducted experiments (it is therefore the safest condition, and at the same time it allows us, due to the geometry of quantum state space, to predict the correlations between the values of the obsevables for distant systems without any physical interaction with them). In such a situation, on the basis of the *supposedly* admissible counterfactual reasoning above mentioned, we in the end obtain the conclusion that the actual states–values of two noncommutative quantities $P$ and $Q$ for two supposedly independent (separable) systems I and II constitute, in line with the criterion, two separable 'elements of reality'. And that is exactly what Einstein–Podolsky–Rosen wanted the reader to conclude, in opposition to Bohr's standpoint. Since those values are real 'states of affairs' independent of one another, and QM cannot supply local observers with their complete and simultaneous description with the aid of one, coherent quantum space of states for all the observables, hence QM is an incomplete theory, which ends the argument. As we can see, the original EPR argument is rather complex and its logical reconstruction may cause difficulties, hence in later years Einstein himself construed its much clearer version, which in a way highlights the whole dilemma and is more interesting to a metaphysician.

In general, we can simply distinguish four basic assumptions of the 'EPR paradox', which at first glance seem logically independent – in Federico Laudisa's [2001] concise formulation for NRQM they are as follows[49]:

$[1]_E$ Reality
If, without interacting with a physical system $S$, we can predict with certainty – or with probability equal to one – result $q$ of the measurement of a quantity $Q$ performed at time $t$ on $S$, then at time $t'$ immediately after $t$, there exists a

49 The 'E's beside the numbers refer to Einstein; further on 'B's will serve to refer to J. Bell's assumptions (which are different, for example they do not require an explicit consideration of separability). Condition $[3]_E$ is with regards to Laudisa's formulation supplemented with separability, in line with the remarks of Einstein himself and his intentions, as will be discussed later. That addition is critical, since the whole EPR dilemma, also in its later forms, is not just a matter of relativistic locality and a ban on superluminal signalling.

property $q$ – associated with $q$ and denoted by $[q]$ – that is actually possessed by $S$: any such property is said to be an *objective* property of $S$.

$[2]_E$ Completeness
Any physical theory $T$ describing a physical system $S$ accounts for every objective property of $S$.

$[3]_E$ Locality (and separability, directly associated with it)
No objective property of a physical system $S$ in state $s$ can be influenced by measurements performed at a distance on a different physical system and any physical state of $S$ is always separable into independent real states of its subsystems in some spacetime regions.

$[4]_E$ Adequacy
The statistical predictions of quantum mechanics are correct.

We can at once notice that the key condition $[3]_E$ can be termed a necessary condition of Einstein's Local Realism – thus if any theory is classically realistic in the physical sense, it has to answer the joint demands of locality and separability, which makes it justified to treat them as two aspects of the same general condition (cf. [Paty, 1995])[50]. Since we consider $[4]_E$ to be undisputable in the light of all the experimental data gathered, hence the conclusion from the reasoning conducted by Einstein, Podolsky and Rosen is the thesis that the conjunction of $[1]_E$, $[2]_E$ and $[3]_E$ is false. So why did Einstein totally relinquish any inclusion of condition $[1]_E$ and the concept of an 'element of reality' in all his later statements on the paradox? It is so because it turns out that it is not of an independent character and in the case of QM it results from $[4]_E$ and $[3]_E$: measurement on system I of quantity $Q$ giving the value $q_r$(I) is to enable, thanks to the theory, a knowledge of value $q_s$(II) of quantity $Q$ for system II with a probability equal to one, and the strong Einstein locality ascertains that measurement on I could

50 The conditions of locality and separability can be treated as two components of a more general condition of locality Einstein had in mind when he spoke of a 'realistic theory' in physics – I shall also assume such a distinction here; also cf. Howard [1985; 1989; 1992]. $[3]_E$ can thus be called a 'strong Einstein locality', which he conceived not so much as a certain negotiable metaphysical assumption, but as a basic postulate of all imaginable physics. The problem with the condition of fundamental 'separability' in the context of the EPR argument is based on the fact that it is concealed in $[3]_E$ and partly overlaps with $[2]_E$, as Einstein in his later discussions with Bohr was not so much concerned with the empty localizability of the mechanical state at a spacetime point, but always with its physical *definiteness* at that point, which makes the discussion more difficult. Condition $[1]_E$ is a *sufficient condition* of local realism.

not have influenced II nor co-created a reality of its state, hence those values have to be objective and independent. In the end, therefore, the whole paradox can be brought down to the diagnosis that $[2]_E$ & $[3]_E$ together with trustworthy $[4]_E$ lead to a contradiction. That being so, we can draw the conclusion that QM is either incomplete, or 'nonlocal'/'nonseparable': $\sim[2]_E \vee \sim [3]_E$, which is finally the proper form of the 'EPR theorem' (cf. [Singh, 2005, p. 2111]). Such a clear and essential formulation of the EPR argument in Einstein's writings appears in several versions, among others in his letters to Schrödinger of the 19th of June, 1935, and to J.L.B. Cooper of 31st October and 18th December, 1949, as well as in his essays: *Physics and Reality* in 1936 [Einstein, 1936], *Quantum Mechanics and Reality* in 1948 [Einstein, 1948], in his *Autobiography* and in the *Remarks* added to Einstein's volume published by Schilpp [Einstein, 1949b]. For example, in the latter case, Einstein presents the 'EPR theorem' in the context of his dispute with Bohr's standpoint and the ideas concerning nonseparability in the case of quantum phenomena:

> … [Bohr] argues as follows:
> If the partial systems A and B form a total system which is described by its $\psi$-function $\psi_{AB}$, there is no reason why any mutually independent existence (state of reality) should be ascribed to the partial systems A and B viewed separately, *not even if the partial systems are spatially separated from each other at the particular time under consideration.* The assertion that, in this latter case, the real situation of B could not be (directly) influenced by any measurement taken on A is, therefore, within the framework of quantum theory, unfounded and (as the paradox shows) unacceptable.
> By this way of looking at the matter it becomes evident that the paradox forces us to relinquish one of the following two assertions:
>
> 1. the description by means of the $\psi$-function is complete.
> 2. the real states of spatially separated objects are independent of each other.
>
> On the other hand, it is possible to adhere to 2., if one regards the $\psi$-function as the description of the (statistical) ensemble of systems (and therefore relinquishes 1.). However, this view blasts the framework of the 'orthodox quantum theory'. [Einstein, 1949b, p. 681*f.*]

Einstein's paraphrase of Bohr's standpoint unanimously reveals that he understood locality and separability (of the 'real factual situation of the system' [Einstein, 1949a, p. 85] or a 'state of reality') as closely bound with each other, not allowing for an acceptance of one and a rejection of the other, although perhaps towards the end of his life his attitude to that became a bit less stringent [Howard, 1985]. The problem pointed out here is simply that if we are to accept the completeness of QM, then we are forced, according to Einstein, to relinquish locality (measurement on system I 'telepathically changes the real situation' of system II, i.e., it violates the geometry of SR) and/or separability ('by denying independent real situations as such to things which are spatially separated from each other') [Einstein, 1949a, p. 85*ff.*]. Let us notice that Einstein's above formulations are determined by the two latter conditions, being only weakly present in the article from 1935, and they marginalize the original 'reality criterion'. The clarity of that statement nevertheless hides a certain ambivalence with regards to the meaning of 'separability', as it cannot in any way be simply narrowed down to a spacetime localizability of mechanical events, but is a metaphysical statement about the *nature of mechanical states* ascribed to points/regions of spacetime: all 'states' are to be *ontically independent/separable* and *noncontextual*, are to possess an 'independent existence' as 'independent situations of things'. Einstein [1948, p. 321] expressed it equally clearly in his famous article titled *Quantenmechanik und Wirklichkeit*, where he offered the classical field theory as an example: 'these things claim an existence independent of one another, insofar as these things lie in different parts of space', calling that existence '*the being-thus*'. In the case of a physicist who – like Einstein – is not ready to relinquish locality and separability, the only conclusion in such a situation is to adopt the thesis that QM is incomplete and does not in any way concern individual quantum events, it is a kind of statistical theory describing whole sets of quantum events, and hence it does not have any say in the matter of individual events in the microworld (analogically to classical statistical mechanics and thermodynamics), thus the description with the help of function $\Psi$ should be supplemented with some set(s) of *local and separable* hidden variables. And it is in that very context that the key moment in the history of the development of research came with the short but extremely influential work by John S. Bell in 1964, inspired by the EPR argument [Bell, 1987a] and a series of significant theoretical and experimental results obtained on its basis – they are all, in point of fact, tests of assumption $[3]_E$, i.e., *a strong Einstein locality* in all its complexity and with its relation to $[2]_E$. Their importance does not merely lie in a blatant non-intuitiveness and a concurrent precision (both formal and experimental), but mainly in the way they reveal a deep structure of QM itself or, as David Mermin put it [1998], 'what is quantum mechanics trying to tell us'. In other words, the Bell theorem gives an insight into the sense of the alternative '$\sim[2]_E \vee \sim[3]_E$' and does that in such a way that the 'nonlocality issue' comes to the fore. But the more general (Bell)–Kochen–Specker theorem mentioned at the beginning of the chapter captures even deeper aspects of that alternative as it probes the problem of the *separability* of quantum states, central to QM and Einstein's standpoint, but in fact hidden in a background of Bell's physical reasoning (it is not explicit there). As we will see, Bohr was right when he consciously and stubbornly downplayed the 'danger' of the 'nonlocality issue' (which worried Einstein all the time) and highlighted the 'nonseparability issue' as really fundamental to QM, as that nonseparability – the core property of QM – is directly associated to the *relativity of quantum states*.

At the foundation of the Bell theorem lies the simple question as to whether it is possible to reproduce all the QM statistical predictions when assuming a familiar, classical local realism, and thus a certain number of locally hidden variables complementing the incomplete state $\psi$, which QM ascribes to every quantum system. Bell's Gedankenexperiment [1978, p. 14*ff.*] envisions a measurement situation on an entangled system, proposed in a convenient form by David Bohm [1951, pp. 614–622] in his QM textbook, and also (together with Yakir Aharonov) in an article on the 'EPR paradox' [Bohm, Aharonov, 1957]. Analogically to the instance of the dimer of mercury described in Ch. 1.2, we are at present analyzing a system of two spin-½ particles in a singlet state, which after interaction move in opposite directions ('to the right' and 'to the left'), until they reach the Stern–Gerlach system of magnets, and next that of detectors. Let us recall that, generally speaking, the basic spin directions set parallel to the $x$, $y$, $z$ axes can be simply represented by the complete basis of traceless Hermitian matrices: $\sigma_x$, $\sigma_y$ and $\sigma_z$, with the unit matrix added to them, $\sigma_0$:

$$
\begin{aligned}
\sigma_1 &= \sigma_x = \begin{pmatrix} 0 & 1 \\ 1 & 0 \end{pmatrix} \\
\sigma_2 &= \sigma_y = \begin{pmatrix} 0 & -i \\ i & 0 \end{pmatrix} \\
\sigma_3 &= \sigma_z = \begin{pmatrix} 1 & 0 \\ 0 & -1 \end{pmatrix} \\
\sigma_0 &= I = \begin{pmatrix} 1 & 0 \\ 0 & 1 \end{pmatrix}
\end{aligned}
\tag{11}
$$

Together they generate the complex algebra ***Mat***(2, ℂ) of 2x2 matrices which fulfill simple relations (where '$i$' is an imaginary unit):

$$
\begin{aligned}
&\sigma_1^2 = \sigma_2^2 = \sigma_3^2 = I \\
&\sigma_1\sigma_2 = -\sigma_2\sigma_1 = i\sigma_3 \\
&\sigma_3\sigma_1 = -\sigma_1\sigma_3 = i\sigma_2 \\
&\sigma_2\sigma_3 = -\sigma_3\sigma_2 = i\sigma_1 \\
&\sigma_1\sigma_2\sigma_3 = iI, Tr\sigma_i = 0,
\end{aligned}
\tag{12}
$$

and thus: $\sigma_i\sigma_j + \sigma_j\sigma_i = 0$. We can notice two things immediately here: firstly, after multiplying the elements of the algebra by $-i$ we obtain the algebra $\mathbb{H}$ of quaternions with three elements: $i, j, k$, corresponding respectively to products: $-\sigma_2\sigma_3$, $-\sigma_3\sigma_1$ and $-\sigma_1\sigma_2$. Secondly, our spin algebra ***Mat***(2, $\mathbb{C}$) is isomorphic to the real Clifford algebra ***Cl***(3) of space $\mathbb{R}^3$, as the basis matrices of ***Mat***(2, $\mathbb{C}$) and their products correspond to the vectors, bivectors and trivectors spanning ***Cl***(3). Algebras $\mathbb{H}$ and $\mathbb{C}$ generate at the same time the eight-dimensional algebra ***Cl***(3), so that:

$$\mathbb{C} \otimes \mathbb{H} \simeq \mathbf{Cl}(3), \text{ and: } (dim\ \mathbb{C})(dim\ \mathbb{H}) = dim\ \mathbf{Cl}(3), \tag{13}$$

where $\mathbb{C}$ is its centre, while $\mathbb{H}$ is its even sub-algebra, $\mathbf{Cl}(3)^+$.

In Ch. 2.2, in order to avoid the language of static objects, we have introduced in the form of a toy model the four-dimensional geometric algebra of the process $\mathbf{Cl}(2) \cong \mathcal{A}(V^2)$, where the basis consists of directed objects, representing fundamental 'elements of action' in an abstract way. Here in turn, as we can see, we encounter a geometric algebra of the process of an even higher order for space $\mathbb{R}^3$, whose sub-algebras are the algebra of complex numbers and the algebra of quaternions. What is extremely interesting is that the whole quantum spin geometry possesses such a simple and elegant representation in $\mathbf{Cl}(3) \cong \mathcal{A}(V^3)$, which means that it has a deep connection with projective geometry of space $\mathbb{RP}^2$ and real multivector algebra $\mathbf{Cl}(1, 3) = \mathbf{Cl}(\mathbb{R}^{1,3})$ of spacetime $\mathbf{M}^4$, where the orthonormal basis of Dirac matrices $\{\gamma_\mu;\ 0, 1, 2, 3\}$ is a relativistic generalization of the basis of Pauli matrices $\sigma_i$. The significance of that fact goes far beyond the problem under scrutiny here, but it is worth noting, first of all, that such a concept of spin geometry enables an elegant account of all of Dirac's theory and a comprehension of its geometric-kinematic sense (see especially the pioneering work by David Hestenes [1975; 1996; 2003]), and also points to deeper connections with the geometry of spinors and twistors (see e.g. [Lawson, Michelsohn, 1990]). The physics of spin, encountered in our B-EPR (Bohm–Einstein–Podolsky–Rosen) experiment, is not simply some singular physical situation, but it immediately refers us back to fundamental QM problems and its connection to the structure of spacetime.

In our B-EPR system, every direction $n$ of the spin measurement in physical space that we may select has a corresponding matrix of the operator:

$$\sigma(n) = n_1\sigma_1 + n_2\sigma_2 + n_3\sigma_3 = \begin{pmatrix} n_3 & n_1 - in_2 \\ n_1 + in_2 & -n_3 \end{pmatrix}, \tag{14}$$

where $\sigma(n)^2 = I$, and its eigenvalues are naturally +1 and −1, with – let us say, in this case – corresponding states (spin orientations) 'spin up' $|\uparrow\rangle$ and 'spin down' $|\downarrow\rangle$. Hence, the experimental question 'is the component of the spin in the direction $n$ up?' is represented by the also Hermitian projective (idempotent) operator

$P(n, +1)$, which projects onto one of the eigenspaces for the corresponding eigenvalue:

$$P(n, +1) = \tfrac{1}{2}[I + \sigma(n)]. \tag{15}$$

In this particular case, we are choosing three random planar directions, $\{n; a, b, c\}$ (with fixed planar angles between them), along which the components of the spin cannot be measured at the same time for a given particle. The singlet state of an experimental system of two entangled particles, expressed by the eigenvectors of the operator for, say, $a$, is given as:

$$|\psi\rangle_{singl} = \frac{1}{\sqrt{2}}|\uparrow\rangle_{1,a}|\downarrow\rangle_{2,a} - \frac{1}{\sqrt{2}}|\downarrow\rangle_{1,a}|\uparrow\rangle_{2,a} \tag{16}$$

We next perform measurement $M$ in one part of the system, where with probability 0.5 we can register one of two possible states for particle 1, for example:

$$|\psi\rangle_{singl} = \frac{1}{\sqrt{2}}|\uparrow\rangle_{1,a}|\downarrow\rangle_{2,a} - \frac{1}{\sqrt{2}}|\downarrow\rangle_{1,a}|\uparrow\rangle_{2,a} \xrightarrow{M} |\psi\rangle_1 = |\uparrow\rangle_a \tag{17}$$

Naturally, in accordance with trustworthy condition $[4]_E$, we can immediately expect the adequacy of QM, as the correlated state of the second particle will be the eigenstate of the operator of value −1 ('spin down'). To formulate Bell inequalities, what is of key importance is to compose a table of correlations in which all the possible combinations of the measured directions for both particles will be exhausted, regardless of the fact that it is impossible to define, experimentally and simultaneously, the values of those observables on *one* particle. As we are dealing with the entangled system of two particles, we shall be *supposedly*, due to the EPR reasoning, in a position to define at one time only two spin components for each particle, which is sufficient to conduct simple statistical estimations (however there is no reason why those particles should not be three or more, such as in the experiment of the GHZ type (e.g. [Bouwmeester *et al.*, 1999], [Pan *et al.*, 2002]). We should start off by calculating the amplitudes $\omega_i$ of transition from state $|\psi\rangle_{singl}$ to specified spin states for random directions, such as, let us say:

$$\begin{aligned}\omega_{a(-1),c(-1)}(|\downarrow\rangle_{1,a}|\downarrow\rangle_{2,c}) &= \frac{1}{\sqrt{2}}\langle\downarrow|_{1,a}\langle\downarrow|_{2,c}|\uparrow\rangle_{1,a}|\downarrow\rangle_{2,a} - \frac{1}{\sqrt{2}}\langle\downarrow|_{1,a}\langle\downarrow|_{2,c}|\downarrow\rangle_{1,a}|\uparrow\rangle_{2,a} \\ &= \frac{1}{\sqrt{2}}\langle\downarrow|_{2,c}|\uparrow\rangle_{2,a}\end{aligned} \tag{18}$$

To conduct further transformations, we should use the Heisenberg operator for the spin in the $y$ direction (we can assume that in this case it is the axis of movement of the measurement apparatus and spin measurement is conducted in the *z-x* plane), which has the form of:

$$\hat{s} = \frac{1}{2}\hbar\,\sigma_y = \frac{1}{2}\hbar\begin{pmatrix} 0 & -i \\ i & 0 \end{pmatrix} \tag{19}$$

Then state $|\uparrow\rangle_{2,a}$ from (18) can be transformed (with the help of the rotation operator and Pauli matrix $\sigma_y$) as being rotated with regards to the axis, given fixed angle $\theta_{ac}$ (we can do so as the above formula '$\frac{1}{\sqrt{2}}\langle\downarrow|_{2,c}|\uparrow\rangle_{2,a}$' refers to one and the same particle):

$$|\uparrow\rangle_{2,a} = \exp\left(-\frac{1}{2}i\sigma_y\theta_{ac}\right)|\uparrow\rangle_{2,c} = \begin{pmatrix} cos\frac{1}{2}\theta_{ac} & -sin\frac{1}{2}\theta_{ac} \\ sin\frac{1}{2}\theta_{ac} & cos\frac{1}{2}\theta_{ac} \end{pmatrix}\begin{pmatrix} 1 \\ 0 \end{pmatrix} = \begin{pmatrix} cos\frac{1}{2}\theta_{ac} \\ sin\frac{1}{2}\theta_{ac} \end{pmatrix}$$

$$= cos\frac{1}{2}\theta_{ac}|\uparrow\rangle_{2,c} + sin\frac{1}{2}\theta_{ac}|\downarrow\rangle_{2,c}. \tag{20}$$

And at this point it is easy to obtain, through the inner product of states, the numerical value of the transition amplitude dependent on the $\theta_{ac}$ angle, being sought by us in (18):

$$\omega_{a(-1),c(-1)} = \frac{1}{\sqrt{2}}sin\frac{1}{2}\theta_{ac} \tag{21}$$

and thus the probability of spin state $\langle\downarrow|_{2,c}|\uparrow\rangle_{2,a}$ taking place is:

$$P(a,\downarrow;c,\downarrow) = \frac{1}{2}sin^2\frac{1}{2}\theta_{ac} \tag{22}$$

Having the distribution of the probabilities calculated for different combinations of states, we can finally write down simple inequalities demonstrating the relations between them, such as, for instance:

$$P(a,\downarrow;c,\downarrow) \leq P(a,\downarrow;b,\downarrow) + P(b,\downarrow;c,\downarrow) \tag{23}$$

Those inequalities can be written down for all the combinations, selected angles between the axes, and also for the number of particles which can be experimentally kept in an entangled state and next measured (e.g. multi-qubit cluster states [Walther *et al.*, 2005]). There also exist many versions of the theorem inspired by Bell's original work, but which differ in their assumptions and conclusions (a good systematization, including proofs, is to be found e.g. in [Placek, 1997]). The first complete experiment with the spin correlation analysis in the B-EPR scheme with photons emitted by the excited calcium atom were conducted by Alain Aspect, Philippe Grangier, Jean Dalibard and Gérard Roger at the Institut d'Optique Théorique et Appliquée in Orsay near Paris [Aspect *et al.*, 1982]. In the third experiment [Aspect, Dalibard, Roger, 1982], now famous, optic switches preceding light polarizers were used, in order to – in line with Bohm and Aharonov's original idea [1957] – enable an effectively quick change in the polarization already after the emission of the photons from the atomic source (Fig. 2). Single polarizers were replaced with pairs of them (in every leg of the setup) with different orientations, and the switch took place every 10 ns, which together with the frequency of the emissions from the source made the detecting events and the optical switching in the other leg of the system remain separated by a spacelike interval. The rapid redirection of the incident beam onto one of the polarizers took place as a result of its acousto-optic interaction with the standing ultrasound wave in water [ibid., p. 1805]. Both in that and in all the later experiments in which the measurement technology has been rigorously prepared, assuring a closure of the potential experimental loopholes (e.g. [Weihs *et al.*, 1998], [Rowe *et al.*, 2001], [Matsukevich *et al.*, 2008], [Ansmann *et al.*, 2009], [Giustina *et al.*, 2013]; see also a concise review in [Hardy, 1998]), *Bell inequalities are violated*, which means that all the QM predictions are correct – regardless of the type of quantum objects, even when they differ in type, such as the atom or photon (e.g. [Moehring *et al.*, 2004]), are massive (e.g. [Hofmann *et al.*, 2012]) and for all quantum properties, e.g. flavour [Go, 2004] – whereas Bell's model, which we use to formulate inequalities, possesses certain assumptions inappropriate to capture the unique physics behind those quantum effects.

*Fig. 2: The setting of the B-EPR type of experiment testing Bell's generalized inequalities (Clauser–Horne–Shimony–Holt inequalities), used by Aspect, Dalibard and Roger [1982], with optic switches ($C_I$, $C_{II}$) at the distance L = 6 m from the source (S), pairs of polarizers (I and II) and an analyzer of the signal from two pairs of photomultipliers (PM). Reprinted with permission from: A. Aspect, J. Dalibard, G. Roger, Phys. Rev. Lett., 1982, vol. 49, p. 1805, http://link.aps.org/abstract/PRL/v49/p1804. Copyright © 1982 by the American Physical Society.*

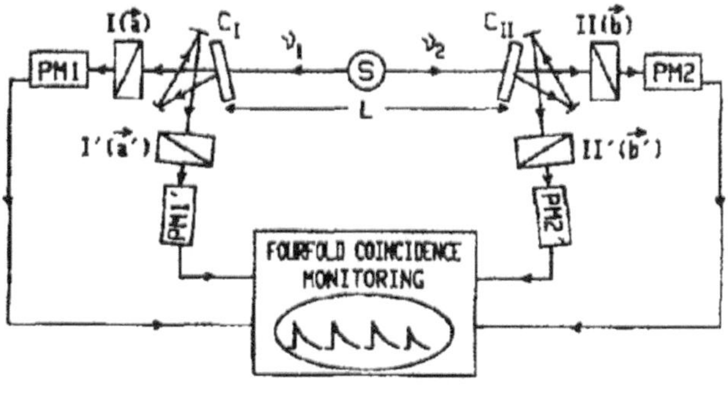

The 'EPR Paradox' (as Einstein called it, beginning with his *Physik und Realität* [1936]) owes, as we have seen, its troublesomeness to two groups of conditions: (i) completeness of the state function as an 'objective' description of reality, and (ii) locality-cum-separability. If we wish to retain the classical local realism and geometry of SR, we resign from condition $[2]_E$ and are forced to introduce additional ontic variables into the quantum description in order to explain the measured correlations in an entangled system, that is we supplement the 'incomplete' state function with 'hidden' or 'postulated' variables which we shall denote as $\lambda$ [Bell, 1987a, p. 15*f.*]. Regardless of the real character of those variables (it may be both an individual physical variable, and a very large set of functions, they could be fully deterministic or stochastic) we can assume that the results of the measurement of the spin direction on each of the two particles, i.e., spin operator eigenvalues: $p_m$ for the 'right' one and $l_n$ for the 'left' one (where quantum expectation values are, respectively, $p = \sigma(m)_1 \cdot \bar{e}_m$, and $l = \sigma(n)_2 \cdot \bar{e}_n$, $\bar{e}_m$ and $\bar{e}_n$ being unit vectors given in the matrix representation and defined for a three-dimensional space, and for distinction we drop here the lower indexes), depend not only on $\bar{e}_m$ and $\bar{e}_n$ but also on the empirically unknown and uncontrollable distribution $\lambda$ so that, after the supplementation, we have:

$$p(m, \lambda) = \pm 1, \qquad l(n, \lambda) = \pm 1 \tag{24}$$

(we shall retain the notation '$p$' and '$l$' for the obtained measurement outcomes in two legs of the experimental setup – the right and the left one, and '$m$' nad '$n$' for arbitrarily chosen axes of the spin measurement). Additionally, the same incomplete quantum state $\psi$ can have various different hidden physical states $\lambda$. We obviously expect that the condition of normalization below is met for the probability density of hidden variables $\varrho(\lambda)$, which adopt the values from the given set $\boldsymbol{\Lambda}$ [ibid., p. 17]:

$$\int_\Lambda \varrho(\lambda) d\lambda = 1. \tag{25}$$

The model of hidden (ontic, objective, noncontextual, but non-operational) variables supplementing the incomplete state $\psi$ of each particle, regardless of the physical nature of λ, has to meet, naturally, the first basic condition: that of empirical adequacy, analogically to $[4]_E$ in the EPR argument, i.e., it has to generate all the predictions of standard QM concerning the probabilities $P$ of obtaining the results of measurements $p_m$ and $l_n$ for given directions and for any state $\psi$:

[1]$_B$ Empirical Adequacy/Objectivity/Noncontextuality (Bell Realism)

$$P_{\psi m}(p_m) = \int_\Lambda P_{\lambda m}(p_m) \cdot \varrho_{\psi m}(\lambda) d\lambda$$

$$P_{\psi n}(l_n) = \int_\Lambda P_{\lambda n}(l_n) \cdot \varrho_{\psi n}(\lambda) d\lambda$$

$$P_{\psi mn}(p_m \,\&\, l_n) = \int_\Lambda P_{\lambda mn}(p_m \,\&\, l_n) \cdot \varrho_{\psi mn}(\lambda) d\lambda$$

The second assumption we made earlier when construing the inequalities had been called a 'vital assumption' by Bell himself, and that was also the 'vital' condition every physical theory had to meet according to Einstein: the condition of locality in [3]$_E$. That is also a guarantee that when two parts of a system, particles $S_1$ and $S_2$, are causally separated, then also the measurements *as well as* the results of the measurements of their states taken at two separate detectors remain totally independent and are not coordinated by any instantaneous interaction (physical signal), i.e., we can expect that appropriate joint probabilities (together with states $\lambda$) will factorize into separate probabilities for each system, which is a proper formulation for all stochastic hidden variable models. The condition, which is not directly identical to Einstein's [3]$_E$, i.e., to the physical locality of interactions together with the *general separability* of events (Bell's condition is apparently simpler and more lucid), shall be called 'Bell locality' for the sake of clarity, which in its original formulation is of a purely formal character: it is a form of Reichenbach's Common Cause Principle, which can next be physically interpreted as safeguarding a relativistic causal structure of events in spacetime by postulating time-asymmetric *spacelike separability* of events (or streams of events). This is an extremely important point as that formulation is the only useful one for the framework of stochastic theories in $\mathbf{M}^4$.

[2]$_B$ Local Causality/Factorizability (Bell Locality)

$$P_{\lambda mn}(p_m \,\&\, l_n) = P_{\lambda m}(p_m) \cdot P_{\lambda n}(l_n)$$

The third and last condition lying at the foundations of the Bell inequalities has been tacitly presumed by us in the equations (24), but could have been considered self-evident and therefore left out – although its importance is as great as that of Bell locality, not to speak of its nontriviality. When we specify the variables in our model which influence the probability distributions coded by state $\psi$, we have to be aware of the fact that the arrangement of the axes of the polarizers on both sides of the system should be independent of the distribution of the hidden states $\lambda$ connected with the sources. In simple words, we assume here a commonsense demand for the state $\varrho_\psi(\lambda)$ of the whole system not to be dependent on *what* measurement will be conducted in the future, or on whether and if at all *any* measurement is made, so one could call it 'timelike separability of events', '(time-asymmetric) independence from history' or just 'autonomy/separability of the source'. That assumption is often called 'Free Will' but I would strongly insist that that should be abandoned as that presumes an awkward anthropocentrism and, more important, is even incompatible with the original motivation of the hidden variable theories to provide some efficient *deterministic* model for QM. We write that down for the distribution of $\lambda$ variables:

$[3]_B$ $\lambda$-Independence/Time-Asymmetric Independence from History (Timelike Separability of Events)

$$\varrho_{\psi m_i n_j}(\lambda) = \varrho_{\psi m_j n_i}(\lambda) = \varrho_{\psi m_i n_i}(\lambda) = \varrho_{\psi m_j n_j}(\lambda) = \varrho_{\psi}(\lambda)$$

The above condition states that there exists a connection solely between the incomplete quantum state $\psi$ and the distribution of the hidden states $\lambda$, while the history of the detecting system and the future 'choice' (switching or singling out) of the axis of the spin measurement are of no effective importance to them (the subscript symbols by the axis indexes refer to those different choices). As the inequalities construed on the basis of $[1]_B$, $[2]_B$ and $[3]_B$ are in fact violated by the statistics of the measurement results, the classic reasoning to be found at the foundation of those conditions has to be false. Thus, the full Bell Theorem (henceforth BT; 1964/1976) can be seen as stating that *no classically realistic (objectivistic), causally local mechanics with hidden, timelike separable variables $\lambda$ (deterministic or stochastic) can fully simulate the statistical predictions of QM.* Contrary to frequent misunderstandings, BT does not speak in favour of the incompleteness of QM and only declares that *if* we manage to construct any reliable realistic QM with hidden variables (stochastic or deterministic), then it cannot be causally local *and* timelike separable, also all the more in the sense of the 'strong Einstein locality' (= $[3]_E$), so, in the end, that cannot be *classically realistic* mechanics in any reasonable sense. That might be seen as disappointing, considering that such a possible theory (cf. e.g. [Dürr, 2001]) – it seems – would not have any particular operational gain over standard QM, as both (even if *in spe* causally local but not fully separable in Einstein's sense) have to violate $[3]_E$ in some way, hence it is difficult to speak *a priori* of any purely physical priority of such a hidden variable theory over the openly operational '(post-) Copenhagen' approach (cf. [Price, 1997, p. 216][51]. Finally, though, BT is extremely important

51 That was in fact the basic reason of Einstein's dissatisfaction with Bohm's [1952] interpretation, who assumed QM to be incomplete but introduced, in the wake of Louis de Broglie's proposals, a radically nonlocal pilot wave. It is that very dissociation into local

as a critical extension of the EPR argument insofar as the latter has had to offer hope for Einstein's 'Local Realism' to be saved. Nevertheless, one should carefully emphasize the popular tale that it was Bell who 'overthrew Einstein's conclusion' about the incompleteness of QM to be incorrect – actually the former was a consistent adherent of the thesis about the promising merits of the hidden variable theories (e.g. [Bell, 1987]). Before we look into the challenge BT poses more closely, we have to become fully aware what the exact sense of the 'realism' mentioned is and what exactly would violating $[3]_E$ imply so that we do not discard ontological realism as such too hastily, when we analyze the different metaphysical scenarios. Travis Norsen [2007] is perhaps partly right in his severe essay when he says that the concept of 'local realism' is so vague and overused that one should remain constantly wary when applying it in a discussion, but that is certainly too weak a reason to – as he advises – completely reject it in the context of BT. The concept of 'realism', even as 'Metaphysical Realism' [ibid, p. 330*ff.*], is neither as unanimous and stiff as Norsen believes, but it is true that a potential redefinition of it has very far-reaching consequences (see Ch. 4 and 5).

We have noticed that in the original EPR argument, the 'reality condition' is not wholly independent, but – in the framework of that argumentation – is possible to be brought out from more primitive conditions $[3]_E$ and $[4]_E$. Although condition $[1]_E$ itself in Einstein–Podolsky–Rosen's article is only a sufficient condition of asserting that a certain 'element of reality' exists, nevertheless the dispute between Bohr and Einstein has revealed a much deeper difference between them, far transgressing the problem of the connection between statistics and the certainty of predictions in QM. Einstein was even prone to accept a certain dependence of the system's behavior on the observer's operations, for the latter could be indeed seen as an active factor whose actions change the physical situation – but he decisively opposed a more radical interpretation of that

material systems and the quantum pilot waves coupled to them that makes possible an exact reconstruction of all the predictions of standard NRQM in Bohmian mechanics, despite a different ontology. Einstein wished to completely eliminate any nonlocality *and* nonseparability, however understood, hence Bohm's proposal seemed to him to be, at best, a Pyrrhic victory (see Freire's [2005] and Myrvold's [2003] historical essays). Bohm's 'quantum potential' is not only nonlocal, but additionally it introduces an extreme contextuality of states, nevertheless he managed to construe NRQM as an externalistic ontological holism with well-defined positions and trajectories of particles. The success of Bohmian mechanics simply depends on the way all that nonseparability of the systems gets transposed from kinematic states of the particles onto a completely new, nonlinear physical potential *in spacetime*.

fundamental dependence, which Bohr in turn tended to. We shall understand Einstein's treatment of realism better when we stop concentrating on the operational sufficient condition $[1]_E$, and, as Einstein himself wished, place an emphasis on necessary condition $[3]_E$, which he himself apparently never left out and which is internally more complex than it looks. Thus, along with the local causality of interaction ('signal locality') – the foundation of the theory of relativity and the whole spacetime geometry as we can currently probe it – what turns out to be of key importance is that definition of physical (mechanical) 'separability', which concerns not only the spacelike separated parts of the entangled system, as in the EPR argument and the experimental tests of Bell-type inequalities, but also all local systems, classical and quantum, including locally interacting systems. Let us spell it out more precisely. Einstein insisted that every physical system should possess the values of all the mechanical observables proper to it, determined by its causally local spacetime history and the laws of physics (complete definiteness) – hence his reference to counterfactual reasoning which irritated his opponents so much: 'what, if we nevertheless measure…'. But in his treatment of realism there is something more than the condition of definiteness (based on the old metaphysical Universal Principle of Definiteness discussed in Ch. 1.2) and it is that very characteristic that is revealed in the EPR reasoning itself. Einstein additionaly requires an ontic noncontextuality of physical quantities, i.e., that such value definiteness be at every moment *noncontextually* ensured, that is ontically independent of all other physical processes, particularly of the whole context of the measurement, including the system extracting the information. That is a demand which is a consequence of strong and general physical separability, which we are going to discuss at greater length in Ch. 3.2, in that case of local separability: the separability of the instantaneous 'objective (full) state' of the measured system and the instantaneous 'objective (full) state' of the measuring system, both of which are to be independently 'real' (:noncontextually definite).

It now becomes clearer why Einstein introduced through conditions $[2]_E$ and $[3]_E$ a decided, or even restricting position which indeed does mirror quite well the way of thinking proper to classical mechanics, as it directly encompasses also the complete definiteness and noncontextuality of objective mechanical states in spacetime. Such a very paradigm corresponds to the expression quoted earlier from Einstein, Podolsky and Rosen [1935, p. 780]: 'reasonable definition of reality'; it is in fact, as Julian Barbour rightly called it [2001, p. 220], 'a naïve atomistic philosophy', and decidedly conservative. What is more, 'Einstein's paradigm is, in its general terms, everybody's commonsense paradigm, the paradigm that we live by. In its operational terms, it is also the paradigm of the present-day physicist' [Malin, 2003, p. 90]. We may thus sum up that Einstein's Local Realism (or

Einstein's 'strict realism') as a metaphysical assumption package is in fact a quite rich conjunction:

| EINSTEIN'S LOCAL REALISM: causal locality & general separability & complete value definiteness & ontic noncontextuality |
|---|

with the latter two being closely bound with the former two constituting Einstein's 'strong locality' (we are writing them all down, so that it becomes clear how Einstein seemed to understand condition $[3]_E$ and its implications for any acceptable mechanics, present or future). We should keep that definition in mind in order to avoid any confusion arising from ascribing to Einstein other, much weaker definitions of 'local realism' (more liberal ones), and at the same time avoid presenting condition $[1]_E$ itself as an allegedly complete Einsteinian 'definition' of reality (though later dropped), which it was not. At the same time, we should probably emphasize that that is not just 'physical realism', but also a very strong version of 'metaphysical realism' in Norsen's sense [2007, p. 330*ff.*] and indeed, as he claims, the foundation of the construction of Bell's inequalities by the affirmation of $[1]_B$, $[2]_B$ and $[3]_B$. We should therefore harbour no illusions: such a radical Local Realism *is certainly being violated by QM* and observed processes of nature, but unfortunately that is another negative conclusion and the problem remains to what degree and what type of realism can be secured in the face of that fact. But it becomes clearer, though, why that Einsteinian realism is utterly irreconcilable with Bohr's philosophical (epistemological) views.

Bohr's 'principle of complementarity', taking aside the issue of signal nonlocality in QM and despite Norsen's protest, remains to be in a blatant discord with Einstein's Local Realism, as it radically violates his demands of general separability and resulting noncontextuality, that is the traditional carrier elements of metaphysical realism in its classical forms. For example, in his famous *Discussion with Einstein* in 1949, Bohr first introduced the distinction between a 'quantum phenomenon' and the classical level of description, and then wrote:

> This crucial point (...) implies the *impossibility of any sharp separation* (*emphasis – M. W.*) between the behaviour of atomic objects and the interaction with the measuring instruments which serve to define the conditions under which the phenomena appear. In fact, the individuality of the typical quantum effects finds its proper expression in the circumstance that any *attempt of subdividing the phenomena will demand a change in the experimental arrangement* (*emphasis – M. W.*) introducing new possibilities of interaction between objects and measuring instruments which in principle cannot be controlled. Consequently, evidence obtained under different experimental conditions

cannot be comprehended within a single picture, but must be regarded as *complementary* in the sense that only the totality of the phenomena exhausts the possible information about the objects. [Bohr, 1963, p. 40]

Bohr stressed the fact of quantum contextuality (let us call that Bohr contextuality), as it was fundamental to, or rather the essence of his principle of complementarity: a definite physical context of the interaction co-determines the phenomenon. Bohr's 'impossibility of any sharp separation' can nevertheless be treated, true to Bohr's very intention, as an unequivocal expression of the thesis that that contextuality of physical states is simply a special case of a more general property of microphysics: *ontologically understood nonseparability.* Despite Bohr's almost exclusively epistemological interests, his understanding of local contextuality as an irreducible form of nonseparability is fully concordant with the counter-Einsteinian metaphysics and he can be safely regarded as a holist (cf. [Esfeld, 2001, p. 232*ff.*]). That means that the *very possibility of composing bottom–up the information about a physical system U from the states of its parts is seriously violated in QM even at the elementary level of local 'measurements' (extractions/transfers of information) successively conducted by system O on that quantum system U, both being parts of another whole (such as, e.g., $U \oplus O$)*, which Bohr fully realized and which he stated at the end of the quoted remark. That also means that quantum nonseparability being the result of an interactional (dynamical) generating of correlations, that is of the entanglement (correlation) of states, plays a fundamental role in all the relational structure of QM and its ontology, *at every level of the physical description.* Opponents of the 'Copenhagen interpretation', such as Einstein or Schrödinger, stressed the problematic nature, or even the unacceptability of that principle and certainly its status would remain to be in a large measure philosophical with regards to the accusations of the 'local realists', were it not fundamental for the Kochen–Specker theorem, being a corollary of the earlier Gleason theorem [1957] as a nontrivial quantum no-go theorem (as different from the famous but in some way trivial von Neumann theorem [1932]).

(Bell)–Kochen–Specker theorem (KST) [Kochen, Specker, 1967] differs from BT, the latter being of a definitely physical character, but it shares at the same time an interest in the existence of hidden, real ('objective') variables $\lambda$. Generally speaking, it states that the conjunction of the two above described, important elements of Einstein's Local Realism, i.e., a complete definiteness and noncontextuality is incompatible with QM predictions, and at the same time QM's empirical adequacy forces us to reject (or significantly modify) either one of those conditions, or both. In consequence no theory with completely definite and noncontextual (Einsteinian) quantum variables, represented geometrically by projective operators, can effectively simulate standard QM, and therefore we can be certain that *no classical or quasi-classical phase space is hidden behind the quantum state space.* To put it more precisely, expectation value $\langle \hat{A} \rangle$, ascribed to a specific Hermitian operator $\hat{A}$ on the Hilbert space $\mathcal{H}$ of a given quantum system, depends on which set of operators $\Omega$ that operator belongs to, that is, the value of the observables measured depends on the context of the measurement in the sense that what is of primary importance is what set of commuting (compatible) observables it belongs to. KST is important insofar as it does not require *any* assumption of local causality (like BT) and 'objects' embedded in spacetime regions, but it refers to the algebraically discussed contextuality in the measurement of observables, which enriches the discussion about hidden variables with a new foundational aspect (in the discussion, one should keep in mind that precise meaning of the 'context' concept and the fact that KST, unlike the older von Neumann theorem, refers to compatible observables represented by operators in a quantum space of states). Let us call that property Kochen–Specker contextuality. A precise formulation of KST goes as follows:

**(Bell)–Kochen–Specker Theorem**

Let $\mathcal{H}$ be Hilbert space spanned by the vectors representing the quantum states of the system, and let $dim\mathcal{H} \geq 3$. Then there exists a set of observables (:operators) $\mathbf{\Omega}$ defined on $\mathcal{H}$ containing $\omega$ elements, so that the following two statements lead to a contradiction:

(1) All the elements $\omega$ of set $\mathbf{\Omega}$ possess at the same time definite values $r(\omega_i)$, i.e., there exist well defined maps into $\mathbb{R}$, $\varphi: \omega_i \rightarrow r \in \mathbb{R}$ (*The Rule of Definiteness*)

(2) The values of the particular observables $\omega$ meet the following constraints:

(2.1) If observables $\omega_i$, $\omega_j$ and $\omega_k$ commute (are mutually compatible), and $\omega_i + \omega_j = \omega_k$, then: $r(\omega_i) + r(\omega_j) = r(\omega_k)$. (*The Sum Rule*)

(2.2) If observables $\omega_i$, $\omega_j$ and $\omega_k$ commute (are mutually compatible), and $\omega_i \cdot \omega_j = \omega_k$, then $r(\omega_i) \cdot r(\omega_j) = r(\omega_k)$. (*The Product Rule*)

KST is interesting, as it is more fundamental than BT in the sense of the limitations placed on all the possible hidden variable theories, and its formulation and experimental testing is independent of the prepared particular quantum state of the system. Earlier on I had suggested that noncontextuality is connected to condition $[3]_E$ of the EPR argumentation and all of the Einsteinian ontology it is based on – but now it is more and more evident that the situation is much more complex, as KST puts significant restrictions on any quantum theory which would also like to save the other components of that ontology: a complete value definiteness for the mechanical observables and a complete noncontextuality of the states as 'elements of reality' represented in the geometry of projections in Hilbert space. As any adherent of Einstein's hidden local variables and Local Realism will naturally wish to have both assumptions of KST intact, just as all three of Bell's conditions, it becomes clear that in effect he/she will never be able to obtain the predictions of standard QM. The KST simply suggests that such hidden variables would have to be contextual, in a strictly *physical* sense of the word. We may treat that as another fundamental violation (in one way or another) of Einstein's Local Realism or Einstein's metaphysics and as proof of its weakness as a certain 'natural' way of reasoning. Einstein, in the light of such of his remarks as the one from *Remarks...* [Einstein, 1949b, p. 682], would have to treat a similar unremovable contextuality of quantum information – even when saving the value definiteness of the observables – as an unacceptable property of mechanics for that very reason that it infringes on the *general separability of the physical states* of local systems: those real states should be ontically independent, intrinsic properties of 'things', regardless of the local frame of reference of quantum action. Once more, therefore, there is a sharp contrast between Einstein and Bohr, which first and foremost concerns non-separability and the resulting contextuality of physical states (and not the locality of interactions, which is of secondary importance here), and hence Einstein remained so steadfast in his judgment on the principle of complementarity. A good example here is Bohm's [1952] mechanics, which is a non-contradictory and empirically adequate NRQM with hidden variables, but is characterized not only by a radical nonlocality, but also contextuality (a non-manifest context of the measurement is always to be found in the hidden variables of the measuring system). In that sense it already contains – despite operating on local particles *in* spacetime – an element of Bohr's nonseparability (holism) and is naturally closer to Bohr's way of thinking than it is generally assumed, succumbing to a stereotype of Bohm belligerently opposing Copenhagen's dominance. The hidden variable theories of that sort are extremely interesting as theoretical models, but it is clear that when we seek them in order to escape from quantum contextuality and nonseparability, we have to expect defeat – both those traits of QM remain to be deeply embedded in its formalism and ontology, which shows that they are a central property of every empirically adequate version of QM. Particular variants of the hidden variable theories may at best differ in a more or less lucid interpretation of that fact, but they cannot recover a causally local *and* noncontextual (i.e., generally separable) realism. From the perspective of classical physics, they suffer a double defeat: not only are they unable to recover it, but on top of that they highlight the radically non-classical properties of QM itself (e.g. by an extremely nonlocal pilot wave and strong measurement contextuality). So the only really foundational issue for ontology seems to be the kind of contextuality we are dealing with. I propose that that is a strictly ontological (not epistemological) problem and the KST itself, as well as Bohr's self-limitation to epistemology, cannot give us any further clues in that respect. Quantum processes are *ontically nonseparable* and the strong local contextuality is a crucial aspect of that property of the physical world.

A mere statement about the inadequacy of Einstein's strict realism is not too satisfactory, as we still do not know what exactly 'quantum magic' depends on and what it tells us about nature. What is more, it is not clear what ontological notion of 'realism' we can save – if it is not possible to present some consistent proposal in that area, then the sarcastic tone of criticism, such as that of Norsen [2007], would be fully justified. The notions of 'contextuality' and 'nonseparability' may sometimes seem too vague and one should strive e.g. to define the way they can be translated onto the conditions in Bell's argument and other similar ones: which of them should be violated and why? But even more important, how does it put in doubt the common views in standard metaphysics of nature? Some of us, habitually presentist realists with a predilection for $[1]_B$ and $[3]_B$, would claim that *local causality* = 'signal (spacelike) locality' (thus the causal structure of physical events in time-oriented $\mathbf{M}^4$) is violated and absolute simultaneity restored, so the stakes are high.

## 3.2 The global nonseparability of nature: state holism, contextual realism and the hidden structure of the physical process

BT contains, as we have seen, two – apart from adequacy ('Bell Realism') – constructive assumptions: 'Bell Locality' and 'λ-independence/timelike separability of events', whose conjunction nevertheless is contradicted by the experimental results. KST, in turn, being much more fundamental, confirms Bohr's intuition in the sense that it eliminates all the hidden variable theories which are not contextual and at the same time provide a classically complete definiteness, such as the one Einstein postulated. A violation of Bell inequalities means that at least one of the conditions has to be rejected or redefined, and that step naturally has to include the special geometry of quantum states, thus the results of KST – it is only then that we can speak of having drawn a lesson from those theorems and that we better comprehend what the physical principles lying at the root of 'quantum magic' are. We have already said that the noncontextuality in Einstein's understanding of 'local realism' is directly connected to a *more basic assumption about a strong general separability*, as it is only the latter which guarantees the local realist that the separate objects and their physical states, also that of the measuring system, will not be ontically mutually dependent: if it fails, that eliminates at least some of their intrinsic/monadic properties. The value obtained in the measurement and its reality should not then depend on the extracting actions (it is that very factor which is taken into account in Bohr's contextuality/holism), nor – all the more so – on any future actions, and hence all the (hidden) quantum history with unknown boundary conditions in the future. But the rejection of separability in any form leads to breaking that local and/or global noncontextuality, and that automatically ruins Einstein's Local Realism, even without the need to violate spacetime (relativistic) locality and to postulate superluminal physical signals as an action-at-a-distance. We therefore have to look more closely at separability itself and recognize what image of nature and types of realism are connected to any of its weakening or rejection.

In QM the unique properties of microphysical processes are mirrored in the construction (geometry) of the state space and that is true regardless of the formulation of the assumptions and definitions in BT. Contrary to classical mechanics, in which the maximum information contained in a complex system is the sum of the maximum information coded in its subsystems, in QM there appears, as we have seen, the phenomenon of entangled systems, whose informational characteristics completely differ from the classical ones – hence what we require there is to analyze the specific arrangement of relations, and not a monadic property [Teller, 1986]. Entanglement is a completely new phenomenon in nature and it can be understood to be a purely quantum resource of information, whose fully classical model is impossible to construe. From the formal point of view, the phenomenon finds an expression in the mathematical structure of the tensor product on Hilbert space $\mathcal{H}$ together with the superposition principle (cf. e.g. [Bruß, 2002], [Belousek, 2003], [Karakostas, 2004]). Let physical state $|\psi\rangle_{singl}$ of the compound system $S_1 \oplus S_2$ of two particles, $S_1$ and $S_2$, be an entangled state, whereas let the state of each be represented in the complex Hilbert space with the orthonormal vector basis $\{|p_i\rangle\}_{\mathcal{H}_1}$ and $\{|q_j\rangle\}_{\mathcal{H}_2}$ – then the appropriate representation of state $|\psi\rangle_{singl}$ is given by a bilinear mapping into another Hilbert space, of the form: $\Phi$: $\mathcal{H}_1 \otimes \mathcal{H}_2 \rightarrow \mathcal{H}_\psi$. The mapping is defined in such a way that the pairwise products of vectors, $|p_i\rangle_{\mathcal{H}_1} \otimes |q_j\rangle_{\mathcal{H}_2}$, create a complete, orthonormal vector basis of space $\mathcal{H}_\psi$, each product vector $|p_i\rangle \otimes |q_j\rangle \doteqdot |p_i, q_j\rangle$ belonging to the image $\overline{(|p_i\rangle, |q_j\rangle)}$ of that $\Phi$. In the instance when we are dealing with a pure singlet state for $N$ quantum objects, the basis vectors of the state space are simply a finite $n$-fold product in the form of:

$$|v_{ij\ldots n}\rangle_{\mathcal{H}_\psi} = |p_i\rangle_{\mathcal{H}_1} \otimes |q_j\rangle_{\mathcal{H}_2} \otimes \ldots \otimes |u_n\rangle_{\mathcal{H}_N}, \tag{26}$$

where $\{|p_i\rangle\}$, $\{|q_j\rangle\}$, …, $\{|u_n\rangle\}$ are vector bases of, respectively, $\mathcal{H}_1$, $\mathcal{H}_2$, …, $\mathcal{H}_N$ (of which dimensions need not be the same) and $\{|p_i, q_j, \ldots, u_n\rangle\}$ is a basis of $\mathcal{H}_\psi$. The tensor product $\mathcal{H}_1 \otimes \mathcal{H}_2$ fulfills two important conditions:

(a) for the scalar product of tensor products the following takes place:

$$\langle p_i \otimes q_j | p_{i'} \otimes q_{j'}\rangle = \langle p_i | p_{i'}\rangle_{\mathcal{H}_1} \langle q_j | q_{j'}\rangle_{\mathcal{H}_2} \tag{27}$$

(b) all linear combinations $|p_i\rangle \otimes |q_j\rangle$ create in $\mathcal{H}_\psi$ a countable dense set (hence the basis in $\mathcal{H}_\psi$ is complete, and $\mathcal{H}_\psi$ is a metric separable space).

In the instance when we choose a statistical-operator formalism for mixed states instead of vector formalism, the states of subsystems will be in general represented by self-adjoint, positive, trace class density operators: $\varrho = \sum_{i=1}^{\infty} c_i \,|\psi_i\rangle\langle\psi_i|,\ 0 \le c_i \le 1$, acting on the product Hilbert space $\mathcal{H}_1 \otimes \mathcal{H}_2 \otimes \ldots$, whose trace is $Tr(\varrho) = \sum_{i=1}^{\infty} c_i \,\|\,\psi_i\,\|^2 = 1$ (the trace does not depend on the choice of the orthonormal basis). In that general operator formalism, pure states, which were earlier on represented by normed vectors in $\mathcal{H}$, will simply be given by positive, idempotent density operators, $\varrho^2 = \varrho$, $Tr(\varrho^2) = 1$, and hence in that particular case by projectors $P_{\psi i} \doteqdot |\psi_i\rangle\langle\psi_i|$ onto subspaces $\mathcal{H}_{\psi i}$ of the product space, for which obviously also: $(P_{\psi i})^2 = P_{\psi i}$. Thus, in the case of state $|\psi\rangle_{singl}$, we can write it down using an operator which is the sum of the density operators: $\varrho_{singl} = \Sigma_i\, c_i|\psi_i\rangle\langle\psi_i| = \boldsymbol{I}_\psi$ on space $\mathcal{H}_1 \otimes \mathcal{H}_2$. The condition of idempotency for the 'coefficients' (weights) of that operator, $c_i^2 = c_i$, together with the condition of the unit trace, $Tr(\varrho) = 1$, make one of the coefficients $c_i = 1$, while the others equal to zero, therefore, for every pure state, we have such one-dimensional projectors $P_{\psi i}$ and also $P_\psi = \boldsymbol{I}_\psi$ on $\mathcal{H}$. We can thus say indeed that 'a system is in state $\varrho$', although $\varrho$ is an operator.

Those quite elementary remarks suffice to capture the meaning of the quantum nonseparability of states regardless of the definitions referring to EPR or Bell's physical argumentation (to which we are still going to go back to). We say that state $\psi$ of the singled out system is s e p a r a b l e only when it can be written down in the form of a vector in $\mathcal{H}_\psi$ such as:

$$|\psi\rangle = c_i|p_i\rangle_1 \otimes c_j|q_j\rangle_2 \otimes \ldots \otimes c_n|u_n\rangle_N \tag{28}$$

for any $i, j, \ldots, n$ (they have to belong to appropriate vector spaces numbered by external indexes) and $c \in \mathbb{C}$, hence there is no quantum correlation between subsystems here. If they are *totally uncorrelated*, even classically, we have a plain product state: $\varrho = \varrho_1 \otimes \varrho_2 \otimes \ldots \otimes \varrho_N$. Alternatively, one can still have a system prepared in a *separable*, but purely *classically correlated state* – then it can be represented as a convex $n$-fold combination of the product states of its subsystems: $\varrho = \sum_{i=1}^{\infty} p_i (\varrho_i)_1 \otimes (\varrho_i)_2 \otimes \ldots \otimes (\varrho_i)_N$, where $p_i > 0$ and $\Sigma_i p_i = 1$ (therefore $p_i$ can be interpreted as probabilities). In turn, we can say that quantum state $\psi$ of the whole system is n o n s e p a r a b l e only when $p_i$ are not real positive and $\psi$ cannot be written down in the factorized form of product (28), as its form has to take into account the superposition of the whole, thus subsystems cannot be in pure states (if they were, they would not be quantum-correlated and $\psi$ would be separable):

$$|\psi\rangle = \Sigma_{i,j,\ldots,n}\, c_{ij\ldots n} |p_i\rangle_1 \otimes |q_j\rangle_2 \otimes \ldots \otimes |u_n\rangle_N. \tag{29}$$

The property is well-established in the above definition of the tensor product of Hilbert spaces $\mathcal{H}_1 \otimes \mathcal{H}_2 = \mathcal{H}_\psi$, as it is *not* equivalent to the Cartesian (topological) product $\mathcal{H}_1 \times \mathcal{H}_2$, i.e., every product of the vectors belonging to those spaces, $|p_i\rangle_{\mathcal{H}_1} \otimes |q_j\rangle_{\mathcal{H}_2}$, also belongs to $\mathcal{H}_\psi$ (they create a proper subset in $\mathcal{H}_\psi$), but *not every one* of its elements can be represented in the form of (28) on account of the unlimited principle of superposition in $\mathcal{H}_\psi$. In the case when we have a situation such as the one in (29), we say that the state of the whole system is *entangled* or *quantum correlated* (the term, *Verschränkung*, was coined by E. Schrödinger [1935a, p. 555]; [1935b] in answer to the 'EPR paradox'). In turn, when it is possible to represent the state of the whole in the form of (28), we can speak of a *product* or quantum *non-correlated* state. In the case of the latter, we treat the states of the systems $\psi_1$ and $\psi_2$ as quantum-independent of each other, that is *separable*, as for any observables $\omega_1$ and $\omega_2$ and – accordingly – systems $S_1$ and $S_2$ we have:

$$Tr(\varrho_1 \otimes \varrho_2)\,(\omega_1 \otimes \omega_2) = Tr(\varrho_1\omega_1) \cdot Tr(\varrho_2\omega_2), \tag{30}$$

otherwise they are quantum-dependent and nonseparable, i.e., none of the reduced density operators for $S_1$ and $S_2$ represents a pure state and the systems are indeed non-classically correlated (see e.g. [Bruß, 2002, p. 4238*ff.*]). The state of the whole composed of separable subsystem states is simply a product of those states (of the physical information contained in them) just as that takes place in a classical case, even if those subsystems are classically correlated; whereas the state of the nonseparable whole does not in any way constitute such a product for – as has already been mentioned – tensor product space $\mathcal{H}_\psi$ is not possible to be decomposed into simple products of product space $\mathcal{H}_1 \times \mathcal{H}_2$. What is more, the situation (28) and (30) is exceptional in QM (it is a singular type of an element of state space $\mathcal{H}_\psi$), and not a typical one, contrary to a classical-mechanical intuition, i.e., it is not entanglement (29) that requires an explanation in QM, but on the contrary: the lack of entanglement is a mystery and requires an explanation of its appearance [Esfeld, 2004, p. 604]. The requirement that every pure state of the quantum system should be a product state is equivocal to the claim that no interaction takes place between it and other subsystems of the Universe (there is no quantum information transfer between them) – the least interaction of the system with the quantum environment forces the product state to be replaced by an entangled one due to the informational entanglement. In that very formalism of QM – be it geometric or algebraic – there is an in-built fact that *nonseparability is a fundamental physical phenomenon* and a novel characteristic of the information transfers which is not accounted for by the classical information theory. The classical (full) separability of properties and events, as represented by logic of the classical phase space, can only be upheld since the quantum entanglement is invisible to classical mechanics.

An issue which seems extremely worrying for an Einsteinian local realist is thus the separability of the particular parts of the whole and the state definiteness of the subsystems, as the above formalism suggests a very non-intuitive situation in that respect – we have come across it in Ch. 1.3 and go back to it here and to the problem of quantum (and Platonic) compositionism. Let us assume that we are dealing, as in Ch. 1.2 and 3.1, with the entangled state of two particles $S_1$ and $S_2$, whose pure spin state $\psi$ is given as:

$$\varrho_{singl} = \boldsymbol{I}_{\psi} \doteqdot \Sigma_i c_i(|p_i\rangle \otimes |q_i\rangle) = \frac{1}{\sqrt{2}}(|\uparrow\rangle_{1,a}|\downarrow\rangle_{2,a} - |\downarrow\rangle_{1,a}|\uparrow\rangle_{2,a}) \qquad (31)$$

where vectors $\{|p_i\rangle\}$ and $\{|q_i\rangle\}$ are the bases of two-dimensional complex spaces $\mathcal{H}_1$ and $\mathcal{H}_2$, $\varrho^2 = \varrho_{singl} = |\psi\rangle\langle\psi|$ is a projector corresponding to state $\psi$, while '$a$' indexes a chosen possible spin orientation in space. It then becomes clear that there *does not exist* any pair of states $|p_i\rangle \in \mathcal{H}_1$ and $|q_j\rangle \in \mathcal{H}_2$, so that: (1) they are completely independent states possible to ascribe, respectively, to subsystem $S_1$ and subsystem $S_2$, and also (2) one can reconstruct from them the physical state of the compound system $S_1 \oplus S_2$. In other words, the vectors $|p_i\rangle$ and $|q_j\rangle$ belonging to vector spaces $\mathcal{H}_1$ and $\mathcal{H}_2$, taken separately, do not represent any states of system $S_1 \oplus S_2$, as *they do not correspond to any eigenvector* of the operator $\varrho_{singl}$. What is more, assuming that the quantum system has been prepared in state $\varrho_{singl}$, then in accordance with von Neumann's 'eigenstate–eigenvalue rule', one cannot say that it is to be found in state $|\uparrow\rangle_{1,a} \otimes |\downarrow\rangle_{2,a}$ or $|\downarrow\rangle_{1,a} \otimes |\uparrow\rangle_{2,a}$, since state $\varrho_{singl}$ simply does not belong to the subspaces spanned by the latter two – the only thing, anyway, that we are able to know about particle $S_1$ or $S_2$, is statistical information contained in the reduced density matrix written down separately for each. That is a s t r o n g  f o r m  o f  n o n s e p a r a b i l i t y: 'separate states of the parts of system $S_1 \oplus S_2$ physically do not exist', which places a grave limitation on the sense of physical statements about quantum systems as, strictly speaking, in the unitarily evolving entangled system its parts $S_1$ and $S_2$ do not at all have *any definite, separate (ontic) states*, even if they are separated in spacetime, as in Fig. 1 and 2 [Karakostas, 2004, p. 288]. That form of nonseparability is in fact a strong statement about the very existence of properties and states and does not need any prior discussion of relations and their structure within some relational ontology. One can nevertheless weaken that conclusion in a way, interpreting the formalism in a gentler form: in accordance with the rotational symmetry of spin state $\varrho_{singl}$ it is irrelevant which axis of measurement the observer selects, as the only thing QM guarantees is the anti-correlation of the spin states for those pairs of particles, i.e., QM predicts that the situation when both of them would have the same spin in some direction will surely not take place, which is in fact written down in formula (31). The weakening of the ontological understanding of nonseparability is thus equivalent to saying that *the state of every part is meaningful solely within the state of the whole and with relation to the states of other parts* – that is therefore a relational form of nonseparability. As we can see, the difference between a strong form and a relational one is not as decisive as a classical local realist might seem to wish, as both stress the ontological dependence between the parts and the whole they compose without any possibility of reducing it to some form of basic separability. Such a relational formula of nonseparability might make us think that the gist of the situation (31) is indeed a non-supervenience of the relation of 'being entangled with' with regards to all the other properties of each particle.

Firstly, we have to have at hand a certain simple definition of the non-supervenience of a relation in the context of microphysics, to be in a position to capture its possible linkage to nonseparability and holism. Secondly, we have to analyze at greater depth whether it is able to characterize the quantum situation (31) and what connection it has to Bell (Causal) Locality of separated states, constituting condition $[2]_B$. We shall realize that the relationship is much more subtle than it seems at first glance and points to a completely new type of ontological realism, impossible to be reconciled with Einstein's Local Realism.

A frequently quoted in that context definition of the supervenience of relations ([Belousek, 2003], [Esfeld, 2001, p. 246], [Esfeld, 2004, p. 606]), one appropriate to our analysis, is the one provided by Carol E. Cleland [1984, p. 25] regarding the relational ontology of space:

Supervenience of relation

A dyadic relation ***R*** is supervenient (supervenes) upon a determinable non-relational attribute if and only if:

1. $\Box (\forall x, y) \sim\Diamond$ [$\boldsymbol{R}(x, y)$ and there are no determinate attributes $P_i$ and $P_j$ of determinable kind $P$, such that: $P_i(x)$ and $P_j(y)$];
2. $\Box (\forall x, y)$ {$\boldsymbol{R}(x, y) \supset$ there are determinate attributes $P_i$ and $P_j$ of determinable kind $P$, such that $P_i(x)$ and $P_j(y)$, and: $\Box (\forall x, y) [(P_i(x)$ and $P_j(y)) \supset \boldsymbol{R}(x, y)$}.

A *determinable property* is – in a somewhat Aristotelian manner – one which occurs, but taken in abstraction from its particular and definite value (expression), for example 'having (being characterized by) a spin'; a *determinate property* is one which has definite, further unspecifiable value, such as 'having a spin up in direction $a$ after measurement at moment $t + \delta t$'. Supervenience condition (1) states that each of the parts (subsystems) entering into relation $\boldsymbol{R}$ has to possess some specific, definite non-relational attributes of the given type, required for $\boldsymbol{R}$ to take place at all; the stronger condition (2) meanwhile, guarantees that the appearance of those particular non-relational properties will automatically mean an appearance of $\boldsymbol{R}$ which fully supervenes on them. The above definition allows us to discern two degrees of non-supervenience: relations that are weakly non-supervenient fulfill condition (1) but not condition (2), whereas relations that are strongly non-supervenient do not fulfill either of the conditions [ibid.]. In the second instance, the given objects (such as electrons, photons or atoms) might even possess some intrinsic properties, but they do not have to be properties of any particular type, as that is irrelevant to the appearance of quantum relations (such as entanglement). In the case of such relations as, for instance, spatial ones, the objects have to possess certain attributes, such as 'shape' or 'extension' (or something alike), though we do not require those attributes to possess any particular values to enter into the relation. As a result we can say that spatial relations, irreducible to certain intrinsic properties of the physical objects, are weakly non-supervenient (they only do not meet condition (2)) [ibid.]. But quantum entanglement relations are diametrically different from spatial relations – they are *strongly non-supervenient* (as rightly claimed in [Teller, 1986], [French, 1989, p. 17*ff.*], [Belousek, 2003, p. 795*ff.*], [Esfeld, 2001, Ch. 8.2], [Esfeld, 2004, p. 606]). Situation (31) seems irreconcilable with the familiar concept of separate, classical or quasi-classical objects, whose ontic independence is manifested by their own definite and intrinsic quantum attributes, and those in turn being the base for some global relations – thus what utterly lets us down here is the externalistic compositionism of the Aristotelian type (cf. Ch. 1.1 and 1.3). Simply speaking, the situation of quantum entanglement itself violates, in a spectacular manner, the ontological atomism of independent substances/elements of reality, which Einstein had expected to be a solid foundation of physics. For that very reason, though the strategy of expressing quantum nonseparability with the help of standard non-supervenient relations seems very promising, unfortunately, after a closer scrutiny, there appear serious problems which touch upon the interpretation of QM as a whole. Let us at present attempt to define more precisely what that quantum 'relational holism' (in contrast to the classical one) might be.

Making our previous anti-Lewisian discussion from Ch. 1 more physical, we can say that *local physicalism* (Paul Teller's [1986, 1989] term) or *particularism* requires that the physical reality be finally composed of independent (separable), physically well-defined (noncontextual) micro-objects and for each relation between them to be finally reducible to their intrinsic properties. Each theory, in turn, which does not meet those requirements, not only effectively but also regarding the foundational principle, particularly when it contains some primary non-supervenient relations, may be generally termed as *holistic* (to a lesser or greater degree). As we can see, that is a great class of potential theories and ontologies, if only because one can point to many different types of non-supervenience and nonseparability, e.g. we can talk (despite historical stereotypes) of Cartesian holism of matter identified with space, which even found its embodiment in the pioneering programme of quantum geometrodynamics in the 1960's [Esfeld, 1999], or about the holistic aspects of contemporary quantized gauge theories with loop representations [Healey, 2004]. The concept of holism as such, insofar as we wish to define it, is of manifold aspects – we can rather speak of 'holisms', although general nonseparability, such as in QM, is certainly a sufficient condition to identify strong holism. In the case of a standard approach to ontological holism, the primary intuition used to come down to the fact that wholes are not made up solely of parts, but that they also contain something more, with that 'something more' not necessarily being ontologically congruent to those elements (and thus does not constitute a 'part' of a different type). Physical relational holism, in turn, states that there exist certain physical relations between objects (or elementary 'states of affairs' in the Einsteinian way of thinking), which do not supervene on their intrinsic properties ([Teller, 1986, p. 73; 1989, p. 214]) – an example of such a relational holism is, according to Teller, offered by quantum entanglement. Nevertheless, such a metaphysical framing of quantum entanglement breeds difficulties when we take into account the assumptions it requires, and above that the issue of its precision, as after all it has to point out the specific nature of quantum nonseparability as such (as compared to, for example, spatiotemporal relations). After all, the physical relationship (31) is definitely *not* like the spacetime relations and that is a decisive aspect of the quantum riddle.

Darrin W. Belousek [2003, p. 797*ff.*] pointed out the first of those difficulties – Teller's argument in favour of his relational holism in QM can be brought down when we set aside one of its three assumptions: (a) the *completeness* of the 'quantum state' given by function $\psi$; (b) von Neumann's '*eigenstate–eigenvalue link*' (A. Fine's term [1970]), which rules that the preparation of a system in the eigenstate of a certain operator is tantamount to that system possessing a corresponding definite eigenvalue; (c) the thesis that the very identification of correlations between the states of the systems implies the *existence of a real, physical relation* between them. What is interesting is the conclusion the author arrives at when reflecting whether replacing the rule from assumption (b) with its weaker version would cause the removal of the non-supervenience of the internal relations in $S_1 \oplus S_2$ on the separate properties of systems $S_1$ and $S_2$. That turns out not to be so, as in accordance with the above definition of the nonseparability of states, it is impossible to fully reconstruct purely global relations in system $S_1 \oplus S_2$, given by the density matrix $\varrho_{singl}$, from the density matrices of its entangled subsystems, $\varrho_1$ and $\varrho_2$, which in the intention of the weaker version of von Neumann's rule were to represent the non-relational, intrinsic properties of those subsystems [ibid., p. 797*ff.*]. In that situation we are still forced to opt for an irreducibility of global relations in system $S_1 \oplus S_2$, only in that case it might be a weak non-supervenience (as we would assume that statistical $\varrho_1$ and $\varrho_2$ nevertheless are somehow 'founded' on some minimally intrinsic properties of both particles). The situation becomes much more confusing, though, when we desist from assumption (a) and accept a quantum theory with hidden variables such as old Bohmian mechanics. In such a case (b) also has to be turned down, but the fact that hidden variables have to be contextual (which *is* a form of nonseparability) to simulate standard NRQM predictions does not nevertheless imply that they have to be at the same time relationally non-supervenient, as some versions of Bohmian mechanics, the so-called minimalistic ones (the school of D. Dürr [2001] and S. Goldstein), do not require any non-supervenience of relations at all, although they are radically spacelike nonlocal [Belousek, 2003, p. 801*ff.*]. Other versions of that mechanics, in turn, with the quantum potential interpreted in a relational way, contain a strong non-supervenience of global relations in multi-particle systems. That does not in any manner determine any principal weakness of the theories with the hidden variables – on the contrary, they could even be a step in the right direction – but Bohmian mechanics in its original form as NRQM is principally not Lorentz invariant, as it contains troublesome absolutes at its core (absolute space as the background to the configuration of the particle system and absolute simultaneity). That is the first serious clue that analyzing quantum holism in the categories of Teller's simple relational (non-supervenience) holism is insufficient, as what escapes us in such an analysis is the essence of microphysical phenomena. The very assumption of the minimally definite, individual 'objects' quasi-classically located in points/regions of space, even if strong local contextuality is explained by spacelike nonlocality, falls prey to the Aristotelian idol – the 3+1-spacetime with a preferred foliation, which cannot be physically determined as that would give the forbidden exact particle trajectories. But quantum non-separability does not need spacelike nonlocality destroying SR and GR, quite to the contrary: as I signalled in Ch. 1.2 and 1.3, it is deeply interconnected with the Principle of Relativity, and that might be obscure when entangled information (31) is treated as a noncontextual 'element of reality'.

There is also a second type of difficulty involved there, though, rightly pointed out by Michael Esfeld [2001; 2004], studying the analogy of the quantum entanglement relations to presumably also non-supervenient spacetime relations. He refers to David Lewis's analysis [1986a, p. 77] in the framework of the metaphysics of possible worlds, in which he wonders whether there could exist any external, non-supervenient relations between individual entities in the world, which (analogically to the spatiotemporal ones) would unify the physical world. One should imagine a certain type of such imaginary relations, such as 'possessing the same charge' (SC) and 'possessing an opposite charge' (OC), which – just as entanglement – would be strongly non-supervenient, i.e., they would not require any determinable intrinsic properties of the individuals and as such would be rival to the entanglement. It turns out, intuitively and in accordance with Lewis's analysis, that such imaginary, purely external charge relations cannot unify the physical world, as contrary to spatiotemporal and entanglement relations there is no reason for exactly the same relations SC and OC not to take place between entities in completely different possible worlds (hence they are in no way specific and they do not differentiate any of them). In the case of entanglement the situation is opposite: we cannot speak of exactly the same quantum relations between 'micro-objects' in a totally different possible world, as specific quantum relations unify only one particular world at a given time. Apart from that, there is another property making SC and OC differ from entanglement: the latter is dependent on the physical (quantum) state of the systems, whereas the former are completely independent from it, i.e., they are imposed on the *ad hoc* postulated objects in an arbitrary way, 'without any reason', as Leibniz would have said [Esfeld, 2004, p. 607]. Hence, we have a further reason to suspect that the very use of the category of 'non-supervenient relations' in ontology does not unfortunately offer an adequate picture of what the specific nature of quantum entanglement is, if it is – as a physical property – so fundamental. It seems that at the core of those problems is the very imagery of separable objects with minimally intrinsic properties (even if they are seen as purely probabilistic, given by the reduced density operators $\varrho$) and non-supervenient relations upon them. Imagining quantum correlations coded in $\varrho$ as ontic 'tendencies' inherent in a subsystem gives nothing (see Ch. 4), since such a rigid scheme leaves off the fundamental Principle of Relativity.

The above remarks draw a picture that has already been sketched in Ch. 1.3 when discussing the Platonic ontology of the whole and two general types of holistic compositionism. For at the starting point of all of Teller's [1986] reasoning lies the conviction that before we introduce a non-supervenient relation, we have to have at our disposal the objectively given, separate and primitive individuals (*relata*), otherwise the whole sense of talking about the 'relational holism' of nature is at risk of being ambiguous:

> By relational holism I will mean the claim that objects which in at least some circumstances we can identify as separate individuals have inherent relations, that is, relations that do not supervene on the non-relational properties of the distinct individuals. Relational holism is free of the incoherence which threatens less clearly stated forms of holism. It is sufficient for an object to be a distinct individual that it have a non-relational property. And it is quite consistent to suppose that two such distinct individuals, each having a non-relational property, should also stand in some inherent relation to each other. ([Teller, 1986, p. 73], also quoted in [Esfeld, 2004, s. 605])

Such a view might be called 'relational particularism' [Belousek, 2003, p. 809], as different from local physicalism and Lewis's classical (mereological) particularism, as despite the introduction of non-supervenient (external) relations, it requires primordial, simple physical elements to be found at its basis, such as particles with individual properties. It is still, in one way or another, a certain form of 'atomistic philosophy' [Barbour, 2001, p. 220]. One has to wonder in these circumstances whether a similar reasoning, leading to relational holism in Teller's version, is not a certain kind of mental stereotype in ontology, which leaves us mystified as to the nature of quantum phenomena. In between the ontologies analyzed above and the types of compositionism pointed out earlier there exist clear links, as each of them offers a generically different perspective on the relations of the whole and the parts. Local physicalism (whose realization was for example Hobbes's materialism mentioned in Ch. 2.2) is reconcilable mainly with extensional compositionism, as in that ontology the relational properties of the world, such as entanglement, should optimally be reduced to intrinsic, non-relational properties of the separable, elementary particulars. Such is, for example, the decided view of David Lewis and of all those who defend 'Humean supervenience' [Lewis, 1986b, p. IX*ff.*; 1994, p. 474] and ontological universalism construed on the basis of mereology, although – one has to admit – that is a very restrictive standpoint and useless in the context of quantum ontology. Teller's 'relational particularism' in turn, as the foundation of quantum relational holism, is naturally connected to (minimalistic) externalistic compositionism, which has its historical roots in Aristotelian holism. Within that standpoint there exist many possible versions, but that type of compositionism simply assumes that the real particulars are non-relationally separate entities as they either possess specific, definite intrinsic properties or – even if that is not so – a certain type of minimal 'thisness' (*haecceitas* in the medieval terminology of Duns Scotus), which supplies them with a discernible individuality. Although externalistic compositionism – as has already been said in Ch. 1 – since the time of Aristotle has seemed very natural and commonsense, it nevertheless gives rise to serious problems, particularly when we attempt to use it to construe a consistent ontology of the microworld.

What proves to be a persistent problem and a weak point of the ontology of 'relational particularism' and externalistic compositionism in a framework of the quantum information theory is the riddle: how is it at all possible to find out what those intrinsic, non-relational properties of the primordial objects (i.e., their 'internal state') are? Everyone who treats that thesis on the existence of such properties in a serious way has to demonstrate how, in the framework of a convincing epistemology and quantum information theory, it is possible for any system to obtain any information about those monadic properties, or at least recognize the primitive *haecceitas* to be a noncontextual attribute of the postulated 'elementary object'? If one acknowledges an operational character of the quantum information processing, that is completely unrealistic – to say the least. The information the external system gains is grounded in the physical (quantum-dynamical) interaction with another system, and any properties are identified *solely* in relation to other properties [Esfeld, 2004, p. 614*ff.*] – that is another aspect of what we have sketched in Ch. 1.2 as the Universal Principle of Relativity, whose violation leads to an introduction of certain imagined internal absolutes independent of the relations, such as those that ground the metaphysical scheme of classical mechanics (e.g., the absolute 'state', absolute time or absolute 'observer'). The discovery, as it is, or rather that philosophical re-discovery lies at the foundation of Bohr's epistemological idea of complementarity and contextuality of the local measurement situation or action: the information gained about some system cannot be separated from the physical conditions it was actively gained in, thus one could say that even the Bohrian elementary 'quantum phenomenon' is intrinsically and praxistically relational. Einstein's Local Realism, 'everybody's paradigm' [Malin, 2003, p. 90], is in turn rooted in the conviction that there exist in nature elementary objects/particulars, possessing states that are independent of the context and the observing system ('*eine vom wahrnehmenden Subjekt unabhängige Außenwelt*'[52]),

52 'The external world independent of the perceiving subject' [Einstein, 1982, p. 159].

about whose intrinsic properties one can, and in fact always does, get objective information. So, after all, if we wish to reject 'relational particularism' and externalistic compositionism, we have to think how otherwise we can express the fundamental phenomenon of entanglement without using the concept of 'non-supervenient relations' specified with regards to such particulars. The possible radical solution is an ontology which is an opposite of extensional (mereological) compositionism, and which we have found in Plato: internalistic compositionism without primitive particulars evolving in time, on which internalistic quantum 'state' holism could be founded. That does not necessarily imply, let us stress once again, that speaking of 'non-supervenient relations' is completely wrong, but only that it does not fully account for the singularity of the situation given as (31), and besides, it forces beforehand a certain compromise with the separable 'particularism' as an ontology of classical mechanics. Thus, perhaps regrettably to some, it is not useful in quantum ontology.

Let us therefore come back to Einstein's key condition of separability connected to the ontology hidden in $[3]_E$ and formulated independently of the quantum-mechanical formula (28), to capture more precisely the essence of *state holism* in QM. Einstein expressed it very clearly for example in his well-known letter to Max Born from 1949:

> ...That which we conceive as existing ('actual') should somehow be localized in time and space. That is, the real in one part of space, A, should (in theory) somehow 'exist' independently of that which is thought of as real in another part of space, B. If a physical system stretches over the parts of space A *and* B, then what is present in B should somehow have an existence independent of what is present in A. (...) However, if one renounces the assumption that what is present in different parts of space has an independent, real existence, then I do not at all see what physics is supposed to describe. For what is thought to be a 'system' is, after all, just conventional, and I do not see how one is supposed to divide up the world objectively so that one can make statements about the parts. [Einstein, Born, 1969, s. 223*f.*] (Transl. by Howard [1985, p. 191])

Howard [1985, p. 187*ff.*] was right to stress that the condition was so important to Einstein as 'it provides the only imaginable objective principle for the individuation of physical systems'. In the context of formula (28) it is worth quoting an even shorter, earlier definition from his letter to Erwin Schrödinger in 1935 (without any reference to the spatiotemporal localization of systems):

> ...The real state of (AB) consists precisely of the real state of A and a real state of B, which two states have nothing to do with one another. (Transl. by Howard [1985, p. 180])

In his later statements, Einstein stressed the status of separability not only as a *condition* (which was the case in the period when the 'EPR argument' was created), but indeed as a *principle* of physics: *Trennungsprinzip*. (There was a similar case with the condition of locality, which Einstein evoked as a universal, physical 'principle of local action', *Prinzip der lokalen Aktion*). Was Einstein therefore a local physicalist/ particularist? Everything points to it being so: relativistic locality and the general, physical separability of states, taken together, require that mechanics be construed upon autonomous spacetime individuals 'existing in reality' ('elementary things', to use the phrase Einstein himself used), while every global state defined on the set of those individuals be possible to be fully reducible to their own separable states. As we have noticed earlier (Ch. 1.3), Einstein might have perhaps allowed for certain non-supervenient relations with regards to the global properties of physical fields (especially the gravitational field), but would not have accepted any violation of separability with regards to 'elementary things' or 'elementary states' in any fundamental mechanics. He declared that clearly in his pronounced opposition to Bohr and pointed out that he understood his 'local realism' as one that was defined jointly by both principles mentioned earlier:

> If one asks what is characteristic of the realm of physical ideas independently of the quantum theory, then above all the following attracts our attention: the concepts of physics refer to a real, external world, i.e., ideas are posited of things that claim a 'real existence' independent of the perceiving subject (bodies, fields, etc.), and these ideas are, on the other hand, brought into as secure a relationship as possible with sense impressions. Moreover, it is characteristic of these physical things that they are conceived of as being arranged in a space-time continuum. Further, it appears to be essential for this arrangement of the things introduced in physics that, at a specific time, these things claim an existence independent of one another, insofar as these things 'lie in different parts of space'. Without such an assumption of the mutually independent existence (the 'being-thus') of spatially distant things, an assumption that originates in everyday thought, physical thought in the sense familiar to us would not be possible. Nor does one see how physical laws could be formulated and tested without such a clean separation. Field theory has carried out this principle to the extreme, in that it localizes within infinitely small (four-dimensional) space elements the elementary things existing independently of one another that it takes as basic, as well as the elementary laws it postulates for them. [Einstein, 1948, p. 321*f.*] (Transl. by D. Howard [1985, p. 187*f.*])

As we have noticed earlier, the original EPR article from 1935 did not precisely present Einstein's own metaphysical reasoning, which was much simpler and put forward those very two principles, that of locality and of separability, as Einsteinian 'strong locality' (:= $[3]_E$) in a direct connection to the concept of the 'state of a physical system'. It has turned out that the formulation using the quite well defined concept of 'state separability' is the best point of departure for capturing the exceptional nature of situation (31). First and foremost, contrary to the case of relativistic (causal) locality, state separability does not depend on specific spacetime relations and does not need them to be defined at all, i.e., it is not significant what particular relativistic interval separates the quantum events at all. What is more, obviously it also concerns situations in which it is difficult to take into account a spacetime localization, such as pairs of electrons on the same orbital of the helium atom [Esfeld, 2004, p. 608]. As Einstein has already noticed, the truly important problem here concerns the fact whether the state of the whole $S_1 \oplus S_2$ *can be fully factorized into separate and independent states* of $S_1$ and $S_2$ and hence it is solely a metaphysical issue of the *classical ontic independence* of parts (elements) of the whole within some relational structure. That points to a certain complication in the analysis of QM ontology, as one has become accustomed – in the wake of Einstein himself – to strictly linking, even unintentionally (as Einstein–Podolsky–Rosen seem to have done), the separability to causal locality within a single strong condition $[3]_E$ and hence the important analysis Howard [1985] proposed is so useful. A good formulation of the definition of nonseparability independent of spacetime relations has been proposed by Michael Esfeld [2004, p. 608; 2001, p. 270], which I am quoting below with the addendum that it concerns a *formal nonseparability* of states (without solving *a priori* the dilemma what kind of global ontology it would demand):

> The Global Formal Nonseparability of States (Esfeld)
> The states of two or more systems are (formally) *non-separable* if and only if it is the joint state of the whole that completely determines the state-dependent properties of each system and the correlations among those systems (to the extent that those are determined at all).

The above definition says that only a joint state of system $S_1 \oplus S_2 \oplus \ldots \oplus S_N$ contains complete information on the structure of relations between its parts, but, contrary to the definition invoking non-supervenient relations between individuals, it states that talking about quantum states/properties of the subsystems–parts is meaningful *only with regards to that whole*. Nonseparability thus defined in a natural manner avoids ontological particularism and essentialism, as it corresponds rather to internalistic compositionism: it is not the primary parts which compose secondary wholes in time, but the primary whole defines the parts simultaneously with the relations between them (cf. e.g. [Karakostas, 2004, p. 295*ff.*]). From the perspective of the ontology of classical mechanics, that kind of compositionism is extremely counter-intuitive, as instead of an atomistic 'bottom–up' thinking, it introduces the holistic way of thinking of the 'top–down' type: the whole order of determination somehow comes from 'above', while the absolute (non-contextual) intrinsic properties are subject to elimination in the same way as absolute base objects. Agreeing with Esfeld's conclusions, we can also agree that holism in QM is a radical state holism – it is founded on the nonseparability of quantum properties of systems, that is, ultimately, all the physical properties: QM formalism adequately represents a certain fundamental property of the microworld, which is revealed by pervasive correlations in the (common) situation of entanglement as a new resource of information. One should keep in mind, nevertheless, that nonseparability thus defined is yet of a *formal* character [Belousek, 2003, p. 809*ff.*] – it is an abstract expression of the fact of the existence of global correlations between the results of quantum measurements on entangled systems, which can in no way be explained assuming the classical complete separability of all states. Its formal character does not exclude beforehand any physical principle of the individuation of the different quantum subsystems – it only ascertains that solely a global quantum 'state' (31) in some action frame is a pure state and contains complete information about the relations between the subsystems, while the quantum 'states' of the subsystems cannot be defined in that frame independently of the whole as pure states. Such a formal nonseparability can also be called 'global', as it concerns the relationships between local physical actions which are independent of all other physical relations, particularly spacetime relations – we shall go back to the issue of the ontology of those relations in Ch. 5. 'Globality' does *not* imply spacelike nonlocality as an action-at-a-distance and absolute simultaneity; but it *does* mean that QM itself, its logical-mathematical structure or geometry, guarantees the existence of holistic relations between processes in nature, which from the perspective of the history of a local, relativistic subsystem may seem 'nonlocal', and hence apparently pushes up against SR and GR, which worried Einstein and many others[53]. That point has been often permanently ambiguous in the discussions, since the conflation of conditions $[3]_E$ and $[2]_B$, and the rich Einsteinian metaphysics lying behind the former one, has led many to state the quantum-mechanical dilemma as simply incompleteness versus spacelike

53 The term 'global nonseparability' seems to be adequate as a rival to popular 'nonlocality', as it does not determine beforehand that relations between quantum states require nonlocal action through the transmission of the physical signals, violating SR and GR. The fact that a relation is 'global' does not mean that it needs spacelike 'nonlocal action' in Einstein's (relativistic) understanding of the term, especially if spacetime itself would be a derivative, quantum-thermodynamic state (see Ch. 4).

nonlocality (setting aside a much more fundamental property of nonseparability not singled out among the Bell's conditions). But that is not so.

It seems natural to ask at this point what the relation of global quantum nonseparability of the state is to the inequalities violated in the experiments and Bell's conditions. As we have seen, the breaking of the formal separability in QM surely causes a serious violation of Einstein's 'strong locality' and the equally strong condition of noncontextuality, which automatically crosses out the possibility of retaining Einstein's Local Realism (or 'strict realism') as an ontology for physics. Nonseparability also implies that there always exists a hidden, global context of local quantum interactions, which is either not accessible to the subsystems in any way as a quantum noise, or it is accessible locally in the form of totally relativized information about the former correlation of actions and represented in a particular action frame by a 'state' function or a density matrix, which the observer has at hand. (The 'state' function is interpreted here as a local representation of the relative knowledge about the external system on the basis of the information about its history, and not as a physical entity in space). But the riddle remains what that 'global context' of every quantum event/phenomenon is, whose physical reality is witnessed by the registered correlations coded in the geometry of QM. In other words, sooner or later, especially if we deflate any action-at-a-distance and treat seriously the Principle of Relativity, the key issue is what the *ontological structure of the quantum universe* $\Omega$ really is, which makes possible von Neumann's condition of consistency to be valid at all. In that case, it becomes particularly important to reflect once again on Bell's conditions. The issue of which of those conditions, $[1]_B$ & $[2]_B$ & $[3]_B$, is inadequate sometimes becomes not sufficiently transparent as, as we have already stressed, 'Bell locality' $[2]_B$, is not a plain correspondent of 'Einstein locality', which is sometimes understood as 'a prohibition of transmitting the signal over a distance with superluminal velocity' (= 'signal locality'). It has become customary to assume that Bell Locality logically contains two other conditions, which might be simultaneously violated or not, as $[2]_B$ can be presented as a conjunction of 'parameter independence' (PI) and 'outcome independence' (OI) [Shimony, 1986]:

$[2.1]_B$ Parameter Independence

$$P_{\lambda mn}(p_m) = P_{\lambda m}(p_m); \qquad P_{\lambda mn}(l_n) = P_{\lambda n}(l_n)$$

$[2.2]_B$ Outcome Independence

$$P_{\lambda mn}(p_m \,\&\, l_n) = P_{\lambda mn}(p_m) \cdot P_{\lambda mn}(l_n)$$

The formal separation of those two conditions was first presented by Jon Jarrett [1984], although he used different terminology ($[2.1]_B$ as *simple locality*, while

$[2.2]_B$ as *predictive completeness*), and included a wider set of parametres than Bell, introducing hidden variables of the measuring apparatus to the indexes (as its quantum microstates). Jarrett's terminology itself suggests that, according to the author, $[2.1]_B$ should be retained, if we wish to avoid a serious conflict with SR, while $[2.2]_B$ can at the same time be violated by QM, which results in a certain incompleteness from the perspective of classical mechanics which is unable to generate the adequate probabilistic predictions. There therefore exists an unresolved controversy as to which of Bell's conditions is indeed violated by QM and what the ontological consequences of that fact are, in particular if QM can indeed be reconciled with SR and GR, in the form they are formulated at present. As I do not wish to discuss here that complex, ongoing debate in greater detail I will just stress at this point that according to some commentators, only giving up OI can indeed be accepted without a conflict with SR, whereas a violation of PI would mean an extreme dynamical nonlocality in the form of action-at-a-distance and a fundamental conflict with the theory of relativity (see: [Jarrett, 1984], [Shimony, 1986], [Howard, 1989], [Teller, 1989]). According to others, giving up the PI condition does not lead to any problems greater than in the case of OI, as Bohmian mechanics, for instance, violates PI, and in spite of that does not allow for the physical transmission of signals with superluminal velocity (e.g. [Butterfield, 1992, p. 63*ff.*]). Moreover, both the breaking of OI and PI *may* be connected to a certain form of quantum holism/nonseparability, but does not have to at all, as it appears that it is possible to theoretically construe such interpretations of QM which will, for example, break the OI condition, but will nevertheless retain the physical separability of the state and will not be holistic in the strict sense (J. Berkovitz [1998] has presented an in-depth analysis of the issue with the above conclusions).

That is a serious warning when we wish to draw physical conclusions from QM formalism and BT: one should always keep in mind that nonseparability and state holism, discussed above in the context of situation (29), are of a *formal, structural* character and do not in any way determine the physical mechanism of their realization in nature. The 'global nonseparability of the state' or 'the global contextuality of the action' is a certain property of nature coded in the geometry of QM, which is experimentally revealed in the form of the global correlations of the physical states of the systems but does not say anything about the physical nature of those relations and even more so of the 'transmission of signals' or instantaneous 'action-at-a-distance'. We have to remember that when we are dealing with the concept of 'nonseparability' or 'state holism', we have in mind exactly that structural property of QM and not dynamical physical processes – the latter require a quite separate analysis. We should in fact not be surprised by that, as the state function simply contains *relative information about the system*, and thus the probability distributions determined by $|\psi\rangle_{singl}$ just contain information about the correlations. Hence, the statement that QM implies a certain 'state holism', does not mean, preliminarily at least, anything more than that it codes in its structure certain purely *global* (i.e., irreducible and independent of spacetime relations) *correlations between the information* possible to obtain, that is some *global constraints* on quantum actions (cf. e.g.: [Rovelli, 1996], [Mermin, 1998; 1999], [Fuchs, 2003], [Bub, 2005], [Auletta, 2005]; see also Ch. 5). On that account, the quantum state holism can also be called a relational information holism, as long as we keep in mind that there is no sense in talking of intrinsic, non-relational states as internal properties of the systems, that is, if one forgets the 'elements of reality' and externalistic compositionism. We shall go back to that issue in greater detail in Ch. 4.2.

Don Howard [1989; 1992], in his analysis of Bell Locality, Jarrett–Shimony's conditions $[2.1]_B$ & $[2.2]_B$, and the EPR condition $[3]_E$, put forward the thesis that the following identities take place:

Parameter Independence $[2.1]_B$ = Einstein's (relativistic) locality
Outcome Independence $[2.2]_B$ = Einstein's separability.

On the basis of that, just as Paul Teller [1989], he argues that breaking Bell inequalities has to be equivalent to breaking OI and Einstein's separability, as breaking PI and Einstein's locality would be, as we have already said, impossible to reconcile with SR and the methodology of construing viable physical theories as such (e.g. [Howard, p. 312*ff.*]). Although such an identification seems tempting with its simplicity, as it orders the plurality of terminological nuances in Einstein, Bell and Jarrett, it nevertheless is rather impossible to maintain, as it contains many doubtful assumptions, such as the above quoted thesis that only giving up PI automatically leads to the necessity of a superluminal transmission of signals ([Laudisa, 1995], [Belousek, 1999]). What is more, doubtlessly the above definition of the (formal) global nonseparability of the state is not solely equivalent to the violation of OI if only because the breaking of $[2.2]_B$, just as that of $[2.1]_B$, is not sufficient for physical nonseparability or holism of the state to take place [Berkovitz, 1998], and hence we may expect the same physical quantum phenomena, represented by formal nonseparability in the QM structure, may be explained without any recourse to condition $[2]_B$. In other words: a violation of OI from Bell's model need in no way express the exact sense of the formal nonseparability of the state given by (29) and (31), and hence identifying it as a proper condition of 'separability' is surely confusing. Moreover, it is unclear in what sense the very violation of OI would be supposed to introduce quantum holism, and the breaking of PI or even condition $[3]_B$ would not. When Einstein's conditions, Bell's conditions and the QM mathematical structure are all analyzed together, a great deal of caution is required, despite frequently identical terminology. Tim Maudlin [1994, p. 94*ff.*] even expresses the view that all of Jarrett's and Shimony's analysis is misleading and does not in any way enable a clear understanding of the character of 'quantum nonlocality'.

Still, when we reflect on how the formal nonseparability of the state can be translated onto the physical nonseparability of systems, the basic fact is that from the perspective of a given subsystem of the Universe extracting the information some *minimal separability (possibility of disentanglement) is locally retained*, which peacefully concords with local Bohr contextuality. Though the fact at first seems trivial, nevertheless the relativistic (causal) locality of interaction, contained also in $[3]_E$, is indeed accompanied by the relational separability of systems being a pre-condition of the local action itself (in line with the thesis Einstein insisted on): every system interacting with spatiotemporally separate subsystem $S_1$ or $S_2$ can extract the relational information about it (and all of $S_1 \oplus S_2$) that is locally accessible to it with reference to a given maximal set of compatible observables (in its local action frame). Let us write down that basic physical fact in the form of a lucid formula, being a slight modification of the definition of that 'trivial' disentanglement separability proposed by Matteo Smerlak and Carlo Rovelli [2007, p. 433]:

> Relational Local Separability of Physical Systems (Rovelli–Smerlak)
> Two physical systems $S_1$ and $S_2$ are *locally separable* if there exists a complete set of observables (in the sense of Dirac) of the compound system $S_1 \oplus S_2$, whose values can be actualized by local action of some system $O$ on $S_1$ or $S_2$ only, resulting in $S_1$'s or $S_2$'s entanglement with $O$ – in such a case those observables are *individual observables*.

The above definition seems trivial as a truly minimal condition for nature to be local in any reasonable sense – otherwise even classical mechanics with its abstraction of separable systems would from the start be impossible, and the world experienced would be, to use Roger Penrose's expression [2004, p. 592], 'an entangled mess'. It is also important to stress the seemingly banal fact that for each observer there exists some set of individual local observables and probability distributions for their eigenvalues, which together with the information that system $S_1 \oplus S_2$ remains in state (31), corresponds to the *locally complete knowledge* (set of information) which each of them possesses. In other words, the relational separability of the physical 'states' of $S_1$ and $S_2$ is locally retained and the fact of entanglement (global correlation) does not lead to any paradoxical phenomena, as it is locally manifested only in the *relational knowledge* each of the observers has about the history of systems $S_1$ and $S_2$ (thus it is, once again, a relational information holism). Such a minimal understanding of local physical separability does not stay in tension with the global formal nonseparability of state, as they concern two completely different aspects of quantum history in the Universe **Ω**. The latter refers to the perfect local consistency between relational information registered by all internal systems in **Ω** (the Leibnizian 'Connection'), while the former to the relational internal events inside **Ω**: local interaction correlating the systems $S_1 \oplus S_2$ (e.g. Bell measurement), local individual (disentangling) interactions with $S_1$ or $S_2$, and the relational information about pure 'state' $|\psi\rangle_{singl}$ of the internal whole. (When that fact is not included the result is such a contradictory statement as [2] from Ch. 1.2, since the Universal Principle of Relativity is at work here). We undoubtedly encounter here the source of a true novelty of QM with regards to earlier mechanics: it combines those two properties in its structure, which are impossible to reconcile in the classical understanding of 'physical reality'. It on the one hand offers knowledge about the possible individual measurements and the statistics of their outcomes, and on the other a description of the holistic, irreducible contexts of those actions in the form of the global constraints on the relations between all the possible outcomes (the information gained), which cannot be simulated by the classical causality and probability theory. QM thus allows us to define the interconnections between the descriptions given by each of the quantum observers and guarantees their consistency exactly thanks to the global nonseparability, analogically to Lorentz transformations introducing a global consistency of all relativistic descriptions for inertial systems, which, as a matter of fact, defines Minkowski spacetime itself (see [Rovelli, 1996] for the same standpoint). But if QM does not conform to Einstein's Local Realism, then what type of realism could it conform to?

We have concluded earlier that 'quantum magic' is the result of structural (formal) nonseparability of state, but we have to admit with regards to that that Einstein's Local Realism, which does not allow for any form of real nonseparability, has to be deeply redefined, as in another case realism is unable to play any role in the ontology of physics. It is extremely interesting that the demand for relaxing the separability condition in the definition of that realism does not require us to discard in advance the local causality condition, as QM as such does not contain anything that might suggest that any physical information has to be transferred in spacetime with superluminal velocity (that is at most a thesis based on the classical-mechanical way of thinking). Clearly, we are not forced to completely discard all the conditions of Einstein's realism under the influence of BT and KST or to opt for rigid instrumentalism, certain exotic forms of ontological 'mind solipsism' or the baroque 'many worlds' ontology. Taking seriously into account the fact that QM introduces structural nonseparability (:'global contextuality') of quantum action/information, directly leads to a special form of local realism, which nevertheless is *not* Einsteinian – let us call that ontology Contextual Realism. Its most important property is a complete relativization of the very concept of 'reality', which always requires the question: 'with regards to what internal system?', 'in what context (quantum frame of action)?'. It seems that such a metaphysical contextualism may be too high a price to pay for taking 'quantum magic' seriously, but it turns out to be an idea with extremely profound consequences enabling one to understand the essence of QM – we are going to deal with those consequences in Ch. 4 and 5. The quantum-mechanical form of contextual realism can be presented preliminarily as the conjunction:

| LOCAL CONTEXTUAL REALISM :<br>causal locality & global nonseparability of action<br>& relative definiteness of observables in a frame & ontic contextuality |
|---|

Each (by definition) local observer describes the physical situation in his/her environment on the basis of the relational information that is accessible to him/her, also the one regarding the past of the subsystem under a measurement action – that is precisely the local informational context for each observer. *QM is a theory which codes in its structure the relations between different local descriptions of situations, i.e., it sets down the formal relations between different sets of information, which different systems can extract in different action contexts.* Contextual realism requires rejecting the assumption of a certain ready-made reality 'out there', to which we are guaranteed an 'innocent', i.e., objective, complete access, independent of the local action context. That is also, I insist, the most important lesson in Leibniz's ontology, in which, as we have seen, there does not exist any absolute point of view of a finite 'observer' as a series of actions. QM suggests a truly radical ontology not because it speaks of some physical nonlocal signalling or mysterious entities being at the same time particles and waves, but because it points to the need of getting rid of two other absolutes from physics: the *absolute observer* and *the absolute state of a physical system* (cf. Ch. 1.2 and 4). Those two absolutes perform a role that is analogical to the absolute time, space and simultaneity in pre-relativistic physics, but unfortunately in the course of the development of physics in the 20th century they were not fully noticed and made an object of fundamental physical criticism, like the earlier ones ([Rovelli, 1996; 2004], [Finkelstein, 2003]). Just as the principle of the relativity of time and space is the foundation of SR, so does an even more profound ontological principle of relativity of the physical 'state' lie at the foundations of the relational (internally contextual) structure of QM. With reference to what has been discussed above, one should also make the following three summing up remarks.

Firstly: a recognition of the central role of the global formal nonseparability of the 'state'/action in the QM structure does not predetermine which of Bell's conditions is being violated by quantum phenomena. Most often it is assumed that QM forces giving up condition OI = $[2.2]_B$, i.e., a violation of Bell Locality, which implies a subtle kind of holism, though it does not enforce a conflict with SR in general. Of course that is neither clear nor convincing, as there does not exist any simple correspondence between a formal nonseparability of QM and the violation of OI. There remains a very interesting possibility, one that is less frequently taken into account, and despite of that contains interesting possibilities in harmony with the ontology of the changeless blockworld: a violation of condition $[3]_B$, 'timelike separability' or '(time-asymmetric) independence from history' (e.g. [Price, 1997], [Aharonov *et al.*, 2001]). Those are ontologies which are too rarely studied in the context of physical theories, but which are neither as exotic and unpromising as it is sometimes suggested by the dogmatic presentists seeking to ensure the time asymmetry at any cost. We shall see at a further stage of our analysis that the ontology of the relativistic block universe possesses deep connections with the relational structure of QM, contrary to the hopelessly popular view that QM requires or suggests absolute time as its physical background not being an observable. A certain terminological confusion between 'complete' and 'partial' observables results in a pessimistic conviction that the problem of the 'passage of time' may be insoluble in the context of the quantization of the gravitational field and the generally covariant quantum field theory. Carlo Rovelli's contribution has been his lucid presentation of that distinction, which is of major importance in the discussion on the *quantum* block universe, not sliding into a many-world ontology.

Secondly: the global formal nonseparability of state is a central property of the QM structure, discovered by Einstein, Podolsky and Rosen who used it to construct the apparent, ill-defined 'EPR paradox', but which was later highlighted by Schrödinger in his concept of 'entanglement'. Its place in microphysics is exactly analogical to the lack of a preferred spacetime foliation in relativistic physics. That might be put in a different way by saying that the nonseparability of the 'state'/action is the ur-phenomenon of mechanics (*Urphänomen* of mechanics in the Goethean sense[54]), just as the relativity of time and space, leading to a spacetime 'block of events', is the ur-phenomenon of classical kinematics. Its importance is also fundamental as it points to a completely different kind of ontological compositionism (holism) than the externalistic, classical-mechanical one: an internalistic compositionism (of the Platonic type). The quantum wholes inside Universe **Ω** are clearly impossible to parcel out into elementary base objects and their independent states (internal 'informational content') and then to compose from those microstates a complete description of the macrostate. Such a summing up of information is rarely possible, as the principle of separability has an extremely limited extent in the quantum world (it is a very special case). On the fundamental level, even the minimal separability of any 'objects' on the basis of commonsense expectations, as a certain type of fiction or idealization, may not be taking place at all (even externalistic compositionism, like the extensional one, thus have to fail completely). In that sense, QM might be perceived as a 'window-theory' on the ur-phenomenal sphere of nature, where ontologically separate 'objects', i.e., autonomous elements of the world with supposedly intrinsic properties and their own open micro-history, do not exist at all and cannot even be reasonably defined in some classical approximation. Quantum wholes are Platonic wholes.

Thirdly: the ur-phenomenon of nonseparability moves the point of gravity in the ontology of nature from the concept of 'substance' to that of structure, as QM as a mathematical theory deals almost wholly, if not solely, with the probability distributions for the observable values and the correlations between them. That justifies the supposition that QM is really a fundamental physical theory of information, or more precisely: a theory of quantum informational structure of **Ω** ([Rovelli, 1996], [Mermin, 1998b; 2006], [Fuchs, 2003], [Bub,

54 The concept of 'ur-phenomenon' (a 'pure phenomenon') was used by Johann W. von Goethe (1749–1832) in his natural scientific studies to denote a hidden dynamics-pattern in nature, which is spatiotemporally disclosing itself in the whole series of orderly and scientifically studied phenomena [Zajonc, 1998, p. 310*ff.*]. The ur-phenomenon operates at the boundary of sensual cognition and has to be 'uncovered' from a series of phenomena, such as – in our context – the quantum events. That does not mean, though – as Goethe stressed – that it is just an abstraction or a theoretical construct. On the contrary; despite being inaccessible in such a way as the particular phenomena are, it nevertheless remains to be fully real as their effective grounding in space and time. Goethe used that methodology for example in his studies on colour (see e.g. [Goethe, 1979]) and in his morphological studies in the realm of botanics and zoology. The concept of an 'ur-phenomenon' with regards to the quantum history as a cosmological 'hidden whole' is used here by me in a way which is close to that of Goethe's.

2005]). Its most astonishing property is – to a much greater degree than geometry of Lorentzian transformations in SR – the coding of the holistic structure of the relations between the elementary informations processed by different subsystems of the Universe. Those relations do not have the character of another absolute: no physical subsystem can find itself outside the quantum universe to reconstruct all the real relationships between the non-contextual states of other subsystems and those of its own ones. That is the first condition of Contextual Realism, which simply realizes the Universal Principle of Relativity from Ch. 1.2, just as in the Leibnizian universe of monads. Not only do intrinsic (immanent) properties of systems not exist in the quantum universe, but also the full history of relational quantum events in $\boldsymbol{\Omega}$, i.e., of the physical relations between all the subsystems and the contextual information possessed by them, always stays hidden. The fact implies a complete relinquishing of the postulate of the 'absolute observer' or (simultaneously) of 'absolute actuality' supporting diverse ontologies modelled on classical mechanics. After all, speaking about 'quantum action-at-a-distance' as a 'physical fact' and proclaiming the breakdown of the Principle of Relativity is a kind of conceptual dead-end situation in theoretical physics. QM does not require that; but, on the contrary, it does require taking the counter-presentist Principle of Relativity in the regime of mechanics much more seriously than some would like to.

# Chapter 4

# '*Esse est percipi aut percipere*': The Quantum Principle of Relativity and Measurement as a Correlation

## 4.1 Quantum frames of reference: relativity, the blockworld and the universal relativization of the observer

The application of the Universal Principle of Relativity to comprehend QM may raise many doubts, arising from the commonsense conviction that both the measured (endo)system and the state of the measuring (exo)system of reference cannot be completely relativized, which nonetheless – as I shall claim – leads to a conflict with SR and GR: the conclusions about nonlocality as action-at-a-distance and the supposed preferred foliation of spacetime as an ill-defined arena for quantum theory. Such conclusions are premature, though, in the situation when nonlocal interpretations are decided in advance, while at the same time the radical contextualization of quantum 'states' is commonly ignored or trivialized in order to make some connection with the phase space states of the more familiar classical mechanics. A more attentive analysis of EPR argumentation has allowed us to discern that the authors assumed – as opposed to Bohr – a strongly 'atomistic' condition $[3]_E$ hiding some conservative ontology and the noncontextuality/separability of each physical action connected to it, which Einstein only later presented explicitly in his programmatic statements. Though the 'reality condition' $[1]_E$ was abandoned by Einstein, as it is not a condition that is logically independent, nevertheless when we analyze it regardless of condition $[4]_E$, it might supply a criterion of 'objective reality' (i.e., independent of the external observer, who is able to gain noncontextual knowledge about it). For that reason the original EPR argumentation is not very clear and what is much more important for the discussion is, as we have seen, treating Einstein's (relativistic) principle of causal locality separately from that of separability, exactly just as Don Howard has consistently proposed [1985; 1989; 1992].

The 'reality criterion' might not be unnatural (for its authors intended it to be commonsense and uncontroversial), but in the context of quantum ontology it contains, as we have seen, a serious flaw: the expectation that quantum 'observer' $O$, possessing a certain particular set of information, may have direct knowledge about the intrinsic 'state' of another system $U$ 'out there', regardless of its own 'state' and the physical context of that knowledge. Putting it shortly, the EPR argument secretly assumes that both the state of $O$ and the quantum state of $U$ are monadic and separable, not relational – the 'observer' can in some way get to find out what the intrinsic (also instantaneous) properties of physical objects are even without any interaction with them and independently of its own perspective, i.e., a local, dynamical frame of reference. That is surely a fundamental epistemological problem, but in the case of QM it generates real, *physical* paradoxes. We are here generalizing the meaning of the 'frame of reference' beyond SR and GR to denote the *dynamical-informational (ontic) setting of any interacting quantum system/'observer' (a subsystem of the Universe)* – hence I shall be using further on, just as in Ch. 2.2, the more general concept of the quantum frame, so that it becomes clear that we are not concerned with the observer in a relativistic (epistemic) sense. The problem with condition $[1]_E$ depends on the fact that the epistemic criterion is rashly rendered also an ontic status, construing thus a certain physical absolute: a 'state of affairs' in the universe, independent of any other internal 'observer'. That is surprising, as it might have been expected of Einstein to have perhaps consistently applied his own, philosophically generalized principle of relativity also with regards to all quantum (which means: just all) 'observers'. On the other hand, though, that is not so very surprising when we realize that an application of that principle in QM leads directly to a truly radical physical ontology, whose consequences transgress far beyond a mere interpretation of QM. In practice, that is an outline of a promising, but deeply non-intuitive (just as nonseparability itself) ontology for quantum cosmology *in statu nascendi*, and also of a new type of realistic ontology based on internalistic compositionism. When we wish to construe that ontology by philosophical means, we first have to understand the fundamental fact, probably most clearly formulated by David Finkelstein [1996, p. IX], that: 'Quantum theory is above all an extended kind of relativity. We quantize by relativizing'.

To present the new realm of problems introduced by quantum correlations in a fuller way, we should begin with a geometric spacetime presentation of Bell's model, which we have already presented in Ch. 3.1. The arrangement of the EPR-like experiment consists of two local subsystems of the system $S_1 \oplus S_2$, separated by a spacelike interval, let us say photons $S_1$ and $S_2$, each of which possesses its own spatiotemporal history and a respective causal past in a regular, globally hyperbolic spacetime $(M, g) := \mathbf{M}$, i.e., a collection of events which could have had an influence on the 'states' of $S_1$ and $S_2$: $J^-(S_1)$ and $J^-(S_2)$ at any time (Fig. 3). At a certain moment $t_0$ for $S_1$ and $t_0'$ for $S_2$, independent measurements of the spin state are performed locally on each of them by respective 'observers', $O_1$ and

$O_2$, in the direction $m$ for $S_1$ and $n$ for $S_2$ (not necessarily different ones), giving results $p(m)$ and $l(n)$. Photons $S_1$ and $S_2$ have been produced in a laboratory as a maximally entangled system $S_1 \oplus S_2$, and thus at that event their zero curves on the past cones $\mathfrak{L}^-(S_1)$ and $\mathfrak{L}^-(S_2)$ intersect (let us assume that time is measured in the frame of reference of that laboratory). In their future, in turn, there exists an event, which we will define as a comparison of the information about the 'state' of $S_1$ and $S_2$ – that information can be transmitted with a certain speed $v \leq c$ and put together in some series of local interactions for example by a third system $S_3$. Let us now designate in the past of $S_1$ and $S_2$ a special submanifold $C^0$ embedded in spacetime **M** – *topological Cauchy hypersurface* $\Sigma$, which is a certain closed achronal set without an edge, $e(\Sigma) = \varnothing$, and thus a so-called *simultaneity slice* of manifold **M**, which we can imagine as a 'global instant of time'. An embedded manifold $\Sigma$ is a topological Cauchy (hyper)surface if every inextendible causal curve (or timelike curve) intersects it exactly once (in fact every spacetime which possesses such a manifold $\Sigma$ is globally hyperbolic). We may now define the *future domain of dependence* for $\Sigma$, $D^+(\Sigma)$, consisting of all events represented by points $p \in \mathbf{M}$, such that every past inextendible causal curve through $p$ intersects $\Sigma$. We can similarly define the *past domain of dependence* for $\Sigma$, $D^-(\Sigma)$, and the *full domain of dependence* as the sum: $D(\Sigma) = D^+(\Sigma) \cup D^-(\Sigma)$, and apart from that the following condition will always hold in **M**: $\Sigma \subset D^+(\Sigma) \subset J^+(\Sigma)$. (We can see from these assumptions that the spacetime slice as Cauchy hypersurface is a type of closed achronal set, for which $D(\Sigma) = \mathbf{M}$, and indeed $e(\Sigma) = \varnothing$.) Such simplifications of that model are convenient, as we can analyze the postulated, maximally deterministic description of the quantum experiment in classical spacetime (with a strong causality and without any naked singularities).

Part of hypersurface $\Sigma$ now codes all of the physical information that is causally relevant for quantum measurements on $S_1 \oplus S_2$, $S_1$ and $S_3$, as any physical signal sent from $D^-(\Sigma)$ to any given point $p \in D^+(\Sigma)$ has to be registered on it as a spacetime event, i.e., knowledge of the complete physical information on $\Sigma$ (Cauchy data) and appropriate dynamical equations can allow us to predict the character of the event at $p$ by defining the so-called data development, unique up to isometry ([Wald, 1984], [Choquet-Bruhat, Geroch, 1969], [Hawking, Ellis, 1973]). That naturally also concerns the reconstruction of the past for all events $p \in D^-(\Sigma)$. All the surface of spacetime slice $\Sigma$, containing information relevant to the history of the EPR experiment, can now be divided into three 3-submanifolds, of which one ($\lambda$) encodes information about the common past of $S_1$ and $S_2$ and thus also about the future production of an entangled state $S_1 \oplus S_2$, the second ($\mu$) all the information relevant solely to the causal history of $S_1$ and the choice of the measurement axis $m$, while the third ($v$) those essential only for $S_2$ and the choice of measurement axis $n$. Causal determination, understood in a classical way, thus should imply the fact that all the information on Σ should be sufficient not only to determine values $p$ and $l$, but also future settings of the axis of spin measurement, $m$ and $n$. Let us remember that we are conducting an analysis here solely with reference to a particularly 'elegant', globally hyperbolic spacetime, in which the condition of causality is satisfied, guaranteeing that there exist no closed timelike curves and singularities in it – it is a stably causal spacetime. Such idealized conditions also lie at the basis of the derived Bell's inequalities for the fully deterministic, local model of QM with hidden variables, while $[2]_B$ and $[3]_B$ immediately arise from the geometry presented on Fig. 3. In practice, we shall of course be dealing with some regions of spacetime and whole series of spacetime diagrams for series of measurements with different random parameters, which is of importance when studying stochastic Einstein locality.

*Fig. 3: The spacetime diagram of the classically deterministic version of the history of the EPR experiment with Cauchy hypersurface Σ as a slice in globally hyperbolic spacetime* **M**, *with such a part of its past domain of dependence $D^{-}(\Sigma)$, which is at the same time the causal future of a certain event cc, and with the causal past $J^{-}$ of each of the quantum measurements on systems $S_1$ and $S_2$ also being pointed out. The area of arbitrarily singled out manifold Σ, which codes all the physical information being causally relevant for the common past of systems $S_1$ and $S_2$ (and also for the joint Bell measurement of entangled state $S_1 \oplus S_2$), is marked as λ.*

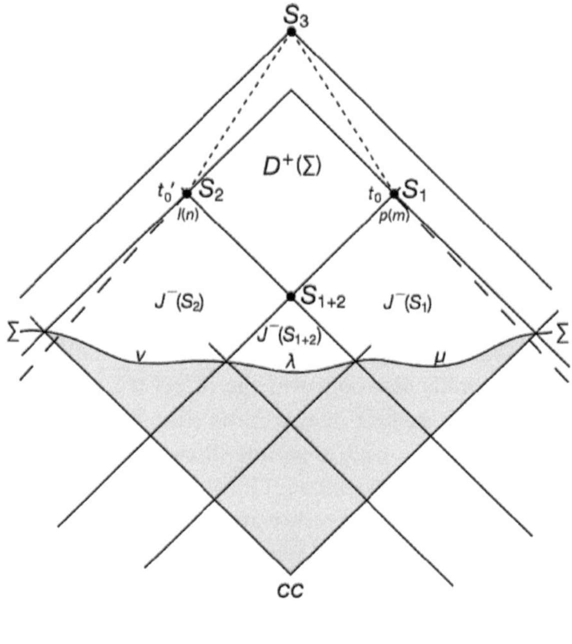

Now, using the spacetime diagram of the EPR experiment it is easier to discern the simple assumptions of Einstein's Local Realism, being a philosophical inspiration of Bell's conditions (of course also in case of stochastic models). The hidden quantum screening-off microstate[55], $\lambda_{S,1+2}$, fuller than state $|\psi\rangle_{singl}$ known to every 'observer' and causally encoded in the content of $J^{-}(S_{1+2})$, for example on information hypersurface $\lambda$, cannot be controlled by any of them, but as the common cause in the past of $S_1$ and $S_2$, i.e., in $J^{-}(S_1) \cap J^{-}(S_2)$, is to be the source determining 'state' $S_1 \oplus S_2$ and the registered future correlations of values $p$ and $l$. Naturally, the quantum-mechanical probability of registering a definite value depends not only on state $\lambda_{S,1+2}$, but also on (effectively) randomly chosen settings of the direction of spin measurement which, though – as we have assumed – are also causally encoded on Σ, are nevertheless independent of one another (they may, for example, depend on physical signals from very distant parts of the universe in the past cone of each observer $O_1$ and $O_2$, registered on $\mu$ and $\nu$). In Fig. 3, those signals determining the choice of the axis orientation have been marked as trajectories passing on to domain $J^{-}(S_{1+2})$ in a very distant past of the experiment producing 'state' $S_1 \oplus S_2$. The postulate of the locality of separated states, lying at the foundation of local quantum theories with hidden variables and Bell–Clauser–Horne inequalities, is thus based on the fact that the probabilities of spin measurements, $P_{\lambda m}(p_m)$ and $P_{\lambda n}(l_n)$, would depend on two distinct sets of data: a common state $\lambda_{S,1+2}$, referring to a common past encoded deterministically on $\lambda$, and settings $m$ and $n$ of the measuring devices, which directly depend on their individual history, also encoded on some surfaces on Σ. Einstein's assumption that all physical processes which take place in the neighbourhood of $O_1$ and $O_2$ should be solely deterministic and local, predetermined by Cauchy data $\{\lambda, \mu, \nu\}$, leads directly to the condition of a factorizability of probabilities $[2]_B$ in Bell's model, based – as has been indicated – on the Common Cause Principle. The condition concerning those two sets of data falls apart into two additional ones, already mentioned: the independence of the local distribution of outcomes from the choice of the settings of the measurement axis for the other spacelike separated system, $[2.1]_B$, and the independence of the registered outcomes from the outcomes obtained in that separated system, $[2.2]_B$. The independence of the choice of the measurement axis from the hidden state

55 In Ch. 3.1 we have used the symbol λ to mark that state in Bell's conditions, but here, so as not to mix it up with the singled out area of hypersurface Σ, we always use the notation '$\lambda_{S,1+2}$'. We can also take into account the hidden states of the measuring device itself, but that complication does not change anything in the model itself and in the argumentation based on it.

of the whole system $\lambda_{S,1+2}$ is found in the commonsense condition of time-asymmetric $\lambda$-independence from history/'timelike separability' $[3]_B$. A hypothetical QM model that would meet all those conditions would have to be completely local in the Einsteinian sense ('strong Einstein locality' := $[3]_E$), deterministic and Markovian, also in accordance with the general condition of Bell Locality for all distributions of hidden variables $\lambda$ (cf. e.g. [Fine, 1982; 1996], [Brans, 1988], [Szabó, 1995; 1998; 2000], [Durt, 1996; 1997]). Such mechanics, no matter how surprising with some pseudo-entangled states, would be classical and would act in a stably causal spacetime, which would satisfy Einstein defending an incomplete, (epistemically) stochastic QM. To construe the above model of the EPR experiment and to formulate Bell's inequalities we do not have to, in fact, depend on such strong limitations defining a globally hyperbolic spacetime – a spacetime having a stable causal structure defined on slices in **M** is sufficient.

For the information given on Σ, we ought to write down the pragmatic condition from the perspective of the observers in such a stably causal **M**, taking measurements in domain $D^+(\Sigma)$:

$$P_{J^-(\Sigma)}(\lambda \,\&\, \mu \,\&\, \nu) \cong P_{J^-(\Sigma)}(\lambda) \cdot P_{J^-(\Sigma)}(\mu) \cdot P_{J^-(\Sigma)}(\nu) \tag{32}$$

A closer look at that condition reveals its obvious idealizing character, though that might not be an obstacle, if we nevertheless intend to treat nature in a commonsense way. That is justified by the fact that when explaining the correlations between the measurement results, we usually do not wish to go further and further back into the past to more and more distant micro-relations between events in **M**. It is obvious that in the evolving universe, even such spatially separated events as the very distant points on $\mu$ and $\nu$ have some common past. All the events in realistic spacetime models possess – as we can expect – converging histories, for example somewhere near the initial cosmological singularity, which means that it is always possible to find some screening-off events for any serial EPR experiments. Fig. 3 presents some event *cc*, which is a distant common cause for all those events on the surface of Σ being essential for the EPR experiment, and hence it could for example be a physical *common* common cause, correlating even many Bell measurements and the hidden states $\lambda_{S,1+2}$ with the selected settings *m* and *n* of the serially working measurement setups. However, when experimentally testing Bell inequalities we can, which has been proposed, make use of signals from the early Universe in a cosmological scale (instead of random number generators in a laboratory), such as streams of photons from quasars positioned in opposite regions of the sky without any causal connection ([Maudlin, 1994], [Gallicchio *et al.*, 2014]). Among the theoretical reasons which refrain us from referring to such extravagant explanations by *cc*, we can of course find resistance to a physically unreliable 'conspirational' superdeterminism: a reluctance to accept a certain type of causal superstructure of events in $D^{-}(\Sigma)$, which would establish on a massive scale the correlations between the most elementary physical processes in the Universe, such as quantum micro- and macro-interactions (those quantum 'elements' as base objects would also have to transmit a full record of their past history, in order for perfect Bell correlations to take place even in a very distant cosmic future of screening-off events). A similar screening-off event could also lie on hypersurface $\lambda$, but that is not the only option, despite the fact that the derivations of Bell–Clauser–Horne inequalities are based on exactly such *common* common causes. The principle of determinism for **M**, treated consequently, may also lead to the idea that the Common Cause Principle and $[2]_B$ (Bell Locality) do not really require *common* common causes, but rather the construction of an extended space of Kolmogorovian probabilities for sequences of correlations and a weaker condition of the existence of sets of different screening-off past events for them, *separate* common causes [Hofer-Szabó *et al.*, 1999; 2002]. It has turned out, though, that the models do not live up to the hopes placed in them: there is an unavoidable conclusion that the experimental violation of Bell–Clauser–Horne inequalities (not to mention GHZ correlations) make it impossible for QM to retain the generalized Common Cause Principle, both in its stronger and in its weaker version (the latter answering the intentions of Reichenbach himself) (see in particular: [Placek, 2002], [Butterfield, 2007], [Portmann, Wüthrich, 2007], [Hofer-Szabó, 2012]). That means that the observed quantum correlations in practice exclude a physically realistic, fully *deterministic* explanation by *common causes in the past* and the *strong Einstein locality* $[3]_E$, even taking into account the finite precision of measurement procedures. From the perspective of local, classical determinism, the Universe with its entangled states is indeed a 'spooky' place. Those entangled quantum actions cannot be replaced by pseudo-entangled classical states.

Nevertheless, why should a violation of $[3]_E$ automatically denote an infringing of solely causal Bell–(Reichenbach) Locality or classical, time-asymmetric determinism? Spacetime **M** as a structure of events is given as a whole (there is no sense in talking of its part being already 'defined' and another part totally 'undefined'), and its symmetries make it impossible to absolutely separate the time coordinates from the spatial ones, hence condition $[3]_B$ could also be called a certain 'locality' but with regards to timelike curves as histories ('timelike locality'). In order not to give rise to misunderstandings, we shall retain the definition of Bell's third condition as 'independence from history' ($\lambda$-independence) or 'timelike separability'. Common sense seems to prompt us to assume that it has to take place as the choice of the settings of the measurement apparatus for measurement on $S_1$ and

$S_2$ should be 'free' in a natural, physical sense, i.e., independent of the state of the observed systems. Were $\lambda$-independence not to be fulfilled, then the history of the system in any part of its past and the hidden state $\lambda_{S,1+2}$ should be globally codetermined also by their future connected to the particular worldlines. That would be a singular 'timelike nonlocality' through *dynamical retroactions* or 'timelike nonseparability' through *global time-symmetric coupling within the quantum history*, which – on account of it being deterministic but totally non-classical – all time-asymmetric models with hidden common causes wish to eliminate. The introduction of a conspirational causal superstructure in very distant parts of domain $D^{-}(\Sigma)$ or on $\Sigma$ *is* indeed physically implausible, but the possible violation of condition $[3]_B$ is much more interesting than is usually supposed, by reason of the causal-temporal nature of the quantum 'paradoxes' (see Ch. 2.3). As the model experiment with the delayed choice quantum eraser, QM does not so much play with space (as the 'action-at-a-distance' terminology suggests), but rather with time and classically orthochronal causality, which makes fully time-symmetric QM a viable option. The classical interpretation of $[3]_B$ and the attempts at an experimental elimination of the loophole connected to that condition are based on a commonsense assumption of the asymmetry of time itself and classical, orthochronal causality. Still, is it possible for *quantum* histories given as timelike curves in **M** to be *globally defined* in a completely non-classical way, violating *orthochronal causality*? And moreover: can such a global definiteness of history have any deeper connections to the ontology of the theory of relativity and the global state nonseparability of QM?

One of the most important properties of SR and GR is the fact that they possess a completely natural ontological interpretation in the form of globally defined histories without the time flow: it is called the ontology of the 'block universe', though we should from the very beginning, for the sake of clarity, call it the 'classical block universe' (henceforth CBU) ontology. Its formulation, originally in the SR context, was first suggested by Hermann Minkowski in his famous lecture titled *Raum und Zeit* given on the 21st of September 1908 during the 80. Naturforscherversammlung in Cologne [Minkowski, 1909]. In CBU ontology, juxtaposed to pre-relativistic presentism, a global present does not exist, as there is no such thing as an absolute simultaneity and the preferred foliation of **M**, but only a four-dimensional block of *classical events* – the postulate was perhaps adequately termed by Minkowski himself as the 'Postulate of the Absolute World' (*Postulat der absoluten Welt*), or simply the 'World Postulate'. Its significance was fully acknowledged by Weyl [1952], who rightly presented it as the chief philosophical component of the relativistic revolution, we can therefore call CBU the *Minkowski–Weyl World*. As the problem of the ontology of classical spacetime as such remains to be marginal to the issues of interest to us here, I will not stop longer at a detailed argumentation in favour of CBU, but shall limit myself to highlighting its development in connection to the by now classic Rietdijk–Putnam–Maxwell formulation ([Rietdijk, 1966], [Putnam, 1967], [Maxwell, 1985]) and Stein's criticism [Stein, 1968]. Contrary to the common view, the Rietdijk–Putnam–Maxwell argument was not refuted by Stein, as the (justified) criticism of some of the conceptual inconsistencies of the former paradoxically additionally strengthens it, which makes the ever-recurring attempts at recapturing a strong classical presentism for SR and GR strained and implausible. My assessment of that discussion is convergent to the excellent analysis recently conducted by Vesselin Petkov [2006; 2007; 2009; 2013] (see also [Saunders, 2002]).

In the shortest variant, the above argument can be presented thus: one should analyze a set of three inertial observers, $A$, $B$ and $C$, whose histories intersect in a specific way (see a concise presentation in [Petkov, 2006]). The histories of $A$ and $B$, approaching each other, meet at some event, say, $p$; worldlines of $A$ and $C$ intersect in the past of that event, while worldlines of $B$ and $C$ meet in its future ($B$ approaches $C$). It then becomes clear that, for observer $A$, event $p$ will be simultaneous with some event $c_p$ from the history of observer $C$ as it lies on $A$'s simultaneity surface $\{A\}_{p/c_p}$ in $A$'s frame of reference. In the frame of reference of inertial observer $B$, in turn, with his/her given surface of simultaneity $\{B\}_{p/c_r}$, event $p$ will be simultaneous with some event $c_r$ from the history of observer $C$, and certainly not $c_p$. In the frame of reference of observer $C$, the event simultaneous with $p$ would be just another event $c_q$ from his/her history, and not the later $c_r$ nor the earlier $c_p$, as $c_q$ lies on his/her surface of simultaneity $\{C\}_{c_q/p}$. All those three events, $c_p$, $c_q$ and $c_r$, are completely different and timelike separated events belonging to the worldline of observer C himself/herself and from his/her perspective they cannot be at the same time 'real' and 'actual'. As the present in SR is always relativized to a kinematical observer with a given proper time, and events $c_p$, $c_q$ and $c_r$ are defined as simultaneous ('actual') in **M** for different observers, then if 'being physically definite' means 'being physically real/actual', then the conclusion is to be made that according to SR observer $C$ should actually exist in all three moments of his/her history, which is in blatant discord with presentist ontology. That is a counterintuitive property of the relativity theory, which is a direct consequence of what I have called in Ch. 3.2 the 'ur-phenomenon of kinematics', that is the relativity of simultaneity. The argument, mathematically trivial when taking into account the Lorentzian geometry, in fact uncovers the key nontrivial property of spacetime physics, but in the above-like formulation it has an important methodological flaw: it wishes to present the radicalism of SR, but at the same time tacitly assumes a pre-relativistic understanding of the present, the past and the future. That in the case of the formulation of physical paradoxes is a rather typical situation, and it in particular occurs also with reference to 'quantum paradoxes' and theses about quantum nonlocality, hence its analysis is quite instructive.

The flaw was accentuated in criticism by Howard Stein [1968], and also by Robert Weingard [1972], according to whom speaking of 'distant present events' is of no use with regards to SR and GR: the only adequate expression is 'the present/now at point/region $p$ of spacetime **M**', that is 'here-and-now'. In pre-relativistic physics the present 'now' is everything that exists at the same time in a certain global moment $t$ (defined independently of the observer), and such a way of reasoning is possible only with a preferred cutting of **M** into topological slices, giving (3+1)-space-time (a preferred foliation and absolute time). Stein was indeed right to accentuate the fact that in SR speaking of 'distant present states' makes no sense exactly because it is based on mixing the language of classical kinematics and a naïve theory of time into the interpretation of SR formalism. That criticism turns in the end against a classical presentism of the 'world in time' and additionally *supports* CBU ontology. A proponent of some form of perspectival presentism (for the three-dimensional world) may claim that for particular observers what is real are only individual pointlike events $p$, but such a standpoint leads to even more absurd consequences, as for each observer the world would then be monadic, zero-dimensional. There is also no sense in claiming that in four-dimensional spacetime **M** one event $p_1$, 'here-and-now', is more real or 'actual' than another event $p_2$ [Petkov, 2006, p. 213]. There is of course another problem with that argumentation – it treats seriously the physicality of such 'point events' and thus interprets the topology of **M** in a naïvely realistic, almost substantialist manner, instead of treating it as an effective geometric model of the relations between certain fundamental physical (:quantum) actions. But there exist no physical, monadic 'points of spacetime', particularly as the means of a desperate escape of a presentist from the four-dimensionality of the world, just as there exists no absolute presence – even the relativized presentism contradicts the relativity [Petkov, 2009, p. 146*ff.*]. Petkov rightly argues that one can – according to the scheme of the generalized Rietdijk–Putnam–Maxwell reasoning – start in argumentation in favour of CBU from the pre-relativistic picture of spacetime and show what modifications of spacetime ontology are required by the *physical fact* of the relativity of simultaneity. It then turns out that one cannot fully explain all the relativistic kinematic effects when assuming the world to be three-dimensional. In their reports, the particular local observers make use of the concept of 'presently existing objects', and thus in practice locally use presentist ontology, defining their surfaces of simultaneity: $\{A\}_p$, $\{B\}_p$ and $\{C\}_p$, but a direct explanation of the symmetries of those reports forces the introduction of the four-dimensional CBU, which means that what is real is only the physical history (worldtube) *as a whole* – and not its particular surfaces/points as supposedly absolutely becoming (sets of) events, gradually actualizing itself only in its own local segment. That can be put even more precisely in a direct analogy to the reflections in Ch. 3.2 in such a way: every three-dimensional object in CBU ontology is formally nonseparable from the whole classical four-dimensional history it is an element of. Everyday intuition fails in this case to the same extent in which it fails when it convinces us that those 'classical objects' possess some noncontextual, intrinsic properties at the quantum level.

The classical (weak) holism of spacetime history is as counter-intuitive as the (strong) state holism in QM, and perhaps that characteristic of theirs arises directly from the fact that even the former might possess some natural connection to the ontology of internalistic compositionism. Petkov himself [2009, p. 141*ff.*; 2006] proposed a thought experiment, worth quoting here to fully demonstrate the strength of the argument in favour of CBU and the ontological holism of history in SR and GR (Fig. 4). Let us take into account two observers in relative motion, *A* and *B* – for whom there is defined a proper time $t_A$ and $t_B$ – and a rod remaining at rest in *A*'s frame of reference (placed parallel to its *x* axis). At the ends of the rod and exactly in its middle lights are attached, which change their colours simultaneously (with relation to *A*), in successive moments of the rod's history: first all of them are green (*g*), then red (*r*), and in the end blue (*b*). When *A* and *B* meet (when their histories intersect at point *M*), observer *A* registers that all the lights are red (that has been recorded as: $t^M{}_A = t^r{}_A$). (The clocks of the cameras registering the light in the frame of every observer were before that synchronized in such a manner as to show $t = 0$ at the moment of the meeting at *M*). After some time, when all the snapshots have been taken, *A* and *B* can compare them and in accord with SR (and contrary to pre-relativistic kinematics) obtain completely divergent data: while *A* in $t^M{}_A = 0$ registers all the lights red ('*r*–*r*–*r*'), *B* in $t^M{}_B = 0$ will obtain picture '*g*–*r*–*b*'. Petkov rightly argues that observer *A*, when asked 'what reality looked like' at moment *M*, would answer that the rod existed then with its red lights, the green lights were a thing of the past (they did not exist at the moment), and the blue ones were as yet an unrealized future. Already merely a perfunctory analysis of Fig. 4 shows that observer *B* does not register *the same* three-dimensional rod as *A*, as the rod observed by him at *M* was composed of elements of what was the past, present and future in the history of *A*. Similarly, from the perspective of *B*, observer *A* registered events–states which were its past, present and future, and thus a different three-dimensional rod (composed of other events–states of the block history as the four-dimensional worldtube). Event *M* is the only common event of the 'presence' of observers *A* and *B* for $t = 0$.

*Fig. 4: Petkov's Gedankenexperiment with the worldtube of the rod with three lights fixed to it (on its ends and in the middle) and two observers A and B, meeting at point M of their history. The lights change at every moment of the rod's history in the following way in the frame of reference of A: first they are green (g) at moment $t^g{}_A$, then red (r) at moment $t^r{}_A$, and in the end blue (b) at $t^b{}_A$. $t^M{}_A$ and $t^M{}_B$ are surfaces of simultaneity of event M for, respectively, A and B. Reprinted with permission from: V. Petkov [2006], Is there an alternative to the block universe view? In: D. Dieks (ed.), The Ontology of Spacetime, p. 219. Copyright © 2006 by Elsevier B.V.*

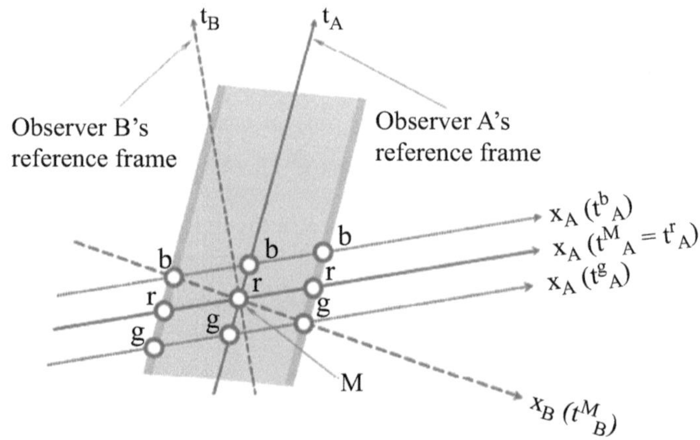


In the light of the above argumentation, and also the analysis of the Twin Paradox (whose explanation directly demands a four-dimensional perspective), CBU ontology is the only natural interpretation of SR and GR ([Petkov, 2006; 2007; 2009; 2013], [Saunders, 2002]). If the observers insist that the real universe is nevertheless only three-dimensional, then it becomes indeed impossible to explain all the relativistic effects. CBU interpretation has an important role in revealing the holism of spacetime (and *not* separate spatial and temporal) relations between the histories, but it does that in such a way that time cannot be considered to be a universal ordering parameter, external with regards to the physical state of the object in a three-dimensional space (the 3+1 perspective is merely a projection of the event collections of the four-dimensional history, the worldtube, in a given frame of reference [Weyl, 1952, p. 180*ff.*]). The global nonseparability of classical-kinematical history in such a CBU is fully compliant also to the ontology of internalistic compositionism applied in kinematics: the relations between the states of motion ('elements of history') of the three-dimensional objects, observed in different frames of reference, can be seen as not external to those objects in the way pre-relativistic time is, but rather as the intrinsic, geometric relations between whole histories composing block universe **M**. That kind of internalistic compositionism of the Platonic type simply says that *the Universe cannot be put together from its primitive base elements, such as 'absolutely present 3-objects', into earlier completely indefinite histories–wholes developing in time (three-dimensionally)*. The hidden spacetime structure is holistic in the sense that the whole of history defines the geometric relations between the events given locally to the 3-observers, who themselves are such spacetime histories within that whole. The principle acts in the same way as in the case of the relational information holism in QM, only here in the realm of kinematics and its classical (non-quantum) symmetries[56]. QM not only does not undermine that ontology, but additionally radicalizes it, though to understand that fact one requires a consistent application of the Principle of Relativity to the quantum systems and quantum actions ('states').

The chief accusation, usually voiced by the opponents of CBU, is not only a violation of commonsense intuition, but first and foremost the presumed incongruence with the QM structure (regardless of its postulated ontology), in which time is not only to be a classical, external background parameter (and not a quantum observable), but also its 'flow' is to be guaranteed by absolute, ontic 'quantum probabilities' (ignoring the fact that they are *not* even a basic ingredient of quantum geometry, but they appear at the output). The suggestion is doubtful and primarily a non-sequitur (cf. e.g. [Petkov, 2009, p. 277*ff.*], [Callender, 2008]), and – as we shall later see – it arises from a mixing up of classical and quantum categories, just as in the case of the presentist attempts at

56 As far as GR is concerned, perhaps one of the most consistent proposals of atemporal physics (apart from Rovelli, see Ch. 5.1) is that presented by Julian Barbour [2001]. An excellent analysis of the assumptions lying at the basis of classical thermodynamics, Boltzmann's *Stoßzahlansatz* and the presumed time asymmetry in cosmology was conducted by Huw Price [1997]. His proposal to consider the violation of condition $[3]_B$ is important in the context of a possible purely quantum version of CBU. A more technical study of the complications which geometry and general covariance of GR introduce to the models of quantum gravity are contained in the by now classic works of Karel V. Kuchař [1992; 1999] and Christopher Isham [1993]. Some of those problems result from the efforts to retain, if only a limited or relativized intuition of presentism and a three-dimensional becoming, yet – I think – those efforts in fundamental physics are from the beginning doomed to fail and are obstructive (cf. [Saunders, 2002]).

undermining the relativity of simultaneity. But, I would repeat, 'quantum theory is a deepening of relativity' [Finkelstein, 1996, p. 36]. If as physical subsystems we live in a four-dimensional blockworld with the causal locality condition satisfied and with a *generalized type of quantum time symmetric determinism*, when, let us say, condition $[3]_B$ is violated in a substantial way [Aharonov *et al.*, 2001], which leads at the same time to a blatant violation of the noncontextual intuitions of the 'elements of reality' from $[1]_B$ *and* a global timelike nonseparability of the states within the whole quantum histories, then such effects as the delayed choice quantum eraser, the breaking of Bell–Clauser–Horne inequalities or Hardy's paradox would not be any problems threatening GR, and should in fact appear as certain spacetime 'quasi-teleological' effects[57] for local observers comparing their measurement outcomes. One does not even need any dynamical retroactive signals/backward causation ('timelike nonlocality') to get that, as a kind of global, time-symmetric coupling or conjugation of the *independent* initial and final boundary conditions within quantum histories ('timelike nonseparability' as a nontrivial feature of the global quantum geometry) suffices *to make any 'element' of the history completely relational in the blockworld.* Such a possibility means a radical deepening of the Principle of Relativity, for, as I shall argue in Ch. 5, it relativizes the local spacetime events ('facts') themselves and the classical asymmetric model of the presentist causality, making it impossible to construe physics and cosmology bottom–up on the basis of noncontextual 'elements of reality' or 'states of affair' accreting into open histories with the flowing time. Will it be possible, nevertheless, *a limine,* without referring to the details of the yet unknown theory of quantum gravity to find any deeper relationship between the geometry of spacetime and quantum geometry, apart from the generally analyzed holism as a metaphysical assumption, which many would prefer to be somehow superfluous?

In the article published in 1981 in the series on the phase space approach to relativistic quantum mechanics (henceforth RQM), Gerald Kaiser made an interesting presentation of the mathematical relation of the relativistic noncommutativity of Lorentz boosts and the spatial translations with the quantum noncommutativity of the operators of position and momentum. That is interesting, as usually one

57 I am using here and further on the expression 'quasi-teleological effects', to make it clear that that has little to do with any classically understood, metaphysical teleology. In the case of a generalized quantum causality, violating condition $[3]_B$ in the case of time-symmetric QM, the microphysical effects should neither be any more 'mysterious' than classical (time-asymmetric) causality.

does not analyze a physical connection between the *noncommutativity* relations of canonically conjugated quantities in NRQM and the *relativity* of simultaneity in SR, while Kaiser – and a dozen years later also Aage Bohr and Ole Ulfbeck [1995], and recently Silberstein, Stuckey and Cifone [2007] – noticed that those relations might be its consequence. To check that, one should take the standard commutators in Lie algebra of the full Poincaré group with the generators of rotations, translations in space and time and Lorentz boosts, and then calculate them for the limit $c \to \infty$ in that algebra. As a result of that operation we obtain those canonical relations for NRQM with the Hermitian operator for momentum (in the case of the generator of spatial translations) and the position operator (in the case of boosts), which means that there would be in the limit $c \to \infty$ a certain geometric remnant of the relativity of simultaneity encoded in the Poincaré group. (Of course, as Kaiser points out at the beginning [1981, p. 706], if we analyze Newtonian spacetime with its Galilean group, instead of **M**, then those relations do not take place and the commutators of the above generators are equal to zero). In the limit $c \to \infty$ there are no Lorentzian transformations any more, as they lack then the relativistic factor $\gamma$, and hence it is not possible to recover the Lorentz contraction and the time dilation arising from the relativity of simultaneity, though at the same time we do not recover absolute time as the time coordinate keeps being transformed with the presence of the expression '$-\frac{vx}{c^2}$'. That means that after moving on to the limit of the Poincaré group we obtain a 'weakly relativistic' regime, in which Galilean transformations do not take place (as we could expect), but complete spacetime transformations for **M** also do not operate. Hence, Heisenberg relations are certainly not connected to the Galilei group, but seem to arise from the relativity of simultaneity connected to Lorentzian invariance. Similarly, Jeeva Anandan pointed out in his interesting article [2003, p. 1945*ff.*] the meaning of the above symmetries and the shift from the Poincaré group to the quantum mechanical inhomogenous Galilean group: the full Poincaré algebra together with the introduction of the Planck constant leads to Heisenberg relations for position and momentum, which means that quantum wave–particle complementarity could have a common source with spacetime operations. But Anandan, contrary to Bohr and Ulfbeck, was right not to treat spacetime symmetries as primitive and did not treat **M** as an autonomous physical entity *per se* after the speculative tradition going back to William K. Clifford. It is rather on the contrary: what is much more convincing is the assumption that spacetime symmetries and Einstein field equations have a purely quantum-thermodynamical origin or even that Einstein equation *is* just a thermodynamic state equation [Jacobson, 1995], hence relational information–action should be treated as more primitive than the spacetime symmetries themselves.

The theoretical result obtained can raise suspicions, as Schrödinger equation is invariant under Galilean transformations, and hence it seems reasonable to claim that NRQM refers solely to Galilean (3+1)-space-time and has nothing to do with 4-spacetime geometry and the Poincaré group (i.e., it assumes an absolute simultaneity). It turns out, though, that the matter is not so evident: suffice it to check what happens to the state function $\psi$ in the case of the operators of the spatial translation by $l$ and of the boost by $v$ acting on it (which satisfy the canonical Poincaré relations), if that action is performed in a reverse order. That simple calculation shows that after the positions of the operators have been switched a pure phase is introduced to function $\psi$, whose mathematical form contains both $l$ and $v$. That phase of course does not influence the calculated quantum expectation values for the system, as the new function offers a description of the physical state that is equivalent to the description of a given initial function $\psi$ (more on spacetime symmetries and their relation with the wave phase is to be found for instance in [Ballentine, 2000, p. 66*ff.*] and the articles by Stuckey, Silberstein and Cifone [2007; 2008]; [Silberstein *et al.*, 2007]). Therefore, that simple operation demonstrates that the action of the relativistic operators on $\psi$ does not lead to a violation of NRQM predictions, as it offers physically equivalent geometric states as computational tools for any local quantum observer in **M**. An even more astonishing result, demonstrating a deep connection between the spacetime geometry and the quantum state space, can be obtained by the derivation of the full density matrix of the quantum system using the spacetime symmetry groups, as Bohr, Mottelson and Ulfbeck have presented it [2004a] together with their proposal of a radical quantum ontology (a popular resume of the ideas: [Bohr *et al.*, 2004b]; a further discussion in: [Stuckey *et al.*, 2008], [Silberstein *et al.*, 2007]). Such deep relations between basic NRQM formalism and the fundamental Poincaré symmetries are indeed a clue that there exists a direct connection between the ontology of four-dimensional CBU (in which spacetime histories are globally defined and nothing 'becomes') and QM ontology with its global, formal nonseparability of quantum histories, even if one is accustomed to the presentist intuitions of becoming 'quantum events'. If the quantum universe is construed upon such histories, then we can in fact expect – as long as we do not treat manifold **M** ontically as a primitive entity – that the macroscopic spacetime symmetries should not so much have some common, deep source with quantum symmetries, but should even stem from them. Statements about the supposed infringement of the Principle of Relativity by QM arise not so much from its serious treatment but – on the contrary – from *a refusal to generalize and apply the Principle of Relativity in microphysics*. Thus one should be careful not to envision a globally nonseparable quantum blockworld as a mere replica of CBU, which it is certainly not.

A frequent accusation against CBU ontology is its counter-intuitiveness manifested in the supposed violation of the physical intuition of the irreducibly dynamical or processual world. Usually that criticism is expressed in such a way that CBU is to postulate a 'frozen universe' being timeless in such a sense that in reality no processes take place in it. That accusation was strongly emphasized already by Popper, who pointed out that it was enough to add one dimension to the universe 'in which nothing ever happens' from the ancient theory advocated by Parmenides of Elea to obtain 'Einstein's block universe' [Popper, 1962, p. 38]. He was even to call Einstein 'Parmenides' in a private conversation [Popper, 2002, p. 148]. The answer to that accusation is of great importance, as it takes us back to the issue of understanding the essence of global nonseparability, which we have discussed in Ch. 3.2. We can turn that accusation down as inadequate simply by nature of the fact that there is no sense at all in the very notion of the 'frozen universe'. Already in 1968, Adolf Grünbaum [1968, p. 22*ff.*] noticed that Popper had commited a tactical equivocation there: one should differentiate changeless ontology negating real ontic becoming (objective coming-into-being in time) from a fully timeless ontology, negating even time itself as some *effective* physical parameter for doing physics *inside* the Universe. Relativistic CBU in fact settles the former (and is able to do that in a consistent way, particularly as the ontology of GR), the latter being perhaps too radical, as it would make impossible even the very formulation of physics for sub-systems locally exchanging information, that is the quantum thermodynamics itself. As we assume, there does not exist any greater enveloping physical universe as an absolute system of reference – just as there does not exist any absolute external time – hence there is no sense in saying that the universe is 'motionless' or 'frozen' (we accept only physical observers/subsystems *inside* the Universe). A pragmatic physicist who avoids metaphysical speculations may even doubt, as Rafael D. Sorkin notices [2007, p. 153*ff.*], if the categories 'dynamic'/'static' have any deeper import for quantum cosmology, as they do not find any crucial reference even to GR mathematics: Einstein equations look the same regardless of the fact if we look at them 'dynamically' or from the Archmidean CBU perspective.

Secondly, CBU ontology for SR and GR encompasses spacetime relations between *classical events* in **M**, rigorously satisfying appropriate symmetries, but the quantum-informational relations, leading to a formal nonseparability of 'states'–actions, are *not* reducible to spacetime relations (cf. e.g. [Anandan, 2003, p. 1947*ff.*]), and quantum relationism is not *a priori* implied by spacetime relationism. SR is a new kinematic theory, whose mathematical heart is the Poincaré group, and indeed it does not say anything about quantum microdynamics and the nature of physical changes, lying at the basis, for example, of the relativistic contraction of macroscopic bodies, though that does *not* imply that one can freely add dynamical postulates to it which violate the ontological Principle of Relativity, openly fixing the preferred foliation of **M** (cf. e.g. the similar standpoint in [Petkov, 2009, p. 277*ff.*]). Interpreting SR in line with the general consensus treating it as a model 'principle theory', which defines the causal geometry of spacetime, one can comprehend it in the sense that it imposes strong formal limitations upon that more basic level of micro-dynamics but, I presume, an opposite view is more justified: the relativity of time, space and simultaneity – similarly to the homogeneity and the isotropy of time and space necessary to introduce them – simply have deep microdynamical sources and that is the perspective I prefer here[58]. Einstein himself, understanding the limitations of his own interpretation from 1905, in the situation when Planck's new quantum hypothesis placed the theory of matter in a completely new light, at least from 1907 occasionally made use of the analogy between SR and statistical thermodynamics ([Brown, 2005], [Brown, Pooley, 2006]) and the same analogy can be used even more interestingly with reference to GR and spacetime covariance as the postulated macroscopic quantum effect (cf. e.g. [Baugh *et al.*, 1998]). I assume, contrary to presentist revisionism and naïve presentist interpretations of QM, that the global dynamics of fundamental quantum processes not only does *not infringe*, but – on the contrary – *generates* macroscopic spacetime symmetries in the thermodynamic regime, even if a local quantum observer has the impression that it is his very frame of reference and measurement action which are physically special or preferred. Spacetime and its geometry are a completely derivative entity, if one can speak of its 'existence' at all – it

58 I do not consider the geometry of **M** to play a causal role and for diffeomorphism symmetries of spacetime to be able to 'act' or 'cause' anything physically to take place (cf. [Brown, 2005], [Brown, Pooley, 2006]). Spacetime symmetries given by appropriate Lie algebras should have their origin in the properties of fundamental, purely quantum dynamics, and hence one should expect a more basic theory to give an insight into the nature of those dynamic effects. When one hears that SR with Lorentzian geometry is a kinematic theory, what we have in mind is the fact that one can use it without the necessity of referring to the details of the quantum dynamics of informational states, but its ur-phenomenon – the relativity of time and space – is certainly not a purely kinematical efect (hence I have proposed the concept of the 'ur-phenomenon of kinematics'), despite it being possible to refer to SR as an Einsteinian 'principle theory'. The quantum nonseparability of state discussed in Ch. 3.2 – the ur-phenomenon of mechanics – should also not be a purely kinematical effect but a dynamical one, connected to quantum actions (the quantum exchange of information) as a primary process.

might be more in place to treat it from the very beginning as a 'glorious non-entity' ([Brown, Pooley, 2006]; also [Rovelli, 2006]). That means that we should consistently employ the Universal Principle of Relativity when construing quantum ontologies, in order to eliminate physical absolutes from it, particularly in the form of objects with intrinsic states, physically privileged subsystems or of the 'present' in an inconsistent quantum presentism. The 'quantum block universe' (henceforth QBU) or 'quantum blockworld', defined by the appropriate noncommutative algebras and the purely quantum statistics generated by them, may have a local, relativized dynamics in a nontrivially generalized sense and local processes of information exchange, but may *not have* any non-relational and noncontextual 'states' of its arbitrarily singled out subsystems, any ontic becoming, and all the more any preferred frame of reference, neither a spacetime nor a quantum one.

If the strategy of construing microphysics on the basis of the Universal Principle of Relativity is an appropriate one, then it seems that also QBU ontology may, by analogy with CBU, be made subject to conceptual criticism of a different type. As we have seen, Minkowski pointed out the 'Postulate of the Absolute World' as the foundation of the new ontology, and thus a critic of QBU may argue that applying the Principle of Relativity in the quantum domain, we get another ontic absolute – the 'block of the quantum universe', that is the noncontextual bulk set of separable quantum histories. That is a misguided argument and it results from both the transference of 'object imagery' [Finkelstein, 1996, Ch. 5] onto cosmology and a literal treatment of the term '*die absolute Welt*', which leads to a quasi-classical conception. The physics of every randomly singled out subsystem inside the Universe is causally local (in the SR and GR sense) and contextual, which means that there does not exist any absolute (non-relational) 'observer' inside this Universe, who would reconstruct a full 'block of facts/events' concerning that system from all perspectives at one and the same time (we have expressed that assumption in Ch. 1.2 by excluding any physical pan-perspectival observer in the realm of cosmology). Resistance to ontologies which globally eliminate ontic becoming has its source in the tacitly accepted prerequisite proclaiming that such a global 'reconstructing system' could exist, that is without entering into any physical interactions it would register the course of quantum events (quantum histories) as a certain cosmological quasi-object. We have encountered that situation when analyzing the problem of the global nonseparability of the state in QM, which is also not an observed phenomenon but a hidden, global structure of relations between fully relativized quantum actions, not satisfying the classical conditions of time-asymmetric causality and noncontextuality. When we analyze physical, finite 'observers' and their relational 'states', no paradoxes take place with any 'frozen dynamics': for all the subsystems entering into quantum interactions a purely relational physics is local and temporal, just as three-dimensional objects in **M**, as certain projections of four-dimensional histories, *seem* to objectively 'come into being' in the relationally (via local clocks) registered 'now'. QBU, which has nothing in common with Everettian multiverse, nevertheless radically differs from its classical version, CBU with classical events, already on the most elementary level – that of *quantum events*, which do not meet the condition of full separability and noncontextuality (taken as a special case of the former). That arises from the fact that quantum (ontic) *frames of action* essentially differ from relativistic (epistemic) frames of reference.

How then are we to apply the ontological Principle of Relativity to understand the structure of QM? Every satisfactory realization of that postulate requires, first of all, a radical relativization of the quantum 'observer' (and hence of the quantum 'state' at the same time), and thus a consistent application of the ontology of Local Contextual Realism to QM. The fact that in the standard approaches to QM the 'observer' is a certain absolute (an 'idol' to use the term Finkelstein [1999b; 2003] borrowed from Francis Bacon), leads at once to a blatantly obvious incongruence with how the relativity principle in SR and GR works, from the very beginning enabling the formulation of the 'EPR paradox', hence the advocates of the preferred foliation in **M** often invoke the measurement in QM as the key argument for its introduction. However, in recent years some interesting, diversely motivated proposals for a deeper understanding of the foundations of QM on the basis of the relativity principle have been put forward, which lead to the idea of 'quantum frames of reference' (cf. e.g. [Aharanov, Kaufherr, 1988], [Finkelstein, 1996; 1999a], [Bene, 1997; 2000], [Bene, Dieks, 2002], [Dickson, 2004]), which I have mentioned in Ch. 2.2 in the context of Finkelstein's ontology. Such quantum frames form a relative context of every quantum action and are a generalization of the frame of reference from SR, but what is the key issue is a precise definition of their character, as they all violate in an extreme way the fundamental pillar of traditional realism: noncontextuality. In line with the perspective adopted in Ch. 3.1, treating Bohr contextuality as a special instance of the general nonseparability, that means that they treat the latter seriously as part of physics and its ontology, and not just instrumentalistic epistemology.

One of the most worked out conceptions of that type, presented by Gyula Bene [1997; 2000], consists in introducing the fundamental concept of the *quantum reference system* to the axiomatics of QM in such a way that ignoring that makes talking of the 'state' and 'measurement outcomes' meaningless. Bene proposes to accept that every quantum state of a certain system *S* is always relative in the sense that it is a certain definite relation between $S$ and a different system $R$, which contains $S$ (or, in a special case, is identical to it) and, at the same time, is its quantum system of reference. When speaking of a 'quantum state of system $S$' we always have to point to a reference system, with regards to which the 'state' is defined, as for different systems of reference *different relations* and *different physical descriptions* are given. Bene formulated that postulate, guided by an interesting property of QM formalism, which is based on the fact that although – according to its predictions – system $S$ as a result of the measurement is to be found, for example, in state $\psi_{S,1}$ or $\psi_{S,2}$, nevertheless the system composed of $S$ and measurement device $M$ remains in an entangled state of superposition $\psi_{S+M}$ (cf. Ch. 5.1), and hence the state of subsystem $S$ with regards to the whole of system $R = (S + M)$ will be given by a reduced density matrix of that subsystem:

$$\varrho_S(S + M) = Tr_M(|\psi_{S+M}\rangle\langle\psi_{S+M}|) \tag{33}$$

The trace operator $Tr_M$ is here defined in the Hilbert space of measuring apparatus $M$, while the bracket notation for operator $\varrho_S(S + M)$ always points to which system the 'state' of the arbitrarily singled-out subsystem $S$ has been defined for. Then the internal 'state' of the system of reference is naturally: $\varrho_{S+M}(S + M) = (|\psi_{S+M}\rangle\langle\psi_{S+M}|)$, and $|\psi_{S+M}\rangle$ is the 'state' of system $R = (S + M)$ with regards to itself [Bene, 1997]. In Bene's interpretation, what is more, the measurement is performed on the system of reference, $R$, which enables the calculation of state $\varrho_S(R)$, dependent on it.

There is nothing surprising in notation (33), as it results directly from elementary QM principles, and yet an understanding of the whole situation is impossible, as long as we insist on 'state' $\psi_S$ being an absolute one, or otherwise independent of the whole which it is an internal part of. Subsystem $S$ has at the same time different physical 'states' $s_1, s_2, s_3, \ldots$ with relation to different (differently singled out) systems of reference: $s(R_1)$, $s(R_2)$, $s(R_3)$, …, as those 'states' are defined with the use of different sets of information obtained through the actions conducted in a particular $R_i$. One should remember that the 'state' of $R = (S + M)$ is also fully relative, thus Bene's designation of $|\psi_{S+M}\rangle$ as 'the state of the system with regards to itself' is problematic and it also should be treated purely relationally (see Ch. 5). The principle, which here finds a direct physical application, is the Universal Principle of Relativity, leading to the Platonic-type internalistic compositionism discussed in Ch. 1: a definition of any 'state' ('part') always requires taking into account the ground of its determination (the whole), while a simple summing up of thus obtained information is not possible. With reference to the earlier analyses in Ch. 2.2, we can say that every quantum action–process has its global and local context, which we cannot abstract from, as that always leads to some base absolutes as Baconian idols (their source lies in the illusion of the independence of the element from the context, the whole). The context which is taken into account in (33) is a local context, accessible to the knowledge of a local 'observer', who knows part of the history of the quantum system and conducts the procedures connected to preparing the measurement. However, there also exists a global, hidden context of every action, to which no observer has direct empirical access, but which is coded in the structure of the general QM formalism. The QM structure as a certain geometry/logic is a set of rules connecting all possible physical descriptions for different quantum frames, and hence it contains the abstract information about the topology of global relations between quantum (nonseparable) histories in QBU. Thanks to that, it is precisely QM which allows us to predict the non-classical correlations between different outcomes and has a global nonseparability of the action built into its formalism, hence it is a strongly contextual theory (both locally and globally). That fact is the reason for the complete incongruity of Einstein's Local Realism to the physical meaning of QM, and also of the serious problems of all non-relativistic theories founded on that realism, which attempt to simulate QM predictions but endemically produce quantum 'paradoxes'.

If we now look at the spacetime diagram in Fig. 3 once again through the perspective of the quantum Principle of Relativity and QBU ontology, it becomes clearer that we do not have to demand that the correlated outcomes of measurements should have their source in any distant past *common* common causes such as event *cc*, or separate common causes from stochastic models. Firstly, we treat the geometric structure of manifold **M** as *in spe* physically and ontologically derivative: the elementary quantum actions do not take place 'in' spacetime (strictly speaking, there exist no real spacetime points, 'in' which anything 'occurs'), neither are they monadic, noncontextual facts of classical mechanics. QM is a theory deeper than SR and GR, as it codes the formal relations between fundamental actions in nature – it is the physics of elementary operations and their statistics [Finkelstein, 1996; 1999a]. Secondly, no physical action in nature is ever fully noncontextual (separable), though in the case of their sufficiently complex combinations it might seem so. The violation of the principle of the separability of action, which from the very beginning Bohr, Heisenberg and others had insisted on, makes it impossible already at the very start to formulate Einstein's Local Realism, despite the relativistic causal locality from $[3]_E$ and $[2]_B$ being fully retained and no superluminal signalling ruining SR and GR. What is more important from the perspective of the discussion on the 'EPR paradox' though, the lack of separable 'elements of reality' and their intrinsic, non-relational properties would make classical Bell Realism $[1]_B$ not to be met – even hidden variables–states $\lambda$ inside the universe $\Omega$ *could not* be defined noncontextually.

A convincing answer to the question why such Einsteinian 'elements' in QBU do not exist is a separate and fascinating problem. With reference to the differentiation between 'principle theories' and 'constructive theories', used in the context of the discussion on the status of SR, we can put Bell's realistic assumptions into doubt solely in the framework of the first of those strategies, by introducing that *state relativity* as a general, *fundamental principle of physics* (that is what, for example, Bene does and – as we shall see – also Rovelli). From the perspective of the local observer, such a violation of $[1]_B$ as a principle is sufficient to ruin Einstein's Local Realism and introduce Local Contextual Realism. A more counter-intuitive and physically even more interesting challenge to that scheme is also possible, arising from the intriguing formal properties of QM ([Aharonov, Bergmann, Lebowitz, 1964], [Aharonov, Vaidman, 1990]) I have mentioned, disclosing its fully time-symmetric structure thanks to the generalized quantum causality or, alternatively, timelike nonseparability for QBU ([Reznik, Aharonov, 1995], [Aharonov, Botero, Scully, 2001]) – then in such a *constructive* approach the relativity itself (or more precisely: relational nature) of each quantum 'state' would have even deeper sources in the purely global properties of histories in QBU. If we were ready to assume the working hypothesis that a fully time-symmetric QM, requiring no passage of time or ontic, absolute probabilities, violates condition $[3]_B$ (as particularly Y. Aharonov, D. Rohrlich, J. Tollaksen and others claim), then QBU can be presented as by analogy to Leibniz's cosmological, quasi-teleological (*avant la lettre* 'block') model, while the relativistic locality of causal nets is still fully retained. The ontology modelled on the Leibnizian one has an interesting asset: it predicts that the more fundamental a physical theory is, the more it is globally contextual/nonseparable. That in fact, the 'Connection', is the leitmotif of Leibniz's ontology, who – as we have seen – incessantly stressed the strong metaphysical thesis on the 'adaptation of all things' (*accommodement de toutes les choses creées*) [Leibniz, 1885a, p. 616]. In such a universe the division into a quantum past, present and future cannot be an absolute one but, true to the postulate of an internal relativization of QM, it has to be a relative one, defined – as the physical 'states' themselves – solely with regards to a given frame of action and a particular subsystem. Taking such a philosophical step meets great resistance in comparison to analogical relativization procedures in SR and GR, as those procedures are transferred onto an even deeper level, a quantum-dynamical one. True to Finkelstein's and Rovelli's postulates, though, I believe that step to be indispensable – without it QM is mechanics that incessantly generates crass paradoxes. In point of fact, it is not QM that is a problem, but the presentist picture of the Universe with the noncontextual (intrinsic), construed bottom–up 'states' of its subsystems. 'Quantum theory is a deepening of relativity'.

## 4.2 Frames of reference, correlations and the breakdown of unitarity

The idea of quantum frames of reference is connected to interesting suggestions concerning the interpretation of peculiar QM properties such as the Heisenberg relations, but perhaps it also sheds light on the issue of local measurements pointed out earlier on. The issue is of a much more universal character and is connected not so much to the QM structure alone, as to the foundations of logic and the philosophy of science as such. Generally speaking, the point of departure for all of that argumentation is the problem of the so-called absolute universal validity of physical theory, which is understood as the thesis that the theory is 'true' with regards to the whole universe without the necessity of referring to the concept of the *mapping/reconstructing system*, i.e., an 'observer' meant as any subsystem collecting, processing and compressing information about that universe. Using rather uncontroversial assumptions, one can prove that no physical theory can be at the same time universally valid, empirically verifiable and deterministic, and hence it makes sense to speak solely of theories *relatively universally valid* [Peres, Zurek, 1982]. Though the thesis, apart from a general philosophy of science, does not necessarily have to be interesting, I nevertheless believe that it has a correlative extension in a much more important general Principle of Relativity, which I have introduced in Ch. 1 in the context of the problem of internal composition and the relativization of subsystems such as 'observers' who perform measurements on other physical subsystems, outside themselves. QBU ontology, in line with Bohr and Heisenberg's original ideas concerning the decisive role of the 'quantum cut', naturally forces a relativization of the 'state of the system' and the 'observer/measuring system', but also a relativization of the completeness of the description/information about the system, which from the classical perspective may be treated with a bias as a spectacular incompleteness of QM itself (cf. [Finkelstein, 1996, p. 71]). Instead of succumbing to that 'classical chauvinism' (as phrased by Ioannis Raptis), though, required by the definition of condition $[2]_E$, we can look at the new mechanics as classically incomplete, but *quantally complete*, which results from the logic of fundamental actions and histories in QBU. I believe that it is simply an immediate consequence of treating seriously the idea that quantum theory is 'a deepening of relativity'.

Michael Dickson [2004] in his article based on the earlier ideas of Y. Aharonov and M. Kaufherr turned attention to the fact that a simple reasoning concerning quantum measurements, described in two chosen inertial frames of reference with an idealizing inclusion of the unitary Galilean transformation between them, leads to a non-classical effect, resulting solely from the procedure of the transformation itself, which allows to relationally explain the quantum formulas of uncertainty [ibid., p. 209*ff.*]. Dickson takes, by way of an illustration, an idealized situation of measuring the position of particle $S_1$ (represented by operator $Q_1^A$) with an apparatus $S_2$ in a laboratory inertial frame $A$ (a momentum – but it could be alternatively a particular component of the angular momentum – of an apparatus is represented by operator $P_2^A$, whereas the position of its pointer by $Q_2^A$). The Hamiltonian of the measurement interaction between $S_1$ and $S_2$ in frame $A$ is then given in general as: $\boldsymbol{H}_{inter}^A = g(t)Q_1^A P_2^A$, where $g(t)$ is a normalized smooth function of time with a maximum ('peak') at the moment of measurement. The shift from frame $A$ to the also inertial, external frame $B$ of a different 'observer' $S_3$ is given by a unitary Aharonov–Kaufherr operator representing pure translation:

$$\boldsymbol{U} = e^{-i\Sigma_{n>0} P_n^B Q_0^B} \tag{34}$$

(index '0' refers here to the laboratory, that is the frame $A := S_0$, whose position is measured in frame $B$), and hence the transformation of any observable $O$ measured in $A$ has the form of: $O^A \rightarrow O^B = \boldsymbol{U} O^A \boldsymbol{U}^{-1}$. For example, in the case of the position of particle $n = S_1$ given in $A$, we have the transformation: $\boldsymbol{U} Q_n^A \boldsymbol{U}^{-1} = Q_1^B - Q_0^B$, for $n > 0$ [ibid., p. 210]. The measuring system $S_2$, which is natural, thus completely vanishes from the description.

Aharonov and Kauherr [1988] used that transformation when reflecting on the problem of the measurement in the instance of a simple hydrogen atom: if we assume that the atom *as a whole* is found in a well defined state of motion, how is it at all possible to give at the same time the position of its electron, as a result of measurement (which QM seems to exclude)? The answer points to the necessity of remembering about the frames of reference of the measurements: strictly speaking, such measurements are possible as it is possible to give the position of the electron with regards to a quite different frame of reference, other than the measurement of the atom's motion (in the case of the former it is the nucleus of the atom, in the case of the latter it is obviously not). When the relativity principle is not taken into account an apparent paradox appears, one we have already encountered in Ch. 1.2. We have to proceed in an analogical way in the case of the measurement of the position of quantum particle $S_1$ in frame $A$, if we would wish – not knowing yet about Heisenberg uncertainty – to simultaneously ascribe to it a momentum or a particular component of the angular momentum. For getting to know that momentum requires, as Dickson was right to point out, ascertaining earlier that frame $A$ is appropriate for its definition, which in the case of the measurement of position of $S_1$ leads to checking if the outcome is very close to the one obtained in external frame $B$, with the realistic assumption that the internal frame is *almost inertial*, i.e., $\frac{dP_0^B}{dt} = i\left[P_0^B, \boldsymbol{H}_0^B\right] \approx 0$, where $\boldsymbol{H}_0$ is the Hamiltonian of the whole system of the laboratory [Dickson, 2004, p. 211*f.*]. (That is in fact the condition for the use of unitary transformation $\boldsymbol{U}$, and also for using the Schrödinger equation in a standard way, to describe the evolution of the state in frame $A$). It seems that such checking is tiresome and does not give much, as in the case of classical mechanics, but it is exactly the analysis of that transformation in the case of quantum measurement which reveals a strongly non-classical effect typical for QM. If we decide to use momentum or a particular component of the angular momentum of particle $S_1$, we of course have to apply a different procedure than the one in the case of measuring its position, but that means that we are forced to also make certain that the chosen frame of reference (laboratory) is appropriate to perform such a measurement, that is that the momentum or the component of angular momentum of the whole laboratory is a constant of motion. In order to do that, it is not possible to limit oneself to frame $A$, as we need external frame $B$, which of course does not appear in the Hamiltonians of systems $S_1$ and $S_2$ and the Hamiltonian of the measurement interaction $\boldsymbol{H}_{inter}^A$, just as there does not appear laboratory $S_0$ in them. It turns out that after calculating the commutators in the external frame, we do indeed obtain a startling result:

$$\begin{gathered}\left[\boldsymbol{H}^B, P_1^B\right] \neq 0 \\ \left[\boldsymbol{H}^B, P_0^B\right] \neq 0\end{gathered} \tag{35}$$

As far as the first equation can be understood by assuming that the measured particle has become involved in a physical interaction with the system measuring its momentum, then the second one is totally counter-intuitive, as the laboratory, as an internal frame of reference of the whole measurement, does not seem to enter into such an interaction, while the inequality suggests that the measurement of the particle's position itself disturbs the momentum of the whole laboratory [ibid., p. 213]. That is immediately significant for the planned simultaneous measuring of the position and the momentum, as the failure to conserve the momentum of the laboratory during the measurement interaction (disclosing the position of $S_1$) makes it impossible for it to be used as a frame of reference for a simultaneous measurement of the momentum of $S_1$. What is most important here is to notice that the quantum effect results directly from the procedure of the mere change of the frames, i.e., the shift from the 'internal' frame to the 'external' one, but for an internal quantum 'observer' that is a real physical effect, which cannot be compensated. There is no possibility to avoid that, as the only way for 'observer' $S_2$ to ascertain that its frame of reference is convenient for conducting the measurement of the momentum (by checking the conservation of the momentum of the whole system $A$) is to shift to an external frame, and that immediately produces physical consequences in the form of quantum uncertainty. Dickson's proposal is quite different from Bene's approach, but both nontrivially illustrate an important property of QM in two different ways: its contextuality in the form of the *irreducible relationality of each quantum action*, which is equivalent to the choice of a certain quantum system of reference. The local context chosen by a given 'observer', as a subsystem of the still greater system, by its nature does not contain all the necessary information to take any other measurement, which in such a case requires changing that context/frame of reference. It is significant that that physical 'incompleteness' of information, accessible in a given frame of reference, is of a fundamental character and acquiring successive pieces of information requires a constant shifting to complementary frames, which are at the same time *a totally new physical situation*. Nevertheless, Dickson's approach, based in a Bohrian style on a classical 'disturbance' viewpoint, is illustrative and instructive but unsatisfactory, as it avoids treating the generalized quantum frame as a fundamental category of the new physics, and that makes quantum systems themselves, though their behavior is analyzed relationally, to be treated in a quasi-classical way. Defining the frames of reference in the way Dickson does, in a strange way singles out position and momentum as quantum observables of objects such as particles, but from the perspective of the full quantum theory that is incomprehensible. The Principle of Relativity works correctly in this case, although it is used in a somewhat conservative way (as in the case of Bohr), which Dickson of course is aware of. Spacetime and its symmetries are treated as the background of quantum actions (Dickson singles out spatiotemporal frames of reference, even in microphysics), whereas the actions themselves are classical measurements of the properties of the material systems in space, whose states are 'disturbed' (in contrast to the standpoint of Bohr, who had never accepted such an explanation, originally preferred by Heisenberg, though he sometimes alluded to it).

We should therefore once again pose the question: what does it mean that the 'quantum state $\psi$' of a given system, such as particle $S_1$ or an atom of mercury from Ch. 1.2, is known to us? In the case of Einstein's Local Realism the answer is that we know some objective property of the particle or atom, being independent of any external system (an 'observer'), and that it has become extracted from the physical reality by the measurement action. The interpretation could even concede in some extent to the subjective character of the partial representation, but state $\psi$ itself is seen ontically as objectively ascribed to an object such as a particle (that results from the complete definiteness and the noncontextuality/separability of that realism). The consequences of that objectivized comprehension of the 'state' are serious for, as we have already observed, it introduces absolutes to the theories in the form of 'objects' which possess intrinsic, non-relational states, and a noncontextual measurement action, creating great difficulties in reconciling QM with SR and GR. Einstein himself, as Carnap remarked [1963, p. 37*f.*], complained about the serious problems stirred by human experience of the 'now', which was not (and could not be, as he stated) encaptured by physics, otherwise the principle of covariance in SR and GR would have to be violated. The introduction of a non-relativized, ontic quantum state to the theory demands that the infamous 'objective collapse' of function $\psi$ takes place at the same time on the whole hypersurface $t = const.$ [Malin, 1982; 1984] and the only solution is to discard the principle of covariance (or treat it instrumentally) by the introduction of the preferred frame/system of reference. The complications caused by such an approach are almost hopeless, as not only do they require an introduction of a nonlocal action in space, but also do not help to understand *state independent contextuality* (in line with KST) and quantum nonseparability as in e.g. (31). But at least since 1927, when Einstein first presented his objections in the debate with Bohr at the V Solvay Conference, illustrating them with the experiment with slits, he expressedly did not take into account the strong generalization (universalization) of the relativity principle and its application to the new mechanics and its 'states', just as he excluded a resignation from the separability of the latter. I claim, though, that in the case of QM we are dealing with exactly such an imperative – *a non-classically holistic type of compositionism requires at the same time a strong ontological relativization of 'states'–actions.*

The quantum 'state' can also be interpreted purely epistemically as a subjective state of the observer's knowledge, saying nothing about the objective, 'noumenal' physical reality, which remains to be mostly inaccessible. Such a Kantian standpoint might seem to be even tempting, as long as it does not become ontologized once again, as then it becomes transformed into a sophisticated ontological solipsism, dissolving physical objectivity in the principles of physical phenomenology (unfortunately its radicalism is not even proportional to its capacity to enlighten the meaning of QM). However, that Kantian-phenomenological approach is peculiarly conservative in the course of post-Kantian ontology itself – what is of interest to us in physics is the run and structure of processes in nature, and not the phenomenology of particular physical observers in their laboratories. The subjectivity of epistemic interpretation reveals its weakness, sliding down sooner or later into a certain psychologization of the observer, who gains knowledge of the 'state' of the quantum system; but what, as Shimon Malin jokes [2006, p. 234], if the physicist had drunk a lot of martini before the experiment? The epistemic interpretation of the above mentioned quantum frames of reference would also stress the fact that they were merely cognitive constructs by analogy with ordinary coordinate frames: but why should therefore the passage from one to the other have real physical effects? It seems that neither the naïvely ontic and noncontextual interpretations of the physical meaning of $\psi$, nor the wholly epistemic ones are sufficiently adequate as being one-sided (they reproduce classical distinctions, which are apparently not applicable here). The quantum theory, as a new theory of information, in a way that is characteristic for it, obliterates the basic philosophical division into what is ontic and what is epistemic (cf. [Heisenberg, 1959]: $\psi$ as '*das objektive Element der "Tendenz"*' and '*das subjektive Element der unvollständigen Kenntnis*'), which is particularly clearly seen in the status of the quantum information itself, problematic for traditional philosophers. It is therefore evident that the 'incompleteness of knowledge' in case of the observer in QM *cannot be interpreted in a purely epistemic way*, as the knowledge itself is in fact always a *physical action*. That is already perceptible in Dickson's reasoning, but because it depends there on classical reference bodies and a classical spatiotemporal background, what still stays in the game is thus some image of the partially relativized 'elements of reality' as internal states of the objects from $[1]_B$, which are dynamically 'disturbed' by quantum operations. To understand QM as a fundamental theory of information–action requires nevertheless a much deeper relativization, without the prosthesis in the form of a classical background and *classical* bodies of reference: the elementary quantum operations themselves should be radically relational, but that is impossible without taking into account the nonseparability of action at the most fundamental level, which Dickson's propaedeutics by the way avoids. I therefore think that that is a unique property of QM and the theoretical imperative – *a physical relativization of the quantum 'state' requires at the same time a completely new ontology: an internalistic compositionism.*

In order to get closer to the answer to the question as to what 'quantum state $\psi$' refers to, we have to go back to the analysis from Ch. 1.3, in which we speculatively interpreted such a state from the perspective of Platonic-type compositionism. It turned out that the standpoint, as different from extensional (radically microreductionist) as well as intensional externalistic compositionism, is the only one able to adequately represent the specificity of the quantum state of a compound system, in which the information about the subsystems does not sum up into a whole and, what is more, the maximally mixed subsystem 'states', taken *in abstracto* independently of the whole they belong to, exist in only a very weak or doubtful sense as a sheer potentiality, which Plato terms 'an undefined plurality', *apeiron*. (We have to remember that Socrates calls in *Philebus* 64e such mixtures of undefined possibilities 'a real catastrophe', συμφορά, or 'an uncombined disaster', ἄκρατος συμπεφορημένη). It was only with the global introduction of a numerical and geometric measure, i.e., an amplitude relation (weights) or geometric phase, that a certain irreducible structure was generated, a 'defined plurality', which rendered identity to the parts by defining only their mutual positions and making them *com*-mensurate (σύμμετρα). Hence, there is no ontological sense in speaking of any '*self*-commensuration' (autometry), as within the whole an arbitrarily singled out part gains any definiteness only relationally. The basis of that procedure of definition, the 'nature' of the particular parts as such (apart from *symmetry*), is completely indefinite and empty, bereft of content, as a sheer potentiality – *apeiron*. As I have made clear, such a perspective raises the objection of the externalist compositionist not merely because it deprives the parts of their own independent individuality, but also because it exhausts the 'essence' of a given whole in the structure of its internal relations without referring to an enduring existence of what enters into those relations. But the definition of 'state' $\psi$ of a quantum subsystem is thus possible only relationally, and in a double sense at that. The 'state' of any subsystem can be defined only with relation to the whole system, whereas the 'state' of the latter solely with relation to some other external system – each of those relations is an expression of a fundamental nonseparability of states, built into a non-classical formalism of mechanics and von Neumann [1932, p. 216*ff*.] was probably the first one to become fully aware of that. Both the systems and their randomly selected subsystems have an exactly identical ontological status being just *subsystems of the Universe* – they can be described *solely* by mutual correlations, which destroys the image of separate 'elements of reality'. Quantum *inter-action*, as an action and at the same time a physical correlation (Platonic 'symmetry'), is something very different from the classical measurement – it does not and cannot possess such a slight significance as in Newton–Euler–Lagrange mechanics. QM is nonseparable in such a sense that it is, as seen from the inside of $\mathbf{\Omega}$, a mechanics of the correlating (entangling) actions, which cannot be reduced (screened-off) to separable/noncontextual ones, hence a *physical* equivalence of dynamical 'action' and adynamical 'relation' in QM – incomprehensible from the point of view of classical mechanics – is natural.

It might not be clear at first how far we can go in such a correlational approach to the 'quantum state', as in the background there always remains a classical image of the phase space 'state of the system', defined for physical substrates (bodies), and, as a last resort, a commonsense object imagery. David Mermin put a consistent stress on the correlative state of the system when he presented his own 'Ithaca interpretation of QM' at a conference titled *Foundations of Quantum Theory* in Bombay in 1996 and developed it in a series of articles in the direction of a post-Copenhagen interpretation [Mermin, 1998a; 1998b; 1999; 2006]. Mermin, perhaps rightly, used an analogy with $19^{th}$ century electrodynamics, which when mathematically modelling the electromagnetic field speculated also about its carrier, to finally conceptually eliminate it. An analogical situation takes place in QM, which when probing the geometry of 'quantum states' is inclined to see them as the properties of some lasting, elementary entities (that is suggested by the terminology itself and the common linguistic *usus* in physics). Still, as we have seen, 'state' $\psi$ does not in fact say anything about the 'essence' of the particular parts–subsystems, whereas it fully informs about the correlations between them. Mermin [1998b, p. 753] even proposed to turn it into a heuristic principle when studying and presenting QM: 'correlations are physically real; what they correlate, though, is not'. The correlative sense of the quantum state has been clearly formulated in Mermin's fundamental *Theorem on the Sufficiency of the Subsystem Correlations* (SSCT), which has an invaluable significance – particularly a didactic one – for the understanding of the exceptional QM formalism presented in Ch. 3.2., but also for the formulation of a fully relational approach to QM. In order to achieve that, it is necessary to eliminate the special status of the laboratory operation of measurement, macroscopic bodies of reference and the knowledge of the specifically human observer. Mermin is right to stress that invoking them when attempting to understand the content of QM is a certain historical relict and the interpretation of QM should therefore not depend on them, but his proposal of the local-realistic treatment of correlations, thus making them an ontic replacement for objects, is unfortunately flawed, which is another instructive lesson in quantum relativity and Platonic compositionism.

Let us take the spin measurement experiment as an example once again, the one from Ch. 1.2, where the quantum spin state of the whole system was described by vector $|\psi\rangle_F$. Of course we are not limited solely to the measurement of the spin in a chosen direction, as we can measure a random quantum physical quantity on every system, represented by an appropriate Hermitian operator with its own spectrum, and for the purpose of greater generality we shall now ascribe to the whole system a density matrix $\varrho_{1+2}$. When we conduct a resolution of the 'state' into states of subsystems we have to remember that it should be physically meaningful (natural) and all the local observables connected to one subsystem should commute with the observables of the second subsystem [ibid., p. 754], which results from the trivial definition of the local separability of physical systems from Ch. 3.2 (there, just as earlier on, we insist that QM is a causally local theory in the sense of SR and GR). The specificity of the quantum state of a composite system nevertheless depends on the fact that it formally codes all its *internal* correlations, i.e., the correlations between the results of any possible measurements performed on its subsystems. The correlations, as Mermin points out, can be understood as the *mean values of all the observables of the system* at a given moment $t + \delta t$, comprising all the products of the particular observables defined on all the subsystems. Among all the particular observables for the subsystems we can also find projective operators onto linear subspaces of their Hilbert spaces, hence the set of the above correlations comprises *joint probability distributions* for the results of measurements on the particular subsystems, which makes it possible in one way or the other to reproduce also the mean values of the observables. Hence, what is to be understood as the 'internal correlations' prescribed by the 'state' of the system are both those mean values of all the observable products, and the distributions of the joint probabilities. Mermin poses a question which can serve as a simple test of comprehending the QM structure: is the 'state' of the whole system defined completely, when we have the mean values of the products of the particular observables on the subsystems (defined for a certain resolution of that system into those subsystems), i.e., without the global observables defined for the whole system? The answer turns out to be unequivocally positive: 'state' $\varrho_{1+2}$ is then completely defined. SSCT therefore states a fact that is fundamental for the comprehension of QM, that *the correlations between the subsystems (for any defined resolution of the physical system into subsystems)* are completely sufficient for a unanimous definition of the 'state' of the whole system [ibid., p. 756]. That means that we have at our disposal a simple and purely correlational or relational definition of the 'quantum state', without the need to refer to the intrinsic properties of entities, *relata*. The quantum 'state' $\varrho_{1+2}$ contains exactly the same amount of information as the correlations between the subsystems – that is an objective piece of information and hence one possessing a strictly physical meaning.

Proof of the SSCT sketched by Mermin ([1998a; 1998b]; see also [Wooters, 1990]) illustrates that simple fact very clearly. Let the states of the subsystems of our system, $S = S_1 \oplus S_2$, which the density matrix $\varrho_{1+2}$ is ascribed to, be measured in four different ways, represented by pairs of Hermitian operators, each referring to a different observable: $\mathcal{A} = \mathcal{A} \otimes 1$(for subsystem $S_1$) and $\mathcal{B} = 1 \otimes \mathcal{B}$ (for subsystem $S_2$). SSCT concludes that in such a case $\varrho_{1+2}$ is unequivocally defined by the trace: $Tr\ (\mathcal{A} \otimes \mathcal{B})$. Finally then, it will suffice to give all the correlations between the observables for the particular subsystems, to have at our disposal complete information about system $S$. Mermin is right to stress that one can choose a number $n$ of subsystems differentiated by a resolution of $S$ and the formalism of QM itself guarantees that, for instance, in the case of resolution $S = S_1 \oplus S_2 \oplus S_3$ it is of no account if the density matrix is constructed from the correlations between the observables on $S_1$ and the global observables on $S_2 \oplus S_3$, or from the correlations between the global observables for $S_1 \oplus S_2$ and the observables defined separately on $S_3$ [ibid., p. 765]. The complete set of marginal probabilities remains to be the same in the case of all the joint probability distributions of the observables, as long as there exists a given one and the same subsystem in those resolutions (it is of no account what measurements are taken on the remaining subsystems) [ibid., p. 759*f*.]. Those properties make the knowledge about the state of system $S$ – as the knowledge about its internal correlations (:distributions of probabilities) – not a subjective assessment of the physical situation, but to a large extent one of an *a priori* character. The structural exceptionality of QM, though, is that our knowledge about the distribution of joint probabilities for the observables and the invariant marginal probabilities of the particular subsystems does not enable us to provide a complete and consistent set of conditional probabilities for all the observables, which makes the assumed physical objects, noncontextual *relata*, on which all those correlations are supposed to be unequivocally defined, *impossible to exist*. That is Mermin's basic argument against the objective reality of quantum 'micro-objects' as such, regardless of their relations, and that is exactly what we have already concluded in Ch. 1 and 3: there does not exist any possibility of constructing an entangled, relational total state from the already defined, primary states of the ontic *elements*, endowed with intrinsic, non-relational properties. Thus, one can safely say that quantum correlations cannot be simulated by any theory, which does not possess some strictly holistic traits and even the status of the notion of 'quantum state' itself is irrelevant at this point. But one should be aware that it is not possible to make such internal correlations $Tr(\mathcal{A} \otimes \mathcal{B})$ the new ontic and local 'elements of reality' or 'building blocks' for realistic, objectivistic ontology (see e.g. [Cabello, 1999]). Shedding *relata* is not sufficient and there is more in global quantum nonseparability and Kochen–Specker contextuality than that, hence a discussion on the non-classical holism in a cosmological context is unavoidable. As has already been said, a new type of compositionism requires at the same time a *strong ontological relativization of the 'states'–correlations themselves*. Quantum internal correlations are not noncontextual quasi-things, exactly as any Bene's reference systems and their 'states' $|\psi_{S+M}\rangle$ are never preferred frames. If they were, such conundrums as the delayed choice quantum eraser or Hardy's paradox would remain essentially intact.

The quantum measurement analyzed canonically is nothing other than a phenomenon of correlating certain micro-processes (such as 'electrons' or 'atoms') with processes on a greater scale, comprising the former and conveniently termed 'classical' by 'observers' [Mermin, 1998a; 1998b]. As the 'observer' is also a physical system, then strictly speaking the result of the measurement is in practice a certain net relation between a directly inaccessible micro-process, the reference object/-s (serving as an indirect 'frame of reference' in Bene's understanding of the term) and the 'observer' in a laboratory. Such a viewpoint on microphysics allows us to push the mystery of the 'measurement' into the background and replace it with the category of net correlations between the different quantum-internal subsystems of the whole of the larger (even cosmological) system, which in turn can be mutual frames of reference for one another giving *different relational states*. The subsystem 'measured', the one the information is gained about, can in the operationalistic context be called a 'specimen', whereas the 'measuring' subsystem, which usually possesses an unimaginably great number of degrees of freedom, can be called an 'apparatus' [Mermin, 1998b, p. 757]; but David Finkelstein [1989] more aptly calls it, respectively, an 'endosystem' and an 'exosystem' as parts of a globally 'nonseparable whole system'. All of the available information about the correlations between the subsystems is indeed contained in an appropriate density matrix of the system, which comprises both of its subsystems as their hidden whole, and it is that holistic information about the structure of the relations that is 'real' (possessing a physical meaning) for another local external subsystem of the universe. However many persistent problems with understanding QM have its source in the fixing of the position of the Heisenberg–von Neumann cut, separating the 'classical world' from the quantum one, but in QM itself there is nothing to suggest such a definite position. To accept the relativity of that cut, in turn – in line with the Universal Principle of Relativity – is a key step towards the relativization of the quantum 'state' on a fundamental level of description. In that sense Bohr's strategy of fixing the classical world as a preferred frame of reference should be treated – as Mermin rightly advises – as a historical relict of the first stage of the development of QM. Shedding the special class of quantum observers cannot mean just making the correlations new building blocks for ontology – it must coincide with shedding the absolute status of the 'state' itself, that is consistently applying the Principle of Relativity.

It is therefore clear that the quantum description has both an epistemic and an ontic/objective aspect (which Heisenberg [1959] was already aware of), but putting it in an adequate perspective requires a complete redefinition of the classical use of both those two terms which support the traditional demarcation line between philosophical epistemology and metaphysics. The quantum 'state' as represented by $\psi$ or $\varrho_{1+2}$ is always a *contextual, maximal piece of information* that one *quantum local subsystem* possesses about another quantum subsystem: it depends both on the local and on a global context (structure) of interactions, that is a network of relations between the arbitrary systems. The structure of the experimental situation, permitted by nature, and the geometry of QM enable a probabilistic prediction of the results obtained by different 'observers', who are always arbitrary parts of a larger quantum system. But every successive action is equivalent to making a successive Heisenberg–von Neumann cut, that is defining a *new quantum system of reference*, with the system not being a correspondent of a passive reference frame, but being of an active character, co-determining the character of the correlation itself. That makes the representation of the 'quantum state' in QBU radically deter from its classical correspondent and always has to be understood as a certain local action, resembling 'force' rather than a 'state', which fact – as far as I am aware – Finkelstein has been the only one to consistently stress for many years (see Ch. 2.2) and which makes the very concept of 'state' in standard presentations of QM rather problematic (that is also why we bracket it all the time). Treating the knowledge of the observer in an Einsteinian manner as a passive reservoir of information automatically introduces a corresponding noncontextual 'quantum state' and an inexplicable dualism which does not enable us to understand why the frame is of fundamental importance in the description of the quantum process, reaching far beyond classically understood parametrization. From the perspective of classical mechanics such a situation would mean that QM is spectacularly incomplete, as every measurement correlation between the subsystems would be connected to the loss of an enormous amount of information earlier objectively contained in the endosystem, but from the quantum perspective speaking of a 'loss of information' of such a kind makes no sense. QBU ontology based on a correlational and relativized understanding of quantum information requires a completely different treatment of 'incompleteness' from the perspective of the finite subsystems of the Universe. The question therefore is: does the procedure of making an arbitrary, local Heisenberg–von Neumann cut have physical effects, arising from its logic as such? One could be tempted to try and capture the singularity of such a quantum 'incompleteness' by purely logical means.

The earliest work wholly devoted to that issue was published by María Luisa Dalla Chiara in 1977, who pointed out the fundamental connection between the quantum problem of measurement in von Neumann's formulation and the logical problem of self-reference and set-theoretical paradoxes, particularly the Cantor–Russell paradox [Dalla Chiara, 1977]. It is assumed that the absolute universal validity of physical theories, mentioned at the beginning of the chapter, is subject to the same limitations as in the case of the logical limitative theorems arising in the fully formalized theories. That in itself is not surprising, as among the philosophers of science such limitations already take root as intuitive ones with reference to the universalistic pretences of fundamental theories. But in the case of QM the situation seems rather exceptional, as the limitative theorems could have an immediate application not so much with regards to the importance of the theory as such, as to the very possibility of representing the states of the objects of the mechanics [ibid., p. 331]. In the wake of Dalla Chiara's article, Peres and Zurek [1982], David Finkelstein [1988] and Hans Primas [1990], for example, embarked on a similar course, but the most general result, concerning both quantum theory and classical theory was presented by Thomas Breuer [1995; 1996; 1997]. Breuer's proof requires no particular assumptions as to the nature of time, it uses the concept of the noncontextual 'physical state' and seems to have a universal application, but nevertheless makes one doubt if it addresses the specific nature of quantum information and to what extent allows us to understand the specific nature of 'quantum measurement'.

The point of departure for the proof presented by Breuer is to analyze the possibly simplest situation of measurement, which requires a 'cut' and the differentiation of an endosystem as an 'observed system' $U$ and of an exosystem performing the role of a 'measuring device/apparatus', an 'observer' $O$. The measurement of $U$ by $O$ is seen in the most natural way as a physical interaction, which produces certain informational correlations (there are many of them in practice) between $U$ and $O$. In general, obtaining information about the state of $U$ ($s_U$) simply denotes an inference about that state on the basis of the registered state of $O$ ($s_O$), with both those states referring to the same moment $t_i + \delta t$. From the formal point of view, the inference in a given experiment can be described by defining the correlation map $\theta$ from the power set $\mathcal{P}(S_O)$ of the set $S_O$ of states of exosystem $O$ into the power set $\mathcal{P}(S_U)$ of the set $S_U$ of states of endosystem $U$, that is that inference/correlation map ascribes to every set $S_O$ a set of states of the measured system $\theta(S_O)$, which is compatible to the information that the exosystem following its interaction is to be found in one of the $S_O$ states. We can now say that in a given experiment with a particular $\theta$ map, the state of the endosystem, $s_U \in S_U$, is *exactly measurable*, if after the measurement there exists such a set $S_O \in \mathcal{P}(S_O)$ of states of the measuring exosystem, which unequivocally refers to state $s_U$ of the endosystem, that is $\theta(S_O) = \{s_U\}$. In turn, we can say that the whole experiment with map $\theta$ can *distinguish states* $s_1$ and $s_2$ of the endosystem only when for $t_i + \delta t$ there exists one set $S_O^1$ of the states of the exosystem referring to $s_1$, but not to $s_2$, and a separate set $S_O^2$ referring to $s_2$, but not to $s_1$, that is: $\theta(S_O^1) \ni s_1 \notin \theta(S_O^2)$ and $\theta(S_O^1) \not\ni s_2 \in \theta(S_O^2)$. Those general definitions by Breuer can be apparently applied to analyze both classical measurement and the quantum one and there is nothing peculiar in them until one introduces the situation of the *observation of the whole from inside*, where the measured system $U$ is to contain a measuring subsystem $O$ (i.e., in general: $U = \mathbf{\Omega}$, as $U$ now encompasses $O$ and the rest of the physical processes in that universum). The situation is depicted in Fig. 5, where a certain part of $U$ is differentiated by a given 'cut' $i$ as an observing system $O$, described after the measurement by internal states $s_O$; the map of inference concerning states that are outside the exosystem is supplied by $\theta(S_O^i)$. Taking into consideration such a scheme introduces at once a fundamental logical paradox – the paradox of self-reference in a general form.

*Fig. 5: The scheme of measurement from inside the whole. $U = \mathbf{\Omega}$ is a postulated physical universe with states $s_U$, in which cut i is defined providing a measuring subsystem (observer) O with states $s_O$ and the measured environment (subsystem U/O). Map $\theta(S_O^i)$ is a function of inference about the states of the rest of the universe from the states of the observer in the experiment.*

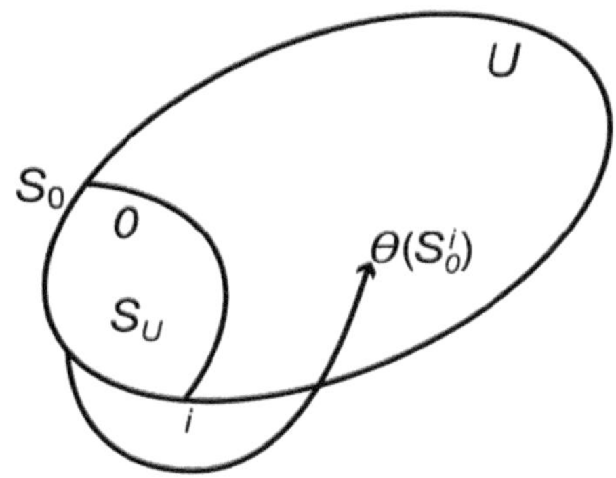


For the proof to work, Breuer introduces two simple and very natural assumptions. Firstly, we assume that all of the system–universe $U = \mathbf{\Omega}$ has more degrees of freedom than the subsystem $O$ being its part, and hence we introduce an *assumption of proper inclusion*:

$$(\exists s, s' \in S_U): s \downarrow O = s' \downarrow O,\ s \neq s', \tag{36}$$

where $s \downarrow O$ denotes the state of the observer as the restriction of state $s$ of system $U$ to subsystem $O$ (hence it is a surjection to the states of $O$). (For that assumption to be met will depend not only on $S_U$ and $S_O$, but also on the function of restriction $s \downarrow O$). If the map of inference $\theta$, which we define for that experiment, is to be consistent, a second assumption is needed, which Breuer calls a *meshing condition* – it comes from the fact that the states of the observer after the observation are self-referential. The condition is that for each state of the observer $s_O \in S_O$ the restriction $s \downharpoonright O$ of the system states $\theta(\{s_O\})$ to which state $s_O$ refers, should also be that state $s_O$ itself, hence:

$$\forall s_O \in S_O: \{s \downharpoonright O: s \in \theta(\{s_O\})\} = \{s_O\}. \tag{37}$$

The key theorem, whose proof Breuer can demonstrate on the basis of the above definitions, concerns a semantic inclosure of each language formulating the description of the global state of the whole $U = \mathbf{\Omega}$ from the perspective of its parts: **[1]** the jointly accepted assumptions about the proper inclusion and the meshing condition lead to the conclusion that not all of the states of the whole system can be exactly measured by an internal observer, hence:

$$\exists s_U \in S_U, \forall S_O \in \mathcal{P}(S_O): \theta(S_O) \neq \{s_U\} \tag{38}$$

The second theorem, which can also easily be proved on the basis of earlier assumptions, is: **[2]** *if $s_1$, $s_2$ are two states of the system U such that $s_1 \downharpoonright O = s_2 \downharpoonright O$, then there does not exist an inference map $\theta$ – and thus neither does the measurement performed by O – such that states $s_1$ and $s_2$ can be distinguished* [Breuer, 1995], thus the following takes place:

$$\forall \theta: \left[ (\sim \exists S_O^1, S_O^2 \in \mathcal{P}(S_O)) : \theta(S_O^1) \ni s_1 \notin \theta(S_O^2), \theta(S_O^1) \not\ni s_2 \in \theta(S_O^2) \right] \tag{39}$$

An analysis of the above formulations easily leads to the conclusion that the state of observer $s_U \downharpoonright O$ is analogical to the Gödel sentence, as it generates an exactly analogical paradox of self-reference regarding the whole of system $U = \mathbf{\Omega}$. Breuer therefore proposed, in the wake of M.L. Dalla Chiara's earlier analysis [1977, p. 338*f.*], a short formulation of a theorem about the incompleteness of the measurement, which refers to the situation presented in Fig. 5: *no measurement from inside the observed system can be informationally complete.*

Every self-reference paradox (including that of Cantor–Russell), possesses a universal logical structure, as it is in fact an inclosure contradiction, that is it realizes an Inclosure Schema. Fig. 6 presents a universal structure of the paradox regardless of its particular realization. The Inclosure Schema is rather easy to identify, though – as we know – that property does not make it as easy to remove it. To obtain that schema and that paradox, we have to have a 'diagonalizing function' (a *diagonalizer*) $\delta$ at our disposal, which is a certain generalization of the measurement inference function $\theta(S_O^i)$, which Graham Priest [2002, p. 156] defined thus:

(1) $\mathbf{\Omega} = \{y; \varphi(y)\}$ exists and $\psi(\mathbf{\Omega})$ — Existence

(2) if $x \subseteq \mathbf{\Omega}$ and $\varphi(x)$, (a) $\delta(x) \notin x$ — Transcendence

(b) $\delta(x) \in \mathbf{\Omega}$ — Inclosure

*Fig. 6: The universal Inclosure Schema lying at the root of the paradoxes of self-reference.* $\mathbf{\Omega}$ *is a set of all the 'objects' y, which satisfy property* $\varphi(y)$*, from among which we single out a subset of objects x satisfying* $\psi(x)$*. Function* $\delta$ *meets in turn the logical condition of transcendence, i.e., it assigns to objects x some objects from beyond their subset, but belonging to universe* $\mathbf{\Omega}$*. If we analyze a special case in which* $\psi(\mathbf{\Omega})$ *takes place, then function* $\delta(\mathbf{\Omega})$*, in line with the closure postulate, projects the whole* $\mathbf{\Omega}$ *into that set, and at the same time out of it. Reprinted by permission of Oxford University Press from: G. Priest [2002], Beyond the Limits of Thought, p. 156. Copyright © 2002 by Graham Priest.*

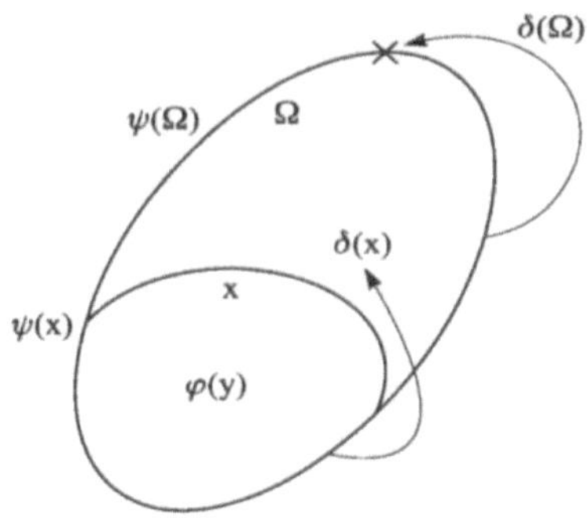


If we wish to take into account the logic of the measurement procedure itself, we are forced to accept that there exists no possibility, for fundamental reasons, to informationally completely represent the state of the whole by parts randomly singled out in it, thus each of those parts also does not have access to a full description of the relations unifying that whole. In order to be able to fully describe the global structure of the whole, it is necessary to shift to a frame/system of reference that is external to it (i.e., an exosystem with regards to $\mathbf{\Omega}$), which – as we have seen – in the case of QM would be specifically bound to produce physical effects resulting from the change of the particular frame of action. In the case of the Universe as a closed system, such a shift is physically impossible. Dalla Chiara [1977, p. 340] pointed to an analogy with the semantic hierarchy of theory and metatheory, writing that '... any apparatus which realizes the reduction of the wave function is necessarily only a metatheoretical object'. But in QM there is more than that. From the perspective of local quantum 'observer' *O*, who performs the measurement at a moment $t_i + \delta t$, one of the states being part of the superposition seems to have become 'chosen' as the one that is 'actualized' with a probability prescribed by the structure of the whole experimental situation, but that only takes place in *the frame of his/her quantum action* defined by cut *i* and some maximal set of the compatible observables, as QM predicts that for another external 'observer' the state of the whole is still described relationally by the superposition of the states of the parts (I am going to deal with this fact in Ch. 5.1.). What seems to be a strange anomaly, an 'incompleteness' of QM formalism, provoking some physicists to seek a mechanism for the 'objective reduction of function $\psi$', can be another strong argument in favour of the contextual realism and the fully relational meaning of the notions of 'state', 'observable' and the 'observer' in the quantum universe. Without providing a given cut and the framework of the observer's injective actions [Finkelstein, 1999a, p. 452] inside $\mathbf{\Omega}$, there is no sense in talking about 'quantum states' and the 'measurement outcome'. Unless they are delineated, there is no sense in giving the bare result of a measurement which leads to a naïvely realistic understanding of the individual, intrinsic 'state' of a particle or atom. (The situation is a radicalization of the one we know from classical physics, where the bare result given without specifying the frame of reference is also meaningless). Dalla Chiara's and Breuer's arguments seem interesting from the perspective of every relational approach to QM, but unfortunately it is difficult to see them as a solution to the problem of measurement without a consistent analysis of the structure of the whole quantum theory, particularly of the fundamental global nonseparability and the status of probability in a local description. Breuer's approach seems to offer nontrivial conclusions concerning the logic of physical measurement, but they still depend on classical notions and mathematical (Boolean) structure, hence, similar to Dickson, he seems to be attempting to solve the problems in understanding quantum theory with the use of classical or quasi-classical means, as in the background there remains intact e.g. the ontology of extensional compositionism working in the set theory. Here again the understanding of the 'loss of information' based on a classical, noncontextual notion of the 'state' and 'measurement action' seems to be doubtful. In short, putting the whole machinery of the inference maps and informational completeness into work presupposes the mechanics with 'elements of reality' being represented by apparatuses. I would agree that the key issue is the internalistic logic of the action itself and of the quantum frames in $\mathbf{\Omega}$, but that is *not* a classical logic of physical states nor its conjugated models of compositionism. Thus the question remains, what can be the source of breaking the unitarity of quantum evolution from the perspective of a given particular, internal subsystem *O* and its quantum frame of action, if any 'state' $\psi$ or $\varrho$ is *not* a noncontextual, registered 'element of reality'? Let us first go back to speculative metaphysics as that seems here to be more than entertaining dialectics.

The abstract, metaphysical scheme of explaining that phenomenon has already appeared in Platonic compositionism. Plato's reflections presenting the so-called second hypothesis of the *Parmenides* seem to be so unclear that even the ancient readers often doubted if they were at all conclusive, but – as we have seen – they contain nevertheless an intriguing conceptual scheme, which C.F. von Weizsäcker uncovered for the philosophy of physics. What is important to note here is that the whole 'closed in itself', the hidden whole, cannot be predicated as 'existing' in any usual meaning; it can to any empirical observer be merely a potentiality, represented in his/her aprioric knowledge, as long as he is aware of the appropriate global symmetries. Such 'plurality (*apeiron*) structured by measure (*metron*)' is given abstractly just as the *hidden structure* of all the *possible relations* between parts/actions which could be randomly singled out/performed in $\mathbf{\Omega}$, but no 'state' $\psi$ or $\varrho$ can represent them as actual. The whole $\mathbf{\Omega}$ is globally nonseparable, which makes Bohr contextuality and von Neumann's consistency condition locally work (and which means that thinking of $\psi$ or $\varrho$ in the categories of 'elements of reality' and externalistic compositionism has no sense), but it is not possible to claim that 'the whole $\mathbf{\Omega}$ as such exists' or is a 'physical object', as it could exist only relationally, in correlation to the other system which would enter into a dynamical interaction with it[59]. Here a kind of internalistic logic of $\mathbf{\Omega}$ overtly comes into play. As we have seen, Von Weizsäcker [1982, p. 490] wrote that if any isolated whole 'comes into being' for a certain singled-out internal observer, that is every time equivalent to its 'destruction': the observer's (exosystem's) local frame of action co-defines the 'state' contextually, at the price of *classically* understood 'incompleteness of information' about the supposed separable 'element of reality' which classical mechanics here demands. So, surely, what is 'destroyed' is *not* some actual whole 'existing out there', exactly as a quantum measurement in laboratory does *not* cause an instantaneous collapse of some full-information wave on the hypersurface in **M**, and what we apply here is the non-classical internalistic logic. In the same vein, the 'paradoxical' wave–particle duality (such as realized in the delayed choice quantum

59 From the perspective of classic objectivistic metaphysics that is a very rigorous requirement, but it really has its source in the ontology of being as 'the other' with regards to 'the One' from Plato's *Parmenides*. That in a way is also a serious adoption of George Berkeley's principle: '*esse est percipi aut percipere*' ('to be is to perceive or be perceived') from his *Three Dialogues between Hylas and Filonous* (1713). That principle also plays a crucial role in Leibniz's earlier monadology, in which there does not exist any such 'state' which would not be a 'perception' of some monad and at the same time the object of the 'perception' of other monads.

eraser) is *not* a manifestation of the 'disturbance', 'destruction' or the classical 'loss' of information concerning some 'elements of reality', but of local Bohr contextuality/nonseparability, i.e., the full relativity of quantum information/reality (the wave–particle duality relations are just the so-called entropic uncertainty relations in local quantum frames [Coles, Kaniewski, Wehner, 2014]). Also any subsystem in **Ω** extracts relational information about a different subsystem in **Ω** apparently at the price of a 'loss of information' connected to its own 'state' as another presumed classical 'element of reality'. It could gain it if it possessed the possibility to *simultaneously* remain at $t_i + \delta t$ outside **Ω** and measure without any dynamical interaction, which is fictitious, as by definition it would then not be a physical system at all, and the very concept of 'simultaneity' would lose all its sense (there exists no absolute physical system, possessing a nonphysical access to all the frames of action in a pan-perspectival way). Such a type of paradox can be treated as a radicalization and *reductio ad absurdum* of a seemingly innocent dualism of the 'observer' and the 'physical world', typical for pre-quantum mechanics. As far as I can assess, Bohr intuitively understood those quantum intricacies as no one else.

Finally, von Weizsäcker's Platonic ontological conviction that 'the Universe can itself *be* only to the extent that it is not one, but many' [ibid., p. 491] leads to the conclusion that existence cannot be seen to be an 'objective existence' as such, but as a purely *relational existence*, which completely changes the meaning of the commonsense concept of 'state', 'fact', 'action' and the 'part'. Perhaps, following Finkelstein and Rovelli (who take that step following Heisenberg), one should completely resign from the concept of 'state' when formulating quantum axiomatics and speak only of *relational information/action* or a *contextual correlation between subsystems* in **Ω**. As I have pointed out, Von Weizsäcker ends the quoted essay by writing that 'when it (the One = unity) is, it is many, but at the price of contradiction' [ibid.]. The plurality of quantum reference systems/frames in **Ω** and their relations is connected to a plurality of classically contradictory descriptions, whose relational coherence in QM is not possible outside the hidden structure–whole **Ω**, which renders them any physical sense (consistency), but which is unconstructible as a classical 'object' in a certain 'state', satisfying the Principle of the Excluded Middle. Although that might be surprising, internalistic compositionism begins to have sense when connected to Contextual Realism, which historically has been seldom associated with Platonism itself.

The above argumentation is expressed in a metaphysical language, but it could possess a definite physical content, particularly when we take into account the meaning of the unitarity of quantum evolution and the above fundamental paradoxes of self-reference, which was pointed out, among others, by Abhay

Ashtekar and Carlo Rovelli [1996]. Unitarity in QM is a formal property, but it has a physical meaning as an invariance under the infinitesimal translation in time for a closed system. Linear operators describing the time evolution of the system, having the general form: $\boldsymbol{U}(t, 0) = \exp\{-i\boldsymbol{H}t/\hbar\}$ in order to be unitary have to meet the condition: $\boldsymbol{U}^{\dagger}\boldsymbol{U} = \boldsymbol{1}$, with the matrices of those generators possessing a simple interpretation in vector Hilbert space as generalized rotation matrices. Such smoothly $t$-parametrized unitary matrices form the Abelian Lie group and at the same time a topological manifold. What is of key importance here is to stress that unitary dynamics has sense only with regards to a closed system, for which we can formulate the condition of state normalization and we can define invariance under infinitesimal translation in time, but the quantum measurement of system $U$ by exosystem $O$ is always a physical interaction (dynamical correlation), hence by definition $U$ is *not* a closed system for $O$, and at the time of the interaction and in the frame of $O$ there is no sense in analyzing the unitarity of the evolution for $U$. Here, once again, the purely internalistic, Platonic-like logic is at work: Carlo Rovelli [1996] postulated that the breakdown of unitarity results directly from the fundamental logic of the measurement itself – it would be based on the Inclosure Schema in quantum universe $\boldsymbol{\Omega}$ and on the phenomenon which can be termed a 'self-referential censorship'. The exosystem $O$, which is performing the measurement inside universe $\boldsymbol{\Omega}$, can use the Hamiltonian of $U$, but it does not know, for fundamental reasons, all the correlations between itself and the measured system, that is it cannot define, at the time $t_i + \delta t$ when a measurement is being conducted, the full Hamiltonian of that interaction. What is more, as Rovelli argues, the whole knowledge that $O$ can obtain is *only* a correlation, but the concept of 'being correlated with itself' does not make sense, hence the description of the whole process of the local interaction between $O$ and $U$ is in $O$'s frame *always* incomplete (that is exactly the physical effect of a self-referential censorship). The effect is thus neither an illusion, nor a consequence of the observer's 'subjective ignorance', but it would result from the very nature of a quantum interaction and the Universal Principle of Relativity, prohibiting the existence of an absolute frame of action. Dalla Chiara [1977, p. 340] proposed in that context a logical interpretation of von Neumann's theorem: if endosystem $O$ is an object of theory ($O \in \boldsymbol{\Omega}$), then it cannot describe the 'reduction' of the state function, i.e., if it describes the process of the 'reduction' of the superposition of states, then it itself is a metatheoretical object (exosystem) with regards to the theory describing the dynamics of the measured object. Thus, in the end, we encounter a deep ambiguity in the notion of 'state' introduced into QM, as seen from the viewpoint of classical mechanics. If that is some superposed state $\psi$ of an isolated system prepared in action by $O$, it cannot be characterized by the definite real values of observables – they cannot be said to 'exist' and $\psi$ cannot be a local 'element of reality' being manipulated in a laboratory, even as a 'set of internal correlations'. If that is a 'state' $\psi$ registered by $O$ in some frame of action, it is an action-relation between systems established at the cost of some information being impossible to extract and some other one being locally established, hence $\psi$ cannot be an 'element of reality' disclosed/inferred by an exosystem. In each case it is *fully contextual*, thus relative, which does away with strong Einsteinian realism. In QM even internal action itself is relative.

The conclusion from the above remarks is that, strictly speaking, it is not possible to talk about the objective 'collapse of the wave function' on the singled-out hypersurface $t + \delta t = const$ in **M**. QM need not remain in any conceptual conflict with the relativity theory, as long as we apply the procedure of a radical relativization demanded by the ontology of internalistic compositionism and contextual realism. That also means that the concept of the quantum frame of action is fundamental for physics and the ontology of physics (cf. also [Anandan, 2003, p. 1944*f.*]). Breuer [1996; 1997], when analyzing the quantum consequences of his theorem, suggested talking about the breakdown of unitarity as a 'subjective decoherence', but the hypothesis, as he himself accepted, did not solve the infamous problem of measurement. The principal reason behind that, in line with the adopted metaphysical perspective, is the fact that the 'interaction' and 'physical state' are still classically seen, and the distinction between what is 'subjective' and 'objective' copies all the basic assumptions of classical realism. The contextual (relational) breakdown of the unitarity of evolution can prove to be a satisfactory solution only in a radically non-classical, Platonic ontological scheme, bereft of existing 'elements of reality' and information interpreted as a noncontextual inference/disclosure. Thus, which I have to stress once more, to speak of a 'loss of information' or 'disturbance' with regards to quantum actions has to be essentially misleading, *if* it silently assumes all that conservative classical ontology. Even in QM – relativity wins, presentism and separable/noncontextual realism lose.

A consistent application of the idea of internalistic compositionism leads, as we have seen, to a very specific form of quantum ontology, a relational block universe, which to many interpreters of QM might be too high a price to pay. On the other hand, the acceptance of 'quantum paradoxes', as well as permanent difficulties with the unification of QM and GR are an even higher price to pay for the attempts to retain the schemes of classical mechanics. A serious treatment of relational, time-symmetric QM and QBU ontology requires nevertheless a total redefinition of the fundamental concepts, with even a marginalizing of others (such as the 'state of the system'), which could help to change the way we think about physics. In Ch. 2.3, I suggested that a presentation of quantum history as a hidden whole could be explained by the violation of Bell's condition $[3]_B$: 'timelike separability' or 'time-asymmetric independence from history'. But even then the classical concept of the 'history' is impossible to be imposed, as the only intuitive physical analogies are representations of empirical wholes, made up of separable objects in spacetime or individual events glued together. For example, the image of a causal 'Schopenhauer sphere' from Ch. 2.3 could be a primitive, metaphysical-speculative version of a relativistic block universe (as a metaphysical interpretation of the spacetime diagram presented in Fig. 3), but it is useless as an image for QBU. In the latter case, all the processes–actions themselves are purely relational (in QBU there do not exist any non-relational 'states of affairs') and the frames are active (dynamical). The additional possibility of the violation of condition $[3]_B$, as a deep physical explanation of those first two properties, makes it impossible to introduce a classical (time-asymmetric/orthochronal) causality and determinism (it is always possible to locally extract *new* information from any quantum subsystem in time-symmetric QM). Also spacetime with its metric cannot be the background, as the measurement of time and distance also have to be treated as quantum actions. Since in QBU an individual event does not have a classically objective character at all (it is physically meaningless independently of its relation with other events), hence there appears here a fundamental discrepancy with Einstein's naïve realism and the standard way of thinking about the relativistic CBU (there is not even any sense in the 'point events' as a physical fiction – their role has to be performed by elementary actions). Thus, the principal theoretical challenge is to construe a quantum blockworld ontology, in which the quantum nonseparability and the quantum time symmetry are compatible with relativistic causality defining the structure of spacetime. Hermann Weyl, who was – as Einstein or Eddington – an early adherent of CBU, wrote in 1928, when analyzing quantum compound systems, that 'in this significant sense quantum theory subscribes to the view that "the whole is greater than the sum of its parts", which has recently been raised to the status of a philosophical creed by the Vitalists and the Gestalt Psychologists' [Weyl, 1931, p. 93]. But discarding Einstein's separability principle and the existence of 'elements of reality' has very serious consequences for cosmology, much more serious than being just an epistemological 'Gestaltist' in a quantum optics lab, and perhaps much more serious than for microphysics itself, which after all might be silently satisfied with the orthodox Copenhagen interpretation (cf. e.g. [Howard 1989; 1997]). Exactly in the context of cosmology, time-symmetric QM and the ontology of internalistic compositionism can perhaps raise the greatest controversies. The relational-atemporal approach to the delayed choice quantum eraser might seem acceptable in a laboratory and its pared-down working metaphysics, but for many not so on a cosmological scale.

One of the issues which automatically raise philosophical objections is naturally a problem of the 'passage of time' in QBU (or rather its phenomenology). One can expect that the source of experience of that passage by the local, relativized observers should be quantum-thermodynamical, connected to the (emergent) geometric time of GR. What seems to be an appropriate direction for a theoretical search is defining the local 'passage of time' as a statistical effect connected to the density of quantum information states (with a certain analogy to temperature or pressure in classical statistical thermodynamics). One should expect such a local 'passage of time', as series of quantum correlations between subsystems, to be an immediate consequence of the fact that the clocks used by local 'observers' in **Ω** can never undergo a unitary evolution, and no local clock can be assumed to be irreducibly classical – as a quantum system it keeps undergoing fluctuations (that is similar in the case of the measurement of length in **Ω** with the help of any 'quantum rods')[60]. The above peculiarities of the quantum universe **Ω** and of quantum histories might make the very use of the term 'block universe' seem to be somewhat misleading, as it overlaps with the picture of the CBU ontology for relativistic physics given by the 'Schopenhauer sphere'. One should therefore remember that if in the QBU the conditions $[1]_B$ and/or $[3]_B$ were not to be met, when there were no events as elementary, non-relational 'elements of reality', and hence **Ω** would possess a *global* quantum (nontrivial) topology enforcing *global* nonseparability, then that would require a completely different ontology and imagery than a relativized 'block of events' meeting Poincaré symmetries, even if relativistic local causality of the emergent GR geometry were intact. Quantum histories as the atemporal Platonic wholes in the quantum blockworld are not just relational modifications of classical-mechanical histories, but are a completely *new entity*, governed by the logic of internalistic compositionism and the Universal Principle of Relativity.

---

60 The idea of the 'decoherent time' in the framework of the relational approach has been suggested, for example, by R. Gambini, R. Porto and J. Pullin [2004; 2007]. It differs from Carlo Rovelli's proposal which I shall discuss in Ch. 5.1. and which I prefer in the context of QBU ontology.

# Chapter 5

# The Relational Structure of Quantum Mechanics and Internalistic Compositionism

## 5.1 The Principle of Relativity, the internal consistency of quantum mechanics and the quantum thermodynamics of time

When discussing the issue of the correlative nature of quantum actions and of the relativity of the endosystem/exosystem division, we encounter one of the greatest peculiarities of QM, resulting directly from the linearity of the state space. It concerns the discrepancies in the description of a physical situation provided by different observers in their quantum action frames: the objectivistic (noncontextual) approach to QM, required by Einsteinian strong realism, inadvertently leads to the troubling conclusion that either QM is internally inconsistent, or that reality itself demonstrates contradictory properties, hence we encounter the paradoxes. Both conclusions are wrong. As I have pointed out in Ch. 1.2 and 4.2, the logical fabric of QM is a very subtle interplay between the Principle of Non-Contradiction and the Principle of the Excluded Middle, which ruins the ontology of classical mechanics with its separable and definite 'elements of reality': in fact, relativity of action and the relativized value definiteness enforce a new ontology of composition without those elements. The basic issue is to ascertain that the application of the concept of local, quantum (ontic) frames solves the problem and that the general contextuality of QM as such can be treated as a very strong argument in favour of its relational structure and internalistic compositionism, which *globally* violates the Principle of the Excluded Middle, hence quantum internalistic logic would enforce non-classical topology. To throw more light on those issues, let us look more closely at the simple way in which Carlo Rovelli [1996] formulated the problem of the 'third observer' in his key paper concerning the relational interpretation of QM. The introduction of a hierarchy of 'observers' (exosystems) possesses the merit of fully exposing the exceptional nature of QM formalism. Such an analysis as that made earlier by Dalla Chiara [1977] is important for a better understanding of the logic of measurement, but places too little stress on those properties of QM formalism which guarantee an astonishing consistency of the quantum theory of relativized information. First, it is instructive to see more closely how that internalistic logic of correlation in the quantum blockworld works, so let us present Rovelli's radically relational approach in a nutshell.

We can consider the simplest possible situation of the measurement conducted by 'observer' $O$ (who can be *any* subsystem inside $\boldsymbol{\Omega}$ conducting the extraction of the information) on quantum system $U$, when the measured quantity is the observable represented by operator $\hat{A}$ on $\mathcal{H}_U$ with two eigenstates, $|1\rangle_U$ and $|2\rangle_U$, associated with two eigenvalues: $a_1$ and $a_2$. $O$ 'measures' the quantity associated with $\hat{A}$ on $U$ at the moment $t_1 + \delta t$ and can obtain two possible outcomes, which are its 'states': $|1\rangle_O$ and $|2\rangle_O$. Next, let us assume the existence of another 'observer', $P$, who is external with regards to both the earlier systems (they are its endosystem), knows their initial 'state' together with their dynamics (and hence is able to construe the equation of evolution for $O \oplus U$) and measures their quantum 'state' only at some later moment $t_2 + \delta t > t_1 + \delta t$ (that is at time $t_2 - t_1$ no interaction takes place between $P$ and $O \oplus U$). We are thus dealing here with an identical situation to the one described in Ch. 4.2: to observer $O$ the 'state' of $U$ prior to measurement at moment $t_1 + \delta t$ is merely a potential (geometrically represented) structure of certain non-actual events as $U$ remains an isolated whole, whereas for external observer $P$ it is the 'state' of $O \oplus U$ prior to the measurement at $t_2 + \delta t$, which is such a potentiality, as $O \oplus U$ remains to be an internally correlated, isolated compound *in a frame of P* (the correlations between subsystems $O$ and $U$ form a nonseparable whole in that frame). Rovelli, just as von Neumann, is perfectly right to stress that that is a fundamental, structural property of QM. Let us now see more closely following Rovelli's example what the correct quantum description of a series $\mathcal{E}$ of 'events' (:actions), called an 'experiment', looks like from the perspective of each subsystem.

At moment $t_1 + \delta t$ a local physical interaction takes place, as a result of which $O$ registers in its local quantum frame the 'state' of $U$ as $|1\rangle_U$ or $|2\rangle_U$. Earlier on, on the basis of the knowledge $O$ has about $U$, that relative 'state' in a frame of $O$ was the whole of all the possible results, defined by the structure of all the purely internal relations between them, that is the 'state function', as: $|\psi\rangle_U = \alpha|1\rangle_U + \beta|2\rangle_U$ (where $\alpha$ and $\beta$ are in general complex numbers, and $|\alpha|^2 + |\beta|^2 = 1$). Thus experiment $\mathcal{E}$ from the perspective of $O$ has to be described as the following sequence:

**[O]**

$$t_0 \rightarrow t_1 + \delta t$$

$$|\psi\rangle_U = \alpha|1\rangle_U + \beta|2\rangle_U \rightarrow \begin{cases} |1\rangle_U \ \textit{with probability} \ |\alpha|^2 \\ |2\rangle_U \ \textit{with probability} \ |\beta|^2 \end{cases}$$

It is quite easy to become accustomed to the algorithm presented above, as long as we are not interested in penetrating the meaning of the 'measurement action' – $\mathcal{E}$

simply refers to the expected outcomes which QM formalism admits, and von Neumann's projection postulate for $\mathcal{H}_U$ can be treated solely as an element of a geometrical language enabling the formulation of those predictions. What is most interesting, though, is to consider the description of $\mathcal{E}$ given by $P$, who knows quantum-mechanical rules and the initial relative 'states' of $O$ and $U$, which come into a dynamical interaction correlating their degrees of freedom. The initial state of $O$ can be represented by vector $|init\rangle_O$, and $P$ knows that $|init\rangle_O$ evolves in such a way so as to save its correlation with $U$ in the whole $O \oplus U$. Hence, in accordance with the general theory of measurement we have indicated in Ch. 4.2, the particular 'states' of $O$ will refer, thanks to some inference map, to the particular 'states' of $U$: in such a sense, to use the Platonic expression, those states are *sym-metrical* (σύμμετρα) within the whole $O \oplus U$, that is *commensurate*, due to a certain holistic measure (μέτρον), without which we would not be able to speak about their correlation at all. In such a situation, if the 'state' of $U$ is given as $|\psi\rangle_U = \alpha|1\rangle_U + \beta|2\rangle_U$, then $P$ in a simple way describes the course of $\mathcal{E}$ as: $|\psi\rangle_U \otimes |init\rangle = (\alpha|1\rangle_U + \beta|2\rangle_U) \otimes |init\rangle$. In accordance with the correlation rule, we thus obtain a local, relative description of the whole experiment $\mathcal{E}$ given by $P$ for a closed system $O \oplus U$:

**[*P*]**

$$t_0 \rightarrow t_1 + \delta t$$

$$\left(\alpha|1\rangle_U + \beta|2\rangle_U\right) \otimes |init\rangle \rightarrow \alpha|1\rangle_U \otimes |1\rangle_O + \beta|2\rangle_U \otimes |2\rangle_O$$

The descriptions of $\mathcal{E}$ given by $O$ and $U$ are totally different, as the simple question, 'does an observable represented by $\hat{A}$ have a definite value at $t_1 + \delta t$?' will receive completely contradictory answers from the observers: for $O$ the value will naturally be uniquely definite, while for $U$ it will not. The contradiction is so clear that the first suspicion to be put forward in that situation is an erroneous application of QM formalism – and still it is easy to find out that no mistake has been made in the whole reasoning and both descriptions have been construed correctly, true to QM rules. (The above record of the course of $\mathcal{E}$ does not result from an abuse of formalism, as exactly that kind of construction is required by the linear rules of the description of $\mathcal{E}$ in QM). The first thing we notice – as Rovelli further argues – is the fact that QM itself does not contain any instructions as to how to make the cut between the endosystem and the exosystem, that is, there is no criterion allowing us to assess whether one of the two descriptions given is more 'adequate' than the other (if such instructions existed, they would certainly not come from QM itself as it is formulated now). All quantum frames of action are equivalent. In description [***O***], the Heisenberg–von Neumann cut is made between subsystems $O$ and $U$ in the quantum universum $\mathbf{\Omega}$, while in description [***P***] the cut is made between $O \oplus U$ and $P$ in the same universum, that is in each of those instances the exosystem (the 'observer') locally performs the logical role of a 'metatheoretical object'. Rovelli concludes that the most important lesson we should begin to learn from that situation is the realization that in QM *different internal 'observers' (subsystems of* $\mathbf{\Omega}$*) can provide radically different descriptions of the internal sequences* $\mathcal{E}$ *of events, as the latter are physically relative.* (That is a situation quite analogical to the one in the case of the registration of collections of classical events in **M**: it seems that different 3-dimensional systems, 'spatial observers', see the rod in space as 'one and the same object', but that is *not* so, which provokes some resistance from the Newtonian three-dimensional imagination, cf. Ch. 4.1). That is hard to accept, unless we do admit the possibility of moving away from strong, noncontextual realism – otherwise we have to assume the incompleteness of QM and seek the sets of *local and noncontextual (:fully separable)* variables $\lambda$, having the sole task of reintroducing the separable 'elements of reality' back into the physical ontology. But that strategy seems irrelevant, apart from the fact that those sets in QM cannot exist, as the structure of the paradox, which becomes visible here, directly resembles the universal Eleatic structure from Ch. 1.2: the introduction of the absolute in the form of an 'objective state of affairs' into the internalistic mechanics of quantum $\mathbf{\Omega}$ forces us in the end to accept the contradiction in the sphere of cognition and/or reality, or to solipsistically negate those very phenomena. However, there exists, as we have seen, a natural counter-strategy pointed out by the firm application of the Universal Principle of Relativity.

To a physicist who uses Bohr's epistemological approach, the dilemma of the contradictory descriptions [***O***] and [***P***] may seem an unnecessary complication, as it assumes that the localization of the Heisenberg–von Neumann cut in $\mathbf{\Omega}$ is not quite random. The 'Bohr cut' singles out measuring device *O* as 'classical' (though also constituted by quantum processes), whereas *U* as a quantum microsystem, which – if it is subject to measurement – has to be finally described in the language of classical physics [Heisenberg, 1969, p. 180*f.*; 1959]. The conclusion is that it is pointless to analyze the correlated quantum state $O \oplus U$ from the perspective of another non-classical system *P*. Though at first such an instrumentalist solution might seem convincing, nevertheless after almost a century of the development of experimental quantum physics it is difficult to retain it in that form: the apparent gulf between microscopic and macroscopic systems is being gradually obliterated, quantum systems on a macroscale are commonly produced and there is no possibility of assuming *a priori* some fixed canonical Bohr cut in the physical world (not to mention the problems in quantum cosmology). As experimental physics is not limited by any type of such a cut, there is no reason to assume that any physical system is by nature classical, which would arbitrarily limit the universal applicability of QM. Every exosystem in interaction is, logically speaking, a 'metatheoretical object' in relation to an arbitrary endosystem, but that by no means implies that QM ceases to be applicable in some domain of **Ω**. Quite an opposite situation takes place, as the fact that with the help of QM one can precisely construct descriptions **[*O*]** and **[*P*]**, constitutes its exceptional strength: it plays the role of a metalanguage in **Ω**. To accept those properties of formalism requires a decided relativization of those descriptions, which is the most natural way to eliminate the contradictions. The question 'does an observable represented by $\hat{A}$ possess a definite value at $t_1 + \delta t$?' is pointless in QM, just as there is no sense in the incomplete question 'what is the velocity of a body at moment $t_0$?': for them to have sense at all, one should provide the frame of reference (frame of action), for which they are formulated, i.e., one has to eliminate absolute action and the 'state' that have no operational and empirical meaning. In the case of QM, such methodically applied procedure leads to a relativization of the actuality as such, as what is said to be real, i.e. given at the moment as the existing 'state of affairs' for *O*, need not be the same for *P*. We should give up the classical complete definiteness required by Einstein's Local Realism in favour of a relativized definiteness of the observables *in the frame of action* introduced by Local Contextual Realism (see Ch. 3.2). Local quantum 'states' as computational constructs encapsulate purely relational and local information (which is never superluminally transmitted), but their full relativity is enforced by the nonseparability of **Ω**, appearing locally in the nontrivial correlation rules.

Thus the adequacy of the relative descriptions given by the particular observers is connected to the coherence of QM itself or, ontologically, to the unique quantum geometry of **Ω** itself. QM meets those expectations and that basic condition, as long as we treat it not as a theory which predicts 'what the objective state of affairs is in the world out there', but as a metatheory of the correlation structure of fundamental actions between arbitrary subsystems in **Ω**. Its most important property is the fact that the formalism guarantees a global consistency of the descriptions in the different frames, i.e., it codes those relations at a very fundamental logical level (through a quantum logic of the actions). Quantum formalism therefore operates in such a way that 'von Neumann consistency' (Rovelli) is met even when the Heisenberg–von Neumann cut is chosen in completely different ways inside **Ω** (it was John von Neumann [1932, pp. 222–237] in fact who first drew attention to that fact). In effect, the local information processed by different subsystems of **Ω**, which enter into interactions, that is which evince correlations between them, is on principle self-consistent by the rules of the quantum universe. In the case above discussed, *P* at moment $t_2 + \delta t$ may perform its own extractive action on system $O \oplus U$ and find out that it will indeed obtain only one of the two possible results, expected by the logic of the correlation: $|1\rangle_U|1\rangle_O$ or $|2\rangle_U|2\rangle_O$, but certainly never $|1\rangle_U|2\rangle_O$ nor $|2\rangle_U|1\rangle_O$. $P$ will therefore not conclude that QM contains an incoherence, as that same simple logic of the correlation works in the case of the prediction of the outcome of the extractive action on $U$ by $O$: when $O$ and $P$ pass on information to each other as a result of the physical interaction, their descriptions will be compatible – we can treat that property as a fundamental property of the relational structure of QM [Rovelli, 1996]. Within that structure, different observers give different descriptions of sequences $\mathcal{E}$ in their *ontic* frames of action, but it is that very hidden whole of $\mathbf{\Omega}$ which guarantees that the structure of relations between those descriptions is self-consistent on a fundamental level. There does not exist any classical 'description-from-nowhere' nor 'absolute action' inside $\mathbf{\Omega}$, but every one of them is conjugated to the given arbitrary, local cut in $\mathbf{\Omega}$, without which the definiteness of the quantum observables and the 'state of affairs' themselves have no physical sense. What results from internalistic relational compositionism – and that is a radicalized antipresentist relativity of SR and GR – is that every such local action–correlation is nonseparable from the global structure of $\mathbf{\Omega}$, i.e., the hidden quantum history, which cannot be retrieved by any subsystem as a cosmological-mechanical 'state', and even less an 'evolving state'. The relational and time-symmetric structure of QM simply becomes much more visible, when we look at QM as a metatheory of the irreducible frames of nonseparable actions inside blockworld $\mathbf{\Omega}$ and of their global, non-classical geometry.

The ontology of internalistic compositionism of the Platonic type, physically interpreted as a nonseparable blockworld, requires the temporal description of quantum processes to be completely relativized, connected solely to local subsystems in $\mathbf{\Omega}$. That of course suggests a rather exotic (from the commonsense perspective), holistic ontology, in which the fundamental quantum actions are more primary than the derivative spacetime characteristics and at the same time nonseparable from the global $\mathbf{\Omega}$ structure. As we have seen, the ontological solution of that type has constituted the basis of Leibniz's metaphysics of nature, where time and space are phenomenal forms (abstractions) of the much more primitive global order of pre-established actions – it is a simple, classical-holistic predecessor of the quantum $\mathbf{\Omega}$. Since Leibnizian 'monadology' is a system that is still rooted in the old tradition of the metaphysics of the changing 'substance' and 'intrinsic nature' (which has already been polemically stressed by Whitehead), and 'monads' are not interactional, it is surely not an adequate model for QBU. Likewise, internal quantum actions, even relativized, are always prone to be understood in a quasi-presentist fashion with some pared-down 'becoming', precisely as many would seek to save some form of the relativized presentism for SR. We find here a problem analogical to the issue of the observers' local time in SR/GR – when we analyze the descriptions of sequence $\mathcal{E}$ given as [$\boldsymbol{O}$] and [$\boldsymbol{P}$], one is confronted with the question from the end of Ch. 4.1: what is the meaning of variable $t$ and what does it really refer to in the formal record of the 'evolution' inside blockworld $\boldsymbol{\Omega}$?

When we pose that question in the domain of theoretical physics, we still remain in the realm of hypotheses and speculative attempts at solving the so-called problem of time, but that should not be a reason to become discouraged, particularly if the metaphysics itself can provide certain working intuitions. Regardless of the philosophical motivation of that question, it is also a serious issue concerning the foundations of quantum cosmology and generally covariant field theory since the stationary Wheeler–De Witt equation was formulated in 1967 for a closed universe, for which the complete energy had to be equal to zero ([De Witt, 1967]; cf. e.g. [Carlip, 1998, pp. 131–142], [Barbour, 2001, p. 238*ff.*]). In such a universe, a pre-relational model of QBU, obviously no global change takes place in the sense that 'change' is locally understood: all the values of the particular variables are timeless and the quantum state of such a universe cannot in any way depend on time for exactly the same reasons for which it is absurd to claim that 'spacetime changes in time'. The Eleatic paradox is evident here, as in the parametrized systems with Hamiltonian constraints the physical observables are defined by operators which commute with all those constraints, and as for the quantum universe the global super-Hamiltonian (functional) $\boldsymbol{H}_{\Omega}$ is defined, itself playing the part of such a constraint, hence all of the Dirac observables are here simply constants of motion with regards to the Wheeler–De Witt constraint operator, i.e., their Poisson brackets vanish ([De Witt, 1967], [Wheeler, 1968], [Brill, Gowdy, 1970], [Kuchař, 1992]). Here spacetime and kinematics are merely phenomenal, and quantum $\boldsymbol{\Omega}$ in a pure state is indeed a hidden Platonic-like whole. When postulating a QBU ontology based on internalistic compositionism, we have to make an important choice, which has serious theoretical consequences for the construction of the quantum theory of gravity: one can either undertake to incorporate (minimally, at least) the time structure and evolution (weak 'becoming') into fundamental formalism, or to wholly eliminate them from it. I think that a consistent internalistic compositionism – holism of the Platonic type – requires a decided adoption of the latter strategy. QBU is essentially Parmenidean, even when its global structure can be interpreted by us as a relational structure of local actions/interactions as *correlations* – speaking of a 'quantum-dynamical universe $\boldsymbol{\Omega}$' by no means excludes the elimination of a real, physical change (a passage of time) from the basic layer of the theory. Time itself cannot be consistently injected into that changeless layer composed of global, atemporal histories. Not only the very mechanical concept of 'quantum state' is problematic; even more so is the concept of the 'quantum state *at time t*', which is perhaps nowhere demonstrated as pointedly as in the experiments of the type analyzed in Ch. 2.3. A hidden 'state' of the Universe would rather be a nonseparable global history as *the* 'law' which cannot be reconstructed or represented from the inside of $\mathbf{\Omega}$.

Thus the key issue concerns the concept of time in QM itself, i.e., the issue of treating independent variable $t$ in the description of experiment $\mathcal{E}$. It seems *prima facie* natural for time to be an observable, and hence to be represented in QM by a certain self-adjoint linear operator $\hat{T}$ on $\mathcal{H}_{\psi}$ with a spectrum topologically isomorphic to the one-dimensional manifold $\mathbb{R}^1$, but unfortunately – in accordance with Pauli's old argument – its quantum conjugate observable, energy, possesses a spectrum that is limited from the bottom, and hence it cannot be represented by any such type of operator $\hat{T}$ and it is not a quantum observable itself [Rovelli, 2002b]. In recent years, the problem has been noticed once again on the occasion of measuring the probability of the time-of-arrival of a particle to a given area of space, where what is being calculated from the state function $\psi(x)$ is the probability density $\pi(T)$ that the particle has reached $X$, when the probability itself is given as: $\int_{T_1}^{T_2}\pi(T)dT$. The simple question posed by the metaphysician is: how is it possible that *time itself is not an observable, but many would claim that it still seems to be measured?* What is worse, if – as we have analyzed before – observables in a quantum fundamental theory are *always* relational quantities (coding particular correlations), then how is it possible in the case of variables apparently dependent on time that we are able to establish such correlations with an unobservable and independent quantity at all? Rovelli and others (T. Thiemann, B. Dittrich, M. Montesinos) point out that a whole series of misunderstandings connected to the role of time in QM and quantum cosmology has its source in a certain subtle, categorial confusion which only seems to be bereft of physical significance. Perhaps a pervasive involvement of the naïve commonsense intuitions is evident nowhere as clearly as in the case of the status of time, hence any QBU ontology, and in general every cosmology based on internalistic Platonic compositionism, automatically raises such general objections. As I have pointed out in Ch. 3.1, at the bottom of the Pandora's box of the 'EPR paradox' lurk the separable (noncontextual) 'elements of reality'; on the other hand, as I have made clear in Ch. 4.1, at the bottom of the Pandora's box of the ideology of 'becoming' lies time as an allegedly special, singled-out variable. But even the machinery of classical mechanics for constrained (gauge) systems does *not* demand that, thus the ideology is surprisingly faint.

Rovelli–Thiemann–Dittrich have probably worked that out most clearly. The key distinction to be made when attempting to understand the 'problem of time' in a quantum cosmology concerns the categorial difference between partial observables and complete observables. Partial observables can be understood as ones where it is possible to associate a certain measuring procedure providing a definite, numerical result, but the theory does not allow for its prediction; they are just *bare* (*non-gauge invariant*) numerical values, which can never be considered to be 'elements of reality', but are a raw base for a kinematical framework. Complete observables are such where the theory enables the prediction of either their definite value, as in classical mechanics, or the local probability distribution for its value spectrum, as in QM; they are *gauge invariant correlations*, or better, *correlation functions* on partial observables, completely defining their dynamics/evolution ([Rovelli, 2002b, p. 124013–2; 2011], [Dittrich, 2007]). All the basic quantities of the empirical theories, without which they would not have any sense and which are directly measured with the help of particular measuring devices in their frames, such as 'position in space' ($q$) or 'clock time' ($t$), are partial observable values, which in principle cannot be predicted, but only registered. The families of observables which are based on them as they depend on the relations between partial observables, are complete observables of a general form $N(x)$, e.g. $q(t)$, whose values can be predicted, but not registered (as a rule those dependencies are supplemented with a certain number of additional parameters present in dynamical equations). From the physical point of view, there does not exist any *a priori* hierarchy of partial observables, hence we have to treat them in mechanics on the same footing, though the way they are habitually treated would rather tend to make one speak of an 'independent' partial observable, such as $t$, and of a 'dependent' partial observable, such as $q$. On the basis of such observables being on the same footing, one may construct an extended configuration space $\mathcal{C}$ – and the extended phase space $\mathcal{V}$ associated with it – in which it is possible to formulate both classical mechanics and QM (e.g. Hamilton–Jacobi formalism, path integrals or propagator theory in the QFT [ibid., p. 124013–5*f*.]). Rovelli is nevertheless right to stress that in nonrelativistic theories the structure of 3+1-spacetime is treated as fixed, which obliterates the distinction between those two categories of observables and instead there is a fixed structure on the space of the observables that is defined. The situation changes completely in GR, where one can no longer speak of an evolution of objects in a fixed, absolute time $T$, measured by an independent external clock, and hence it is no wonder that a disavowal of the distinction of those categories of observables leads also directly to confusion in the canonical quantization of gravity. In GR the systems are described with the help of variables which 'evolve' relationally (with regards to one another), hence there is no sense in an *a priori* distinction of 'independent' partial observables from 'dependent' ones – e.g. in a certain relation $f$ of readings $t_1$ and $t_2$ of clocks in two different accelerating frames, none of them will beforehand play the role of an independent or dependent observable. In GR, complete observables in turn are functions on an extended phase space $\mathcal{V}$ of the system, commuting with all the constraints of the first class, including the Hamiltonian one.

In QM, the situation becomes somewhat more complicated, as much depends on the analyzed picture: in the Heisenberg picture all the operators (as ones that evolve in time) represent complete observables, while in the Schrödinger picture, the theory requires one to define the external time as an independent partial observable for the whole system having a 'state' evolving in such a time. In accordance with the previous conclusions, we should not expect QM to provide a quantum description of 'time', as value $t$ is only a partial observable and it cannot have any operator representation on $\mathcal{H}_\psi$ – QM simply does not deal with the variables $t$ or $q$ themselves (just as classical mechanics), but with correlations between them, which are provided as complete observables by operators on $\mathcal{H}_\psi$. The categorial confusion, which is perhaps not a serious one in classical physics, on a fundamental level of the description of nature thus proves to be the source of grave misunderstandings. We might expect that the relational description of quantum time should be possible to construe, but only if time is treated as a complete observable $T(t)$ – the clocks whose readings we are comparing are after all also quantum subsystems in $\mathbf{\Omega}$, between which a certain correlation is being established subject to purely quantum fluctuations. And indeed, if we construct a certain complete observable in GR of the general form: $T = \tau_1(\tau_2)$, where $\tau_1$ and $\tau_2$ are proper times (and partial observables), it turns out that it is represented as a function on the extended phase space $\mathcal{V}$ of the system, it does not depend on the parametrization of each of those readings taken separately and that it commutes with all the first class constraints (including the Hamiltonian ones) [ibid., 124013–4*f.*]. The first relational theory of that type was proposed by Don N. Page and William K. Wooters [1983] already in the 1980's: a clock could be a physical quantity, which serves as an argument of the function relationally describing the 'evolution' of a different quantity. The standard objection an opponent of relational thinking may raise boils down to highlighting that if $T$ is an observable, then so are $\tau_1$ and $\tau_2$, and hence they should be represented by appropriate operators and should commute with Hamiltonian constraints, and in the general covariant theory for $\mathbf{\Omega}$ also with an operator of Wheeler–De Witt's type. As we can see, though, such an accusation results from a mistaken treatment of the observables themselves, as $\tau_1$ and $\tau_2$ are *not* complete observables, that is mathematical objects of the theory capturing the patterns of constrained correlations (just as neither are the chosen coordinates and gauges), hence it is pointless to require them to commute with any operator. Quantum clocks are *fully relativized clocks inside blockworld* **Ω**, and they should be defined purely relationally by locally entangled relational actions. Such an idea has been recently realized experimentally on a very simple quantum system of two photons [Moreva *et al.*, 2014] – a beautiful piece of the experimental physics of time.

Putting it shortly, the problem of understanding time in a quantum universum **Ω** results from a reluctance to the Universal Principle of Relativity and to relational thinking with regards to certain saved absolutes. If we treat that principle consistently, though, we should remove any fundamental differentiation between independent and dependent variables among the partial observables, and accept that there is no sense in the concept of an intrinsic 'objective state', independent of the observers' action frame in **Ω**. The phenomenological 'time flow' should only be local and be defined on a fundamental level solely with regards to arbitrary Heisenberg–von Neumann cuts and with regards to the dense correlational (entangled) 'states' connected with them, that is, to put it differently, it should be possible to reconstruct it completely in the framework of elementary quantum thermodynamics (as is the case of e.g. the wave–particle duality [Coles *et al.*, 2014]). That has far-going consequences for the understanding of QM, as both classical Hamiltonian mechanics and Schrödinger's wave mechanics are habitually interlinked with the intuition of absolute time, and sometimes even with presentism, further moulding the classical understanding of the probabilities as the actual dispositions, 'propensities' or 'objective opportunities to come about in the future' (cf. [Rovelli, 1991, p. 453*ff.*]), or naïve interpretation of correlations as changing 'elements of reality'. If we accept the fact that defining time as a preferred observable not only at the Planck scale, but also in all of quantum **Ω** loses its physical sense (in particular, we abandon using the time-dependent Schrödinger equation in the description of quantum-gravitational states), then QM has to be interpreted in the Platonic manner as a *theory probing a global structure of* **Ω**, and not mechanics of the world 'coming into being' in time, which allows us to understand many of its startling, at first glance, properties. That Parmenidean physics, a quantum extension of the Rovelli–Thiemann–Dittrich approach, requires nevertheless a reinterpretation of the meaning of probabilities (their generalization in the context of noncommutative quantum measure theory and the complete divergence from any presentist intuitions) and a fully time-symmetric formulation of QM (different from the time symmetry of GR and requiring no modification of the standard formalism operating for local quantum frames). Quantum probabilities cannot be a reflection of the intrinsic 'dispositions' of the fixed micro-objects, as the latter simply do not exist and cannot behave or 'become' due to those inherent 'propensities' as if they were micro-stones or micro-cats from the Aristotelian or Popperian cosmos. That makes the very formulation of the 'timelike separability'/'time-asymmetric independence from history' condition $[3]_B$ even more suspect as has already been clear in the case of the quantum eraser experiment. With the joint use of Finkelstein and Rovelli's mottos with regards to mechanics one should say: *quantizing is a deepening of relativity, and the deepening of relativity is forgetting the flow of time.*

Proposals going in that direction, earlier considered to be speculative, are extremely intriguing from the perspective of a metaphysican of nature, if only for the reason that they overtly or tacitly assume the ontology of internalistic (structural) compositionism, and what is more, aim in the direction of a unification of GR and QM exactly on the basis of procedures of a fundamental relativization. When mentioning in Ch. 2.3 certain modelling conceptions of the blockworld as a 'hidden whole' of the structure of actions, I have stressed that they introduce a fundamental atemporality: the temporal parametrization is here always connected to the local (ontic) perspective and results solely from relations between the singled-out actions *defining* such a perspective. Leibniz, for example, analyzed – as primary – the global relations of a mutual 'representation' of actions in the framework of their 'hidden harmony' (timeless 'ichnography' described by global symmetries), and was only secondarily interested in the empirical phenomenon of time from traditional kinematics. The conclusion is that if we have at our disposal any quantum 'algebra of the process', representing an atemporal logic of actions–relations in blockworld $\boldsymbol{\Omega}$, then it should contain an elementary formal property which can be physically interpreted as a *purely statistical* representation of the local 'time flow' for a singled-out frame. The idea has been stirring a certain interest in quantum thermodynamics for the past dozen or so years, as the fundamental von Neumann $C^*$-algebras $\boldsymbol{\mathcal{A}}$ of observables, which I have mentioned in Ch. 3.1 in the context of AQFT, possess exactly such properties. The formal, statistical representation of 'time flow' in $\boldsymbol{\mathcal{A}}$ has become the cornerstone of the thermal time hypothesis, put forward in the 1990's by Alain Connes and Carlo Rovelli ([Connes, Rovelli, 1994], [Rovelli, 2004, p. 140*ff.*; 2011, p. 1486*ff.*]), which might be worth presenting here in greater detail from the technical side as an illustration of how any quantum 'algebra of the process' or algebra of operators can at all generate a statistical representation of the local time (strictly dependent on an arbitrary relational 'state' of the quantum system, provided in general by a density matrix $\varrho$). The hypothesis is perhaps the key component of every future fundamental theory which, by rejecting presentism, accepts a certain variant of QBU ontology and internalistic compositionism. A thermodynamical reconstruction of the phenomenology of the 'passage of time' and the geometry of time is also a necessary condition to demonstrate the effectiveness of the Universal Principle of Relativity in the framework of the quantum gravity theory. In QBU ontology, the key issue is thus a Platonic strategy: an algebraic reduction of that phenomenology to fundamental, strictly quantum (noncommutative) pre-dynamics without time, which on the one hand extremely relativizes, more deeply than GR, all the physical 'states' and Heisenberg–von Neumann cuts, and on the other is global and nonseparable (with reference to the quantum $\boldsymbol{\Omega}$ the scheme of externalistic compositionism, built upon separable base states with open future, completely breaks down). Putting it shortly, the Platonic strategy is based on an attempt to reconstruct the elementary algebra of internal actions–correlations in $\boldsymbol{\Omega}$, but it requires neither any passing time nor, even more so, *any* temporal background.

Generally speaking, the idea is based on extending the concept of 'time flow' onto generally covariant field theories, but in such a way that the passage depends on a thermodynamical characteristic of the quantum system and is described by a one-parameter unitary group of *-automorphisms on the observable algebra in the generalized Heisenberg picture. In the case of a non-generally covariant theory, 'passage of time' $\alpha_t$ is statistically coded in formalism as the automorphism of the Hamiltonian flow in phase space $\Gamma$ by nonrelativistic Hamiltonian $\boldsymbol{H}$ and by equilibrium thermal states $\varrho$, which are Gibbs states: $\varrho = \exp(-\beta \boldsymbol{H})$, where $\beta$ is a constant. In the case of a generally covariant theory, which cannot single out any preferred local Hamiltonian and partial time variable, we presume that the 'flow of physical time' $\alpha_t^{\varrho}$ (i.e., *thermal time* for $\varrho$) should be naturally generated by an arbitrary statistical 'state' $\varrho$ itself and the corresponding Hamiltonian flow, but in such a way that with the shift to Gibbs states the equations should become reduced to a limit case, i.e., to Hamiltonian equations for $\Gamma$ [Connes, Rovelli, 1994]. In CBU, which is globally described by a certain generally covariant geometric structure, the 'flow of time', registered locally by observers using some convenient clocks inside it, will be for instance the result of the observed statistical state of the thermal environment, in particular of the cosmic background radiation, which would give the so-called Friedman–Robertson–Walker time [Rovelli, 1993]. The property of von Neumann $C^*$-algebras, which is of key importance for QM and expressed by the fundamental Tomita–Takesaki theorem, is the connection between every positive operator $\omega$ on $C^*$-algebra $\mathcal{A}$ (e.g. the density matrix on $\mathcal{H}$ as a quantum 'state') with a strongly continuous, unitary one-parameter group of *-automorphisms $\alpha_t$ of such an algebra (being, strictly speaking, a so-called modular group for real $t$). State $\omega$ on $\mathcal{A}$ is given as a normalized positive linear functional on $\mathcal{A}$, $\omega(A) = Tr[A\omega]$, where $A \in \mathcal{A}$. It is exactly that connection between the algebraic morphism of the Tomita flow and the 'flow' of thermal time that lies at the basis of that conception in the AQFT context. For any operator $A \in \mathcal{A}$ we have a general equation of the Tomita flow (modular flow):

$$\alpha_t A = \Delta^{-it} A \Delta^{it} \tag{40}$$

where $\Delta$, the modular operator for the pair $(\mathcal{A},\psi)$, is a certain self-adjoint, positive operator dependent on the choice of a particular $\psi$, called a cyclic and separating vector in $\mathcal{H}$ for a given $\mathcal{A}$ ($\psi$ automatically generates a faithful normal state in $\mathcal{A}$), and real $t$ is *Tomita time* (formalism and more details e.g. in: [Bratteli, Robinson, 1987, p. 83*ff*.], [Connes, Rovelli, 1994]). In the quantum-field case, after making some assumptions ($\omega$ on $\mathcal{A}$ should satisfy the so-called KMS condition [Haag *et al.*, 1967] and we have a system with finite degrees of freedom) and for a given nonrelativistic Hamiltonian $\boldsymbol{H}$, we obtain a statistical Gibbs state as a matrix: $\omega = N \exp(-\beta \boldsymbol{H})$, hence knowing Gibbs state of the system we are able to reconstruct its dynamics. If we are interested in a generalization to a generally covariant case for AQFT, in which – as we assume – no universal time nor any local dynamics could be singled out as preferred, then we can use the Hamiltonian: $\boldsymbol{H} = -ln\varrho$ for the thermodynamical distribution of microstates $\varrho$, which also generates Tomita flow $\alpha_t$ interpreted as a flow of quantum thermal time for KMS state and inverse temperature $\beta = 1/T$ [Rovelli, 2004, p. 142].

In the general thermal time hypothesis one therefore assumes that the phenomenal local 'flow' of physical time has a purely quantum-thermodynamic source and is represented in the quantum algebras of observables by a modular flow $\alpha_t$ for an arbitrary thermal state $\varrho$. The modular flow, represented by an external automorphism in $\mathcal{A}$, is – as Radon–Nikodym theorem says – a canonical property of the algebra itself (regardless of the particular proper states $\omega$), which in that sense, as the logic of quantum observables, is 'pre-dynamical', i.e., it codes on the elementary level quantum-statistical dynamics in $\boldsymbol{\Omega}$ [Connes, Rovelli, 1994]. We can therefore speak sensibly of a generalized dynamics in blockworld $\boldsymbol{\Omega}$, which in local, relativized quantum frames gives a purely thermodynamic effect of a 'time flow' for *statistical* states. In that perspective it is pointless to talk of an absolute 'passage of time' for any generally covariant quantum theory, just as there is no sense in the assumption that nature prefers some particular 'state' $\omega$: local physical time depends solely on a relational quantum 'state'/relative correlations, i.e., if a system in a certain quantum frame is in state $\omega$, then its quantum proper time is given by flow $\alpha_t^{\omega}$ (in a special case, equilibrium Gibbs states are such whose modular groups are simply time translation groups) [Rovelli, 2004, p. 143]. That means that all the information about the observed quantum dynamics, which is normally coded by a Hamiltonian, is also contained in a thermal quantum state of the observer's environment and one can thus purely statistically 'recover the time' in QBU [Rovelli, 2011, p. 1484], completely eliminating its assumed fundamental character: 'Time is, that is to say, the expression of our ignorance of the full microstate' [ibid., p. 1486]. What is more, in line with the 'Platonic strategy', in the context of a generally covariant theory and QBU as its ontological interpretation, that thermal time is *the only one* that we can in fact have at our disposal.

We can now consider a uniformly accelerating observer $O$ in $\mathbf{M}$ moving along a certain spacetime trajectory and thus demarking Rindler wedge $\hat{\mathrm{R}}$ (a globally hyperbolic space), for whom one can define an appropriate $C^*$-algebra of local observables $\mathcal{A}_R$, defined solely by a trajectory ($O$ is then the so-called Rindler observer, and the edge of the wedge is a *causal horizon* demarking $\hat{\mathrm{R}}$ as an independent spacetime). It turns out that there is a direct connection between its geometric proper time $\tau_O$ in $\mathbf{M}$ and the thermal time given by the modular group and its geometrical action [Bisognano, Wichman, 1975], which are bound together by a proportionality factor, which is temperature $T_v$ of quantum vacuum measured by $O$, that is Davies–Unruh temperature[61]. That relation depends solely on the geometry of $\hat{\mathrm{R}}$ and an appropriate quantum algebra $\mathcal{A}_R$, which is a restriction of the complete algebra of quantum fields in $\mathbf{M}$ to wedge $\hat{\mathrm{R}}$, $x > |ct|$, ([Connes, Rovelli, 1994], [Martinetti, Rovelli, 2003]). The Fulling–Davies–Unruh effect depends on the fact that the stationary state of the Minkowski vacuum of the inertial observer is for a uniformly accelerated Rindler observer in $\mathbf{M}$ (with an appropriate restriction to $\hat{\mathrm{R}}$) a KMS thermal state $\varrho$ for inverse temperature $\beta$, when the time evolution and time $\tau_O$ are parametrically connected to an appropriate Lorentz boost, i.e., with translation along the accelerated worldline. Putting it shortly: the Rindler change of coordinates generates a maximally mixed vacuum state $\varrho$ and Tomita flow in $\mathcal{A}_R$, defined for it, which we then interpret as the local thermal time of observer $O$ strictly bound to its geometric proper time $\tau_O$. Davies–Unruh temperature $T_v$ measured by $O$ is a *relation* of those two times. Just as in the case of the inertial observer, we can reconstruct the dynamic state of the Rindler vacuum, given by field Hamiltonian $\boldsymbol{H}_R$ (for subspace $\hat{\mathrm{R}}$), when we know the thermal quantum state in $\hat{\mathrm{R}}$ given by density matrix $\varrho$.

The very idea that the 'flow' of physical time for the arbitrary subsystem inside $\boldsymbol{\Omega}$ should be completely determined by the flow generated by statistical quantum states is profound and of a great importance for the research at the meeting-point of QFT, quantum thermodynamics and cosmology. It also allows to interconnect

61 On Rindler spacetime see R.M. Wald's introduction [1984, p. 149*ff.*], while on the Unruh effect and the thermodynamics of spacetime with a horizon e.g. Wald's [1994, pp. 105–132] introduction or T. Padmanabhan's [2005] review.

some different conceptualizations of time (particularly the purely geometric one from GR and the information-thermodynamic one from quantum theory) with emphasis on the thesis that, globally, the universe as a closed system is a stationary quantum blockworld, whereas the phenomenal 'flow of time' has neither an absolute, nor even a fundamental character – it is therefore a truly consistent realization of the Parmenidean-Platonic principle in fundamental physics. Empirical 'observers', i.e., subsystems having physical clocks at their disposal which are subject to quantum-thermal fluctuations and whose environment always has a certain non-zero temperature, could in fact replace in their local frames of action all the information about quantum dynamics with information about the thermal state, and because of fundamental measurement limitations and relativity they will never be able to gain knowledge about the global 'state' of **Ω**, i.e., see the universe in a thermal state as 'frozen' or 'unchanging', just as they will never register the quantum unitary evolution in QBU in their frames. Information about the correlations inside **Ω** is for the Rindler observer fundamentally restricted by the causal horizon and its quantum thermodynamics (he/she loses quantum information, hence the value of von Neumann entropy as a measure of entanglement will generally be different than in the case of inertial observers), which has a direct connection with measured Unruh temperature $T_v$ and the phenomenon of the local 'flow' of time. The Unruh effect in itself is very important from the perspective of the Principle of Relativity in QFT, as it also presents the relativity of the 'particle' concept in reference to the movement of observers in **M** ([Fulling, 1973], [Wald, 1994, p. 116]), but that principle shows a much deeper aspect here: no local quantum time of an arbitrary observer can be singled out as preferred, as there is no preferred thermal quantum state $\varrho_0$ of the vacuum in **M**. The Universal Principle of Relativity from Ch. 1.2 and the general internalistic compositionism also work consistently in AQFT: just as it is not possible to obtain a quantum 'state' of universe **Ω** by the summing up of separable states $\varrho$, so it is impossible to get some global quantum time from local thermal times in local action frames (externalistic compositionism/holism of an Aristotelian type fails completely in this case, as always). The hypothesis of thermal time is also a strong argument for treating GR as a constructive theory in Einstein's sense and in tune with his intuitions (cf. Ch. 4.1), and thus of modelling spacetime **M** itself as an emergent entity – a purely quantum-thermodynamic phenomenon. To use Socrates's phrase, 'what is astonishing' (*Parm.* 129b1) in the fact that geometric Minkowski time in CBU cannot 'flow' in any sense, if the noncommutative microdynamics of the fully relativized quantum actions of the subsystems inside **Ω** does not need any temporal background at all, and geometry of **M** would be a global statistical effect. If Einstein field equations are seen as thermodynamic equations [Jacobson, 1995], the century-old worries about the alleged conflict between QM and SR/GR should be put to rest, but then QBU with relational histories as Platonic wholes is a necessity. 'Quantum theory is a deepening of relativity' (Finkelstein).

For historical Plato, 'a moving picture of eternity' (*Tim.* 37d5: εἰκὼ κινητόν αἰῶνος) inside the cosmos was a special physical clock in the form of planetary cycles (*Tim.* 38c–d), presenting 'forms of time' (εἴδη χϱόνου), while for us it is a generalized clock, i.e., a quantum thermal state (a dense informational state) of the environment of any correlated subsystem in **Ω**. We can repeat after Plato that when we speak about a 'becoming/changing universe **Ω**' or 'the past and future of **Ω**', we are committing a mistake being misled by the delusive subject-predicate structure of natural language: 'that what *is*' – authentic whole **Ω** – 'is not subject at all to any of the conditions which becoming has bestowed upon the things moving in the realm of senses' (*Tim.* 38a5–6). That is an old Parmenidean metaphysical intuiton, in which the universe is ἀκίνητον μεγάλων ἐν πείϱασι δεσμῶν (B 8 DK 26: 'unchanging in the limits of its mighty bonds'), i.e., it resembles a quantum version of the Leibnizian bundle of relational histories–'ideas'.

## 5.2 Carlo Rovelli's relational quantum mechanics and internalistic compositionism in fundamental physics

The most important conclusion from the analysis of the 'third observer problem' seems to be the necessity of taking into account the fact that there does not exist any unique, noncontextual description of sequence $\mathcal{E}$ of events on a fundamental level: QM requires an acceptance of a radical contextualization of observation right up to the *relativity of the actuality* itself. The introduction of quantum frames, with regards to which the actuality can be at all defined, is the best solution in such a situation, which additionally points to an intriguing relationship between the QM structure and relativistic physics. QM finally reveals its universality as a certain metalanguage, which offers an opportunity to maximally precisely define the formal relationships between descriptions of any $\mathcal{E}$ given by different exosystems in their quantum frames of action. We have also established that relational state $\psi$ (*not* being a noncontextual 'element of reality') of the compound quantum system $U$ is completely exhausted in internal correlations for all its possible resolutions into subsystems and for all the possible local observables defined on those subsystems: from the viewpoint of QM all the information about system $U$ is only that of a certain correlation, either 'internal', *relatively potential*, given *a priori* on the basis of a relational knowledge of the local history of $U$ and of QM rules, or 'external', *relatively actual*, established as the result of a physical interaction between $O$

and endosystem *U* in a given quantum frame[62] (hence those are two types of the observer's knowledge about the structure of the whole quantum situation, and the 'potentiality' and 'actuality' in Platonic **Ω** are always relative, i.e. quantum-contextual, which makes traditional, post-Aristotelian use of those notions useless, cf. along the same lines [Gambini, Porto, 2002]). QM is thus – as I have mentioned in Ch. 3.2 – not only a fundamental information theory, a theory of correlational interactions (information transfers) inside the closed Universe (cf. e.g.: [Fuchs, 2003], [Auletta, 2005], [Bub, 2005]), but first and foremost a relational and internally contextual theory. All those findings are a point of departure for the interpretation of QM put forward in the 1990's by Carlo Rovelli ([Rovelli, 1996; 2004], [Smerlak, Rovelli, 2007]; see also: [Brown, 2009], [Laudisa, 2001], [van Fraassen, 2010]). Its naturalness is a consequence of such an approach to QM which does not force it to confirm any commonsense expectations and intuitions about nature, but looks at the formal structure of the theory and the way it works, attempting next to comprehend them in simple and philosophically consistent categories (cf. also a similar original motivation in [Mermin, 1998b; 1999], or [Bub, 2005]). The strategy is worth attention, even if we are finally not ready to accept the present QM as a *principle theory* (in Einstein's understanding of the term).

Rovelli [1996], when creatively referring to the Copenhagen tradition and at the same time in a way simplifying it, begins with two assumptions, which contrasted with the labyrinthine development of QM philosophy seem rather simple and natural. The first one of them [ibid., p. 1644] results directly from the reflections on the 'third observer problem' and the meaning of the Heisenberg–von Neumann cut (which is a specific, physical realization of the Universal Principle of Relativity, analyzed by us in Ch. 1.2, in this case for quantum universe **Ω**):

> The Universal Equivalence of Quantum Systems
> All quantum subsystems of **Ω** are equivalent – there is no *a priori* physical difference between quantum and classical (macroscopic) systems. If system *O* can give a quantum description of system *U*, then it is always possible for another subsystem *P* in **Ω** to give a quantum description of endosystem $O \oplus U$ relative to *P* inside **Ω**.

---

62 Mermin's analysis, as we have seen, did not take into account the strong *ontic* relativization of the latter, which means that he did not consistently apply the Universal Principle of Relativity. That was also a case of Heisenberg [1959] himself, who was not ready to fully relativize his 'Tendenzen'/'Möglichkeiten' without any classical reference bodies and time asymmetry.

That postulate turns against the hypothesis of Bohr's 'canonical cut' (put forward by Bohr himself), which – as long as it is treated dogmatically – leads to enormous complications, as the history of the interpretations of QM in the 20$^{th}$ century reveals, including the mystery of the observer's consciousness (e.g. E.P. Wigner's ideas, Albert–Loewer many-minds interpretation etc.). The postulate of equivalence states that it is pointless to introduce *a priori* a canonical cut in universe **Ω**, to differentiate with its help certain internal absolutes in **Ω** (in Finkelstein's sense), particularly those concerning human observers. In that sense it is a quantum-mechanical generalization of Einstein's principle of relativity, encompassing the whole spectrum of internal physical actions in **Ω**. The postulate is natural, as it does not require going beyond the very effective QM formalism, which does *not* contain any clues as to the localization of the Bohr cut. Additionally, as we have seen, its consequence is a broad relativization of all the physical phenomena connected to the quantum 'state'–action of a system, e.g. the statistical 'time flow' (defined for any state $\varrho$) in the framework of the thermal time hypothesis.

The second postulate [ibid., p. 1650], giving rise, as we know, to many controversies is:

> The Local Completeness of Quantum Mechanics
> Quantum mechanics provides a complete and internally consistent framework for the local description of the physical world **Ω**, in full accordance with all the operations available and all the data accessible to local subsystems ('observers') inside **Ω**.

Its controversiality does not even depend on the suspicion that it would be in some way contradicted by future experiments (it states what is a fact, as the QM algorithms do not fail in any of the known instances), but rather on the abandonment of the pre-quantum mechanical reasoning which demands a supplementation of QM with a series of local-separable *classical variables* $\lambda$ which would simulate quantum statistics. That as such does not negate the existence of any $\lambda$ but only states that without their inclusion QM as a set of algorithms is self-consistent and complete in its present formulation, as it perfectly encaptures the fundamental properties of *local* microphysical processes, which cannot be, however, reduced to time-asymmetrically causal processes with separable states. One can imagine reformulated QM (as a constructive theory in Einstein's sense, without any revisions in the present QM structure) with the generic variables $\lambda$, yet they would be of an extremely non-classical and global (i.e. cosmological) character: in line with KST, internal variables $\lambda$ in **Ω** would have to be also strongly contextual, just as all physical 'states', which makes it impossible to recover completely classical states of any systems for QM. In accordance with the postulate of completeness, QM is already in its present form a kind of metatheory, which enables any physical 'observers' (subsystems) within the quantum universe **Ω** to formulate maximally complete descriptions and predictions in *their local frames of action*, whereas it does not enable us to obtain complete information about the physical 'state' of nonseparable whole **Ω**, which is inaccessible to its subsystems for fundamental reasons. The hypothesis also suggests that it is pointless to try and make QM similar to classical mechanics, for – as the example of old Bohmian mechanics shows, for instance – it will nevertheless then force us to accept the radically non-classical effects in the description of **Ω**, particularly the spacelike 'signal nonlocality' understood as an instantaneous action-at-a-distance ruining the causal structure of GR.

In Ch. 4.2, we reflected upon the meaning of the quantum 'state' and underlined that it had a double nature: an epistemic one (as none of the 'states' has any sense without providing the action frame/context in which it is defined) and ontic in the sense that it is not completely arbitrary, but defined by the physical structure of the whole quantum situation (in that sense *it is not* an individual 'state' as the property of some separable 'object'). Rovelli puts great emphasis on eliminating the objectivistic understanding of a particular 'state' $\psi$, at the same time remembering about the objective character of the *global context* of every measurement, which QM codes with its formalism. The quantum 'state' $\psi_U$ *is not* the same as a mechanical state in classical theory and *has to be replaced* by $\psi_{U/O}$. As the local representation $\psi_{U/O}$ depends strictly on the frame of action, one should take into consideration the fact that $\psi_{U/O}$ always models local information $O$ (as an exosystem) has about $U$ (as an endosystem) inside nonseparable **Ω**, and it is dependent on the information $O$ has about the past of system $U$ and the general rules of constructing the description in QM. Such a remark is perhaps trivial, because naturally that knowledge of the context is indispensable for any statistical prediction of the interaction result; but what is not trivial, though, is the fact that knowledge about the past is also not of an absolute character, but depends only and solely on the actions of $O$ itself (injective and extractive actions of $O$ as an internal subsystem–history in **Ω**), i.e., it is also purely relational[63]. If we neglect those simple assumptions, the result will not only be an absolutization of the particular description $\psi_{U/O}$, but also that of the past of $O$ and $U$, which in three different ways violates the Universal Principle of Relativity. If the given representation $\psi_{U/O}$ itself is purely relational *and* relative, then it does not make sense to state that the internal histories of $U$ and $O$,

63 Cf. Rovelli's crisp slogan: 'facts are interactions' [Rovelli, 2004, p. 215].

which are constructed solely on the basis of local information gained by $O$, are independent from the physical actions of $O$. For that very reason function $\psi_{U/O}$ allows $O$ to formulate relational predictions [$\boldsymbol{O}$] in its quantum frame, predictions concerning the possible result of the future interaction–correlation with $U$ with the use of the formalism of the unitary evolution in QM. One should of course clearly differentiate the 'description' from the 'measurement/interaction': the former is the maximum amount of relational information $O$ can get about $U$ (or $P$ about $O \oplus U$) without any physical extractive action on $U$ (or of $P$ on $O \oplus U$), accessible solely on the basis of the locally known contextual history of $U$ and QM rules, while the latter is always connected to the actual physical interaction of $O$ and $U$ (or $P$ and $O \oplus U$), that is to the establishing a dynamical correlation between them. Since quantum 'states' as purely relational, contextual actions are extremely far-removed from what is represented by the states of classical mechanics, hence in line with Finkelstein's approach [1999a, p. 449] one can even avoid calling them 'states' altogether, so as not to provoke equivocation (if one does not accept the classical Leibnizian approach to the relativization of action/observation, as in Ch. 2.2, then it might be advisable to completely desist from calling functions $\psi$ 'states' as a relict of the old wave-mechanical approach to quantum phenomena [Finkelstein, 2003a, p. 483]). In any way, *they cannot serve as base elements of an ontology of externalistic compositionism* and exclude the modelling of nonseparable $\boldsymbol{\Omega}$ while using that kind of metaphysical imagery: hence internalistic compositionism (holism) with extremely relativized internal actions is, I believe, an appropriate *metaphysical* reformulation of Bohr's (epistemological) standpoint in opposition to Einstein's metaphysics.

Such a relational treatment of description $\psi_{U/O}$ leads to further consequences, arising from the way in which we reflected upon the problem of time in the previous chapter. The information $O$ has about $U$, coded each time in $\psi_{U/O}$, can be updated in an incontinuous way only by successive physical interactions with $U$, that is, strictly speaking, without those local interactions between the subsystems of the Universe, $\psi_{U/O}$ does not evolve in time. That suggests that the Schrödinger picture, though formally it is equivalent to the Heisenberg picture, is not necessarily equally appropriate to understand microphysical processes. The Heisenberg picture seems much more natural, in which it is the operators which evolve in time defined for the local subsystem, i.e., their expectation values are connected to the evolving probability distributions, which $O$ can calculate on the basis of the locally known history of system $U$ (cf. e.g. [Gambini, Porto, 2002]). (Also Dirac considered that description to be more fundamental: Smerlak and Rovelli [2007, p. 432] anegdotically recall on that occasion the last of Dirac's seminars in Sicily, during which he presented only one transparency, titled 'The Heisenberg picture is the right one'). Such a solution is very important for the way QM is presented, as it emphasizes the primary role of the operators, which can be treated as part of Dirac's heritage, with reference to his terminology of generalized $q$-numbers ('queer numbers') and their rings, in place of $c$-numbers, which he had introduced in the early days of QM [Dirac, 1926]: the appropriate point of departure to formulate quantum theory is not so much Hilbert space itself, as von Neumann operator $C^*$-algebra $\mathcal{A}$, where 'states' are the simply defined functionals on elements of $\mathcal{A}$. That is in full accordance with the philosophical thesis from Ch. 2.2, stating that what should be treated as primary are the quantum relativized 'actions' in a blockworld and not mechanical 'states' from which we would next have to 'compose' a physical reality. We may therefore postulate an ontology in which those elementary actions are the basis of a derivative definition of relative 'states' and the properties of 'time' (as a relativized variable in a quantum frame) associated with the statistics of the former. As we have seen, when analyzing the problem of complete observables in QM, the Heisenberg picture seems to be more fundamental, as it allows us to understand the purely relational nature of all the observables in the theory and points to the way in which such an operator of time can be construed, one that would be both relational and meaningful in the generally covariant theory ([Rovelli, 2002b]; more in [Reisenberger, Rovelli, 2002]). The treatment of the 'state' as a secondary category and the contextual action as a primary one also has an extra-physical meaning for philosophical epistemology itself. The categories of '(individual) substance', 'essence', 'things' and '(intrinsic) properties' exerted a great influence in Western philosophy not only on the theories of cognition, but also on the admissible metaphysical options. Even if one accepts the active nature of cognition and takes into account the different forms of the principle of relativity, then still some basic construction elements of ontology remain untouched, which results from the dualistic schemes historically incorporated into the structure of philosophical thinking. The complete relativity of physical states and the need to treat quantum action frames as a fundamental physical concept imposes at the same time a completely new conjugation of epistemology and ontology, which in modern philosophy has been realized extremely rarely, and thus it requires a certain speculative courage.

To see how Rovelli's relational QM works when applied to the 'EPR paradox' [Smerlak, Rovelli, 2007], one can now make a start with a modification of the assumptions that lay at the basis of that paradox, from a new perspective. One can change them in such a way so as to remove the absolute elements initially preferred by Einstein, Podolsky and Rosen as 'natural'. The ones that were most troublesome concerned that very physical objectivity/actuality of the 'state' and the absoluteness of the 'observer', as well as the noncontextual completeness of any mechanics offering a description of reality. In a modified version of Federico Laudisa's formulation [2001, p. 125], the relational assumptions will be thus:

> $[1]_R$ *R*-Reality (or *R*-Actuality)
> If a system *O*, without interacting with a physical system *U*, can predict with certainty – or at any rate with probability one – at a time *t* the value *q* of a quantity *Q* measurable on *U* in the known state *u*, then, at a time *t′* immediately after *t*, *q* corresponds to a property of *U* that is *objective/actual relative to* *O* (in its quantum frame of action).
>
> $[2]_R$ *R*-Completeness
> Any physical theory *T* describing a physical system *U* accounts for every property of *U* that is *objective relative to some other system O* (in its quantum frame of action).
>
> $[3]_R$ *R*-Locality (or *R*-Causality)
> No property of a physical system *U* that is *objective relative to some other system O* (in its quantum frame of action) can be nonlocally influenced by measurements performed on a different physical system in a spacelike separated region.

As the *adequacy* condition $[4]_E = [4]_R$ is retained in its original form (we know that the local statistical QM predictions are fully correct), we are not repeating it again: *R*-Adequacy concerns the aptness of quantum algorithms in every frame of action, which in particular implies that the global, formal nonseparability of action is a completely new, non-classical property witnessed in global correlations (or von Neumann consistency – see below) – the ur-phenomenon of mechanics. As we can see, the strongest modifications we make concern the concepts of 'reality' and 'objectivity'/'actuality', which become ontologically relativized to a quantum frame of action with a simultaneous retention of locality in a relativistic sense. Condition $[1]_R$ as such is included only with reference to the reflections in the original article by Einstein, Podolsky and Rosen, but strictly speaking, it could be by-passed as a result of $[3]_R$ and $[4]_R$ (it could solely help to highlight the argumentation removing the paradoxical nature of the EPR-type experiments). Generally speaking, the definitions can be directly inferred from the general formulation of the Local Contextual Realism I proposed in Ch. 3.2: $[3]_R$ denotes 'causal locality', $[2]_R$ the 'relative definiteness of observables in a frame', $[4]_R$ also comprises the 'global nonseparability of action', which constitutes an immanent property of QM formalism, while they all are irreducibly contextual ($[4]_R$ also means that the quantum algebra of actions meets the constraints of KST). By local *R*-Separability we shall simply mean the *local relational separability of physical systems* or the 'disentanglement separability' in Smerlak and Rovelli's definition (cf. Ch. 3.2), which in fact is quite trivial, as it concerns the possibility of the dis-entanglement and re-entanglement of systems: measuring local observables on physical, arbitrarily singled out subsystems in **Ω**. Such a reformulation of the assumptions of the analysis of the original EPR experiment is no longer merely a change in the initial hypotheses, but leads to a completely new treatment of the whole situation described in experiment $\mathcal{E}$. Though conditions $[1]_E$–$[4]_E$ and $[1]_R$–$[4]_R$ can at first glance seem to be quite similar, yet the latter ones introduce a completely different ontology, as they result from a principal change in the way the concept of the 'quantum state' and that of a physical 'fact/event' are conceived. At a deeper level, surpassing purely physical redefinitions, the difference between them results from the difference between two ontological schemes: (Aristotelian-like) externalistic compositionism and (Platonic-like) internalistic compositionism, as $[1]_R$–$[4]_R$ automatically impose a completely different type of cosmology. That is a cosmology without the non-relational base states and with the purely global structure **Ω**, which we have called a 'Platonic whole'.

In order to fully see the physical naturalness of relational argumentation, let us go back to the spacetime diagram of the EPR experiment, which is presented in Fig. 3 from Ch. 4.1. Two systems separated by a spacelike interval perform extractive actions on subsystems $S_1$ and $S_2$ of the entangled system $S_1 \oplus S_2$: the first one, $O_1$, measures the spin along axis $m$ and gets outcome $p$ in its quantum frame, while the second one, $O_2$, measures the spin along axis $n$ and obtains in its local quantum frame a certain outcome $l$. According to the relational approach, the key mistake which Einstein made when formulating his version of local realism, was his too strong definition of realism itself, which cherishing classical 'objectivity' ignored the principle of relativity on the level of elementary events. Each of the systems performs extractive actions in its own quantum frame and there is no physical access to information about the result or even about the measurement itself taken by the other 'observer' – if they wish to obtain such information, they can do it *solely* by way of physical interactions in their future cones. The fact of conducting the measurement on $S_1$ by $O_1$ thus obtaining, for example, the result: $p(m) = -1$, does not generate, in full accordance with SR, any immediate spacelike action over a distance, which would mechanically change the state of $S_2$; the only thing that 'changes' is the information $O_1$ has about the spin measured along axis $m$. Any stronger statement in that case is unjustified, as there is no causal, immediate connection with the help of some spacelike signal between $O_1$ and $O_2$ or $S_2$: the only thing they can do is exchange a signal in the future, to confirm a consistency of their local results. Putting it briefly, nothing in QM suggests that condition $[3]_R$ would have to be violated in any way. The problem with Einstein's reasoning results from his failure to take into consideration quantum frames and, instead, of treating the whole situation with the help of the concept of 'elements of reality', i.e., he tacitly assumes that there exists in **M** and inside the quantum universe **Ω** some third 'observer', who performs the role of the absolute and, regardless of its own frame of action, registers the outcomes of actions performed by $O_1$ and $O_2$. But all the 'states' of subsystems $S_i$ and quantum events in **Ω** are *always* relative ones. As we have observed at the end of Ch. 4.2, such a classical illusion is sustained by the spacetime event diagrams themselves, which we look at as if from an absolute perspective. In the case of QBU that results in a violation of the relational structure of QM: there exists no such absolute 'observer' inside **Ω**, possessing information beyond the internal contexts (frames) of action. Were it to exist, then it would have to be found inside **Ω** (be its subsystem) and not be found there at the same time (be its exosystem), but by definition, nothing can exist outside **Ω**. Such a type of situation created by the physical absolute generates generalized (quantum-)Eleatic paradoxes, which are impossible to solve without the procedure of relativization.

One should become aware of the importance of the relativized condition $[1]_R$ for Einstein's reasoning, who for those very reasons could never accept Bohr's physical intuitions concerning nonseparability in any guise. The information about $S_1$ that $O_1$ updates in the course of a dynamical interaction pertains solely to his own local quantum frame of action and has nothing to do with actuality referring to $O_2$ and $S_2$. That means that the extractive action does not nonlocally 'create' any 'element of reality' at a distance at moment $t_0$, but that it only modifies the relative knowledge $O_1$ has about $S_1 \oplus S_2$ *in its own local frame.* From that point of view the relational approach not so much makes the EPR argument more difficult to apply, but – as Smerlak and Rovelli [2007, p. 437] have rightly pointed out – *it makes it impossible to formulate it in any reasonable way right from the beginning*, by rejecting $[1]_E$ and its metaphysical background. The Einsteinian type of realism ignores such a possibility, as it does not treat the option of a relativization of the actuality in a serious way. Yet, if we nevertheless take that step, then we can assume that what is actual in the quantum frame (i.e. with regards) to $O_1$, certainly is not and cannot be actual in the frame of $O_2$ (the same applies to the *potentiality* of the outcomes of interactions). The operation of comparing their results *before* their interaction in the future cone simply does not have any physical sense (an observer who could simultaneously conduct such an operation simply does not exist in **Ω**). Hence, the fact that the EPR correlations exist has to be taken into account in a completely different way, such that does not violate causal (relativistic) locality. Let us now, to follow Smerlak and Rovelli's example [ibid., p. 438*ff.*], mark the local spin measurement taken by $O_1$ on system $S_1$ in the direction $m$ at a moment $t_0$ as: $Sp^{m}_{S1/O1}$, while the outcome obtained as $p = \epsilon_{S1/O1}$; the spin measurement, in turn, taken by $O_2$ on $S_2$ in direction $n$ and at a moment $t_0'$ as: $Sp^n_{S2/O2}$, while its outcome as: $l = \epsilon_{S2/O2}$. It is to be inferred from what we have said above that there is no sense in comparing values $\epsilon_{S1/O1}$ and $\epsilon_{S2/O2}$, treated naïvely as noncontextual, before a real dynamical interaction of $O_1$ and $O_2$. The outcome obtained by $O_1$ is a relationally actual piece of information that had earlier been predicted by it as a contextual possibility, as a result of it earlier registering the entangled 'state' of system $S_1 \oplus S_2$ (on account of the interaction in the past: $J^-(S_1)$, $O_1$ knew that $\psi_{S1+2/O1}$ was a maximally entangled pure state). Similarly, the possibility obtained by $O_1$ of further predictions referring to the 'state' of system $S_1$ in the future depends solely on its frame of action and is relativized with regards to it. As QM permits, though, it has at its disposal at a time $t > t_0$ some additional information regarding subsystem $S_2$. That can thus be put down in the form of a simple reasoning, conducted solely on the basis of quantum formalism:

$$\psi_{S1+2/01};\ Sp^m_{S1/O1} = p = \varepsilon_{S1/O1} \Rightarrow Sp^m_{S2/O1} = -\varepsilon_{S1/O1} \tag{41}$$

One should note that the conclusion $O_1$ draws concerns solely his own frame of action, i.e., the outcome he would have obtained were he to have measured the spin on $S_2$ in the direction of $m$, but does *not say anything* about the distant space-time event as the action conducted (or not) independently by $O_2$ (at time $t_0$ in the frame of reference of $O_1$'s laboratory). What may seem pedantic in a macroscale is of great significance for the avoidance of a paradoxical absolute description, one not respecting the Universal Principle of Relativity. The most astounding property of QM, revealed from the analyses made in Ch. 3.2 and 5.1, becomes evident precisely at the moment when we are deliberating the situation of the interaction between $O_1$ and $O_2$ in their future and the full accordance of their quantum descriptions in $\mathbf{\Omega}$. (In the relational interpretation, there is no need to introduce a huge number of Everettian quantum worlds beyond $\mathbf{\Omega}$). The property depends on the fact that formalism automatically guarantees that consistency globally (and formally), but does *not* allow us to reconstruct at any stage the histories of systems and their action frames in a classical (noncontextual) way. QM captures that characteristic of the global structure of $\mathbf{\Omega}$ as a hidden whole, thanks to which it can code formally (for example by the observable algebras $\mathcal{A}_R$, connected to the trajectories of the observers in $\mathbf{M}$) the relations between the pieces of information processed by arbitrary subsystems of nonseparable $\mathbf{\Omega}$. The whole content of that consistency gets exhausted exactly in the global correlations, which cannot be reconstructed bottom–up by the summing up of information about elementary 'states' of the subsystems. There is no need for any object in $\mathbf{\Omega}$ to pass a signal with superluminal velocity about its 'state' to another object in order to conform it to 'states' in other regions of the Universe: such an expectation is the result of using a naïve version of externalistic compositionism in QM, which is often conjoined to a customary presentism. True to internalistic compositionism and QBU ontology it is much more natural to accept that the internal correlations in **Ω** simply globally 'are' rather than 'come into being' by the incessant spacelike nonlocal effects breaking the relativistic causality, thus a fundamental feature of **Ω** (the Leibnizian 'system of correlations') is nonseparability, not dynamical spacelike nonlocality[64]. In the case when $O_1$ and $O_2$ get into an interaction in the causal future, enabling them to compare the results of the measurements that had been performed locally, QM formalism guarantees that the results of that comparison will be in accord with the abstract model of correlations between the subsystems of the whole $S_1 \oplus S_2$, i.e., *the quantum history of each subsystem is consistent.* Let us assume, just as Smerlak and Rovelli have done [2007, p. 439], that both $O_1$ and $O_2$ have performed spacelike separated spin measurements along the same axis $m = n$ (although they could not have been certain of that before the future interaction via e.g. $S_3$, even if they had earlier agreed to do so). The answer $O_2$ gives to $O_1$'s question, 'what spin value have you got in the direction $n$?', is going to be described as $Sp^n_{O2/O1}$. Now, one can assume that there is a third system $O_3$, which in the future interacts with both $O_1$ and $O_2$, and gets information on the outcomes of their extracting actions, which is described as: $Sp^n_{O1/O3}$ and $Sp^n_{O2/O3}$. The formalism of QM referring to the measurements in each of those action frames and – in particular – to the spin operator, now guarantees that simple relationships take place between them. Firstly, if system $O_1$ first conducts a measurement on $S_2$, and next on $O_2$ (true to the universal equivalence of quantum systems, neither $S_2$, nor $O_2$ are physically privileged in **Ω**), it is bound to confirm the accordance of their states, i.e., $Sp^n_{S2/O1} = Sp^n_{O2/O1}$. Secondly, system $O_3$, after its interaction with the former two, will surely find an anticorrelation: $Sp^n_{O1/O3} = -Sp^n_{O2/O3}$. Both those relations between the quantum descriptions are guaranteed by the *von Neumann consistency condition* itself ([von Neumann, 1932, pp. 222–37], [Rovelli, 2004, p. 216*f.*]).

An important requirement introduced by relational QM from the very beginning is the clear specification as to what frame of reference (with regards to what action/'observer') the 'quantum state' or 'quantum whole' (such as entangled state $\psi_{S1+2}$) is defined in. One can do that by adding information about the action

64 David Mermin [1999, p. 583] in his lecture on the occasion of Daniel Greenberger's 65th birthday, used verbal play to propose replacing the expression 'action at a distance' and Shimony's mysterious 'passion at a distance' with 'fashion at a distance'.

frame to the description, such as e.g. [***O***] or [***P***], or aligning appropriate indices to the 'state' symbol itself, e.g. $Sp^{n}_{S2/O1}$. Neglect to delineate that leads to great confusion in the quantum description associated with the misconception of the 'state' itself, which quickly undergoes a classical or quasi-classical interpretation and becomes an absolute of mechanics (e.g. R. Gambini and R.A. Porto [2002] also arrived at similar conclusions when analyzing the problem of the 'wave-function collapse' in RQM). That, in consequence, leads further on to a series of troublesome concepts, such as an immediate spreading of correlating signals in space, the absoluteness of dynamics and time, or even ends in statements on the 'paradoxical nature' of QM itself. The latter appears when we analyze for instance such seemingly contradictory judgments as [2] from Ch. 1.2:

State $\psi_{S1+2}$ is both an indivisible whole and is not an indivisible whole.

Strictly speaking, we are dealing here with quite different wholes and, additionally, those not properly indexed wholes are defined by a certain singled-out noncontextual 'observer', which automatically generates contradiction. The entangled 'state' $\psi_{S1+2}$ of a certain system *S* is an 'indivisible whole' when it is an isolated whole in the frame of another system *O* (it is 'potential' with regards to *O*), while it is not such a whole when *O* enters into a dynamical (entangling) interaction with any of *S*'s arbitrary subsystems – then the full structure of the relations between systems *O* and *S* is a 'hidden (inaccessible/potential) whole' to *O*. The whole of Universe $\mathbf{\Omega}$ is always hidden, as each internal interaction assumes an arbitrary Heisenberg–von Neumann cut *i* inside $\mathbf{\Omega}$, in turn introducing not only an irremovable contextuality and relativity of the correlations between those subsystems, but even a full relativity of 'actuality' and 'potentiality' (always with regards to the cut and the quantum action frame). By definition, and regardless of concrete cut *i*, there always exists within $\mathbf{\Omega}$ *a certain irreducible hidden whole* encompassing arbitrarily differentiated subsystems *O* and *S*, which are part of it – the fact can be seen as an openly ontological generalization of Bohr's original doctrine, as opposite to that of Einstein. The differentiation of arbitrary subsystems, such as $O_1$, $O_2$, $O_3$, $S_1$, $S_2$ and $S_1 \oplus S_2$, guaranteed by $[3]_R$, is possible thanks to their local, relational separability, but at the same time their quantum histories, as the series of correlations, *are consistent thanks to the nonseparability of actions* in $\mathbf{\Omega}$. Thus, the confusion concerning wholes and parts in QM results from a permanent refusal to see nonseparability as the ur-phenomenon of mechanics and acknowledge the internalistic logic operating in quantum ontology. The problem does not have a solely pragmatic or epistemological character, but an ontological one: one can claim that it is a consequence of ignoring the Universal Principle of Relativity, i.e., an ontological absolutization of parts within an ontologically absolutized whole. But in the relational paradigm there seems to be no sense in talking about the 'evolution of Universe **Ω**', as there is a physical meaning only in the evolution of (arbitrary) subsystem *S* with regards to another subsystem *O*, which is, for given cuts, an ordered series of quantum actions, defined *only as relations with other actions* within the global structure of **Ω**. The quantum 'perspective' of *O* (its relational 'state') should single out the relative parameter used to describe local evolution, i.e. 'the quantum time of *O* (for cut *i*)', and we expect the quantum thermodynamics of time in a fully covariant theory to have exactly such a form. The cosmology, which should be possible to be constructed on the basis of such a scheme for a globally nonseparable **Ω** as QBU, would simply be a radically relational cosmology with a kind of internal (canonical or built-in), relativized dynamics in the Leibnizian spirit. One should ask, what concrete shape the programme of relational cosmology should take, if we are to assume that internalistic compositionism is the only viable option for quantum theory?

One of the interesting options for the generally covariant, relational quantum theory has been developed, among others, by Louis Crane (cf. [Crane, 1994; 1995; 2005; 2009], [Crane, Yetter, 1993], [Barrett, Crane, 2000]) since the beginning of the 1990's, on the basis of the topological quantum field theory (henceforth TQFT) and the categorical approach. TQFT was formulated at the end of the 1980's by E. Witten, M. Atiyah and G. Segal (see the axiomatic formulation in [Atiyah, 1990]) and is a class of theories which – in general terms – are background independent, global (i.e., deprived of local degrees of freedom) quantum theories (e.g. John C. Baez [2001] offers a presentation accessible to philosophers). They do not assume beforehand any given topology of space or spacetime, nor even less so require any predetermined background metric, but allow the choice of a manifold of arbitrary topology as a convenient representation of the physics under study. What is an extremely important fact is that TQFT possesses a complete and precise formulation in the language of the category theory: putting it technically, TQFT is a functor from the category **COB** of $n$-oriented cobordisms (with a certain type of tensor product)[65] to the category **VECT** of finite-dimensional vector spaces with a usual tensor product (the functor refers objects, morphisms and composition in both categories, with the composition of cobordisms being noncommutative)

65 Generally, $(n+1)$-cobordism in algebraic topology is a triple of manifolds ($W$, $M$, $N$), where $W$ is $(n+1)$-manifold with a boundary, which in turn is a disjoint union of both $n$-manifolds $M$ and $N$. Such cobordism creates an equivalence class of those two manifolds (we say that they are cobordant then, $M \equiv N$), i.e., $W$ is a cobordism between $M$ and $N$, or otherwise, from $M$ to $N$ (cobordism can be seen as a certain topological map, such that it meets e.g. the condition: if $M \equiv N$, then $\partial M \cong \partial N$).

[Crane, 1995, p. 6182]. TQFT in the original formulation by Atiyah [1990] assigns to every ($n$–1)-dimensional manifold $S$ a space equipped with a *-structure, i.e. Hilbert space $Q(S) = \mathcal{H}_s$, whose vectors represent the states of the universe with a space topology given by $S$, while to each $n$-dimensional cobordism $K$: $S \rightarrow S'$ some linear evolution operator on $\mathcal{H}_s$, i.e., $Q(K)$: $Q(S) \rightarrow Q(S')$, with the composition of cobordims corresponding to the composition of operators. In the case of such a type of formalism, what is extremely surprising is the elegant linking of the very different mathematical apparatuses of QM and GR and that example proves the force of generalizing the theory of functions to the categorical theory of morphisms, which we have mentioned in Ch. 2.1: categories and functors enable a fundamental reference of descriptions which possess seemingly different properties. Nevertheless it is of key importance to assume that the generally covariant quantum theory for $\mathbf{\Omega}$ cannot be, strictly speaking, a quantum theory of the field (a continuum) in such a form as the one in which we construe QFT and AQFT, hence the technique of Feynmann path integrals should be rejected as inadequate to the description of processes at the Planck scale. That is of great importance, as originally 3D TQFT was constructed by Edward Witten from a topologically invariant Lagrangian precisely on the basis of that technique and it was only secondarily reformulated purely algebraically by other theoreticians, for example in the category of modular tensors [Crane, 1995, p. 6182]. That assumption in itself is not sufficient: as we have seen in Ch. 4.1 and 5.1, the relational quantum Universe $\mathbf{\Omega}$ with its vanishing Hamiltonian should not evolve in time, as it does not in particular possess any physical exosystem which could register it (a lack of evolution-in-time, just as the global nonseparability of quantum 'states', is the ur-phenomenon of the hidden whole $\mathbf{\Omega}$). Thus that second assumption is connected, for example, with Witten's suggestion that the so-called Chern–Simons–Witten gauge theory as the topological quantum field theory (CSW TQFT of 1988) simply gives the topological phase of the quantum-gravitational state of the Universe [ibid., p. 6186]: $\mathbf{\Omega}$ would be found, which should not be surprising, in a non-evolving state described by CSW functional. We may – like Crane – provisionally assume that the CSW TQFT theory is a toy model of the quantum gravity theory for three dimensions, allowing us to imagine what its full abstract structure could look like: we of course lack here a full generalization to 4D TQFT and an introduction of local 'observers' (subsystems of $\mathbf{\Omega}$) with their local clocks. We can assume that the quantum gravity theory cannot be similar to standard QFT, as space $\mathcal{H}_s$ (or the appropriate operator algebra $\boldsymbol{\mathcal{A}}$) for a random finite subregion of $\mathbf{M}$ has to be finite-dimensional, whereas the appropriate geometrical representations for the generally covariant model of quantum spacetime should be sought with the tools offered by singular homology.

With a consistent quantum-relational approach to TQFT for QBU we are forced to abandon talking of a certain 'quantum state' of the whole of **Ω**, as such a non-relational 'state' is undefinable (it has no physical meaning): from the perspective of each internal 'observer'–subsystem the CSW-type 'global state' of quantum Universe **Ω** always remains an ur-phenomenon of all physical interactions, whereas the relational 'states' make sense only as *correlations* between those subsystems (they are thus rather *relative actions* defined for pairs of arbitrary subsystems). The ontic probabilistic interpretation for the 'quantum state' of **Ω** is in that situation pointless and leads to an absurd multiplication of universes as elements of superpositions, while fundamental theory should be referred solely to *internal relations* within **Ω** (there are no observers who would render an empirical meaning to the superpositions of the universes). We should therefore insist, when construing relational cosmology, on the simple idea that there is never any quantum interaction (dynamic correlation) in **Ω** without a quantum extracting subsystem, which is its proper part – there therefore does *not* also exist any Hilbert space which could be associated with all of **Ω**. The full relativization of the geometrical 'states' in QBU also at the very outset makes us abandon the idea of introducing any preferred 'evolution of states' (a global clock) as in classical mechanics. The relational approach also assumes that there does not exist anything like 'distant measurement' – it is always a physical interaction based on a strictly defined local topological Heisenberg–von Neumann cut within **Ω**. In the category-theoretical approach of Crane and others we can thus provisionally think of a quantum observer within that whole as an embedded, finite and oriented topological 3-manifold $O$ with boundary $b$, which refers to a particular cut $i$ (boundary $b$ is of course also a manifold, only of a lower dimension, $b^{d-1}$). Fig. 7 schematically illustrates the topological cut and observer $O$ thus defined together with quantum environment $U$ in **Ω** 'measured' by it. The topological closed space $b^{d-1}$ or its finite subspace can also be supplied with additional data – the space of quantum-gravitational states (construed for example combinatorically or by a quantization in the framework of Chern–Simons theory), which refer to the quantum action frame of $O$ in a quantized spacetime, i.e., to elementary, relational geometric information on $b$, defined by an appropriate set of operators. Relational 'states' in **Ω** can be modelled with the help of $n$-simplices and their complexes, where unitary representations of the Lorentz group are *elementary (quantized), oriented dynamical objects such as bivectors* (see Ch. 2.2) for which appropriate quantum operators should be defined (e.g. area operator). In that case, it is of course necessary to generalize the operation of integration over states in order to have appropriate sums over topological states. In a concrete physical model, each surface $b$ can be described by, for example, a certain quantum spin network $\Gamma$ (full information about the quantum-gravitational states of $\mathbf{\Omega}$ on $b$ for $O$), which induces a finite number of punctures or, otherwise, of points labelled by spin numbers, where the edges of network $\Gamma$ cut through $b$[66]. Crane proposes to call such a closed, oriented surface with labelled cuts the *skin of observation* [Crane, 1994, p. 123*f.*], which is a properly generalized concept of Heisenberg–von Neumann cut in the context of quantum cosmology (vector spaces $\mathcal{H}$ are ascribed to those skins). Thus, there is a given quantum 'state' associated with any arbitrary observer $O$ in a globally unchanging $\mathbf{\Omega}$: a certain very complex, yet finite graph with vertices in labelled points of $b$. Assuming that the fundamental relational states of nature are states of $q$-spin networks at the Planck scale, independent of any physical background, and that the classical (non-relational) description should emerge from them in certain bulk approximation, one can construe simple models with numerical predictions (cf. e.g. [Poulin, 2006]).

*Fig. 7: The scheme of observation in the categorical TQFT: its precondition is to make a cut in blockworld* $\mathbf{\Omega}$*, i.e., to make a division into an 'observing system' (exosystem) O and 'observed system' (endosystem) U. From the formal perspective it is described with the help of a topological manifold (more precisely: cobordism)* $K_\Omega$ *with boundary b given by a disjoint union, and manifolds O and U. The Universe as a certain 4D simplicial complex* $\mathbf{\Omega}$ *constitutes a 'hidden whole' for the operations of cuts defining particular 'observers'* $O_i$*, with whom skins of observation (L. Crane) are associated. Every skin can be ascribed with spaces of quantum-gravitational states* $\mathcal{H}$*, which are at the same time quantum-informational states. The 'observing system' itself is defined in a quantum-informational way (it establishes a correlations solely in region b) – observation at a distance is physically meaningless.*

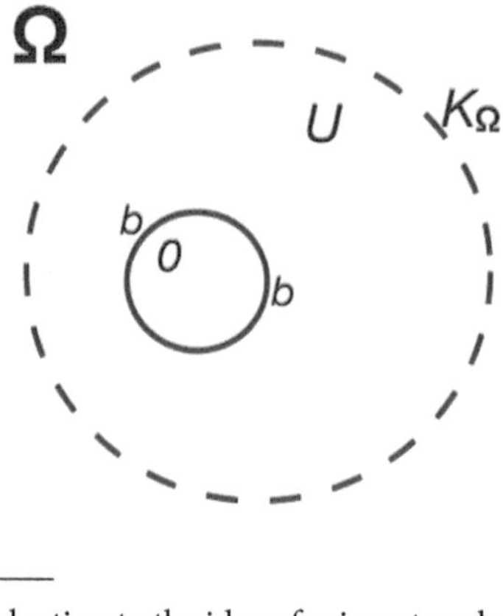


66 A nice popular introduction to the idea of spin networks and quantization on topological surfaces is offered by L. Smolin [2001, pp. 125–45] or R. Penrose [2004, pp. 934–57].

The whole logical simplicity of that notion of 'measurement' depends on the fact that the cut described, which is a common boundary for two different manifolds, can serve to supply the fundamental definition of relational observation in universe **Ω**, which forms a category. The objects in that *category of observation* **OBS** are regions of interaction, that is boundaries $b_i$ (skins of measurement), while its morphisms are three-dimensional cobordisms $K_\Omega$ on those boundaries, referring to the process of information extraction/'inspection', with appropriate conditions of pairing the manifolds and the lattices of projection operators on them [Crane, 1994; 1995, p. 6185*ff.*]. Hence, the 'state' of such a relational-topological theory of quantum gravity can be formally defined as a functor from the category **OBS** to the category **VECT** of finite-dimensional vector spaces (or, alternatively, finite C*-algebras $\mathcal{A}$ of an appropriate type). In accordance with our assumptions we can see that TQFT for **Ω** cannot give some unique state space $\mathcal{H}_\Omega$ from the perspective of any arbitrarily singled out *O*. That is impossible as there is physical sense only in speaking about a category of observations which are associated with all the possible skins. The language that is able to refer those descriptions to one another has to be extremely general and has to capture only the abstract relations between them, that is – as we have postulated in Ch. 2.1 – has to be based on a universal generalization of the simple concept of function (in a more general case it even has to allow in TQFT to create for $b_i$ vector spaces from the superposition of other vector spaces). In a categorical formulation, CSW TQFT is exactly that type of fundamental theory in which no description with the help of a definite representation of the algebra of actions (projections) $\mathcal{A}$ is singled out as physically preferred: the only thing that matters (and which at the same time constitutes the physical content of TQFT), is *the topology of the relations between finite, commutative C*-algebras $\mathcal{A}$ (quantum action frames) for all the possible boundaries $b_i$, that can be defined for a given area of the universe.* The fundamental consistency of the individual descriptions supplied by different 'observers' ('parts of **Ω**') can be expressed solely with the help of category morphisms and the functors on categories, and it is the *n*-categories which are an appropriate mathematical arena to express such criteria of consistency in a non-classical QBU, where there is no possibility of gluing the descriptions together in a Boolean way: 'If A watches B watch C watch the rest of the universe, A must see B see C see what A sees C see' [Crane, 1994, p. 123]. Putting it briefly, the global topology of those relations is a strictly mathematical form of the idea of the Platonic 'hidden structure' or Platonic whole, regardless even of the astonishing fact that the concept of, for example, the quantum causal triangulation of spacetime [Ambjørn *et al.*, 2012] and of construing fundamental physics on the basis of simplicial complexes and their topological symmetries is deeply Platonic *per se* (cf. Plato's famous geometric speculations in *Tim*. 53c–56c [Harte, 2002a, pp. 235–247]). Perhaps only a categorical approach enables an effective, consistent implementation of fully relational mechanics and the physics of spacetime, and in that sense that is very Leibnizian (cf. a similar opinion by Crane [1994, p. 129]). There is another, deeper fact behind that statement: the categorical approach is mathematically ideally tailored to construct QBU theories, when it is global internalistic compositionism that induces their local logic, while physics has to be formulated from the inside of **Ω** from the perspective of arbitrary subsystems. For the universe of externalistic compositionism (with its noncontextual, atomic base elements and at least *minimally* non-relational states) a categorical formulation is at best redundant.

The non-trivial, global topology of relations in quantum universe **Ω** is responsible for the fact that QM as a metalanguage referring interaction results in different quantum frames to each other is at all adequate, i.e., despite the radical relativity of local 'states' for cuts *i* the condition of *R*-Adequacy $[4]_R$ can be met at the same time. As a matter of fact, the latter expresses the global nonseparability of action in **Ω** and the appropriateness of the representation of quantum actions given for example in formula (31), which generates quite new, non-classical statistics for every *i* in **Ω**. But how is *physical R*-Adequacy grounded, though, or – in other words – what is the mystery behind the global nonseparability of action in QBU? As already stressed, the option assuming any direct spacelike interactions (spatial nonlocality) is on account of the causal structure of GR excluded as physically unacceptable (and thus nonseparable QM is a theory which is completely causally local in the sense of $[2]_B$), while QBU does not require such effects, and does not even suggest them, exactly as the relativistic 'Twin Paradox' is not a paradox at all. For condition $[4]_R$ to be satisfied, strictly quantum correlations have to occur regardless of particular Heisenberg–von Neumann cuts and the proper time/simultaneity surfaces of local observers, and hence that formalism enforces – as we have already stated – not the presentist world of becoming states of affairs, but the atemporal structure of *whole histories of relational actions* in a unique **Ω** (statistics for arbitrary subsystems, coded by density matrices for any *i*, are secondary). The only promising way out of the 'paradoxes', such as those as in Ch. 2.3 – if only the baroque metaphysics of many worlds is not taken seriously – is thus a strategy naturally associated with the ontology of internalistic compositionism for physics: the assumption that base elements cannot be non-relational individual events or noncontextual 'states of affairs', but a global quantum history **Ω** generating a spacetime geometry in a thermodynamic regime (as is suggested particularly by Jacobson's [1995] astounding theoretical result). QM, and quantum cosmology in particular, require Platonic wholes (particularly **Ω** as *the* Platonic whole), not familiar Aristotelian wholes with dispositions–potentiality or the extensional collections–mosaics of microproperties *à la* Lewis.

A physical explanation of $[4]_R$ and the global nonseparability in **Ω** can in fact be attempted in a number of ways and to a great extent depends on the interpretation of the status of QM itself. It is possible to treat QM as a *principle theory* (in Einstein's sense), at least *in spe*, analogically to the standard 'postulate formulation' of SR, just as Rovelli [1996] insists. In such a case, nonseparability is a consequence of the von Neumann consistency condition, interpreted as a fundamental postulate of the new theory of information: the acceptance of $[1]_R$–$[4]_R$, which are the implementation of the Universal Principle of Relativity, will suffice to eliminate every quantum 'paradox' and a further metaphysical interpretation of the effectiveness of QM formalism is unnecessary. One can nevertheless treat QM as a *constructive theory*, i.e., an effective information theory from the perspective of local subsystems of **Ω**, and construct quantum theory (without the slightest disturbance to the cohesion of the actual structure of QM), in which the local nonseparability of 'states' mirrors even more fundamental, global properties of **Ω**, which can prove important from the perspective of quantum cosmology.

One possible tactic, still somewhat faithful to the 'principle approaches', is to eliminate any trace of dynamics, even its generalized atemporal form, from fundamental physics and to reject the Common Cause Principle in any form, and thus to opt in favour of purely relational, *adynamical physics*, in which the correlations of the EPR- or GHZ-type result from global constraints in **Ω** (among the proponents of such an approach are, in particular, M. Silberstein, M. Cifone and W.M. Stuckey [2007; 2008], quite in Rovelli's spirit as far as they reject any hidden-variable constructive models). One can attempt to regain global nonseparability of events in **Ω** due to violation of $[1]_B$ and/or $[3]_B$ (while retaining Lorentzian invariance and relativistic locality), without engaging any dynamical effects (backwards causation), by the introduction of nonlocal constraints in a globally hyperbolic **M**, e.g. restricting the degrees of freedom of systems such as particles and 'measuring apparatuses' (that is a singular 'nonlocality without nonlocality' [Weinstein, 2009]). In such a case, there is no nonlocal action in **M**, only the *global, 'Parmenidean' consistency conditions* for quantum **Ω**. The construction of such Lorentz-invariant theories is possible and interesting, but – I think – they tend to be a bit too conservative, as they too strongly depend on modelling NRQM on the basis of the physics of spacetime, its symmetries and classical events (which, I claim, together with their causal structure should rather be treated as a quantum thermodynamical phenomenon). That is a version of QBU ontology still based on the principal CBU intuitions and the spatiotemporal relations, but it endorses both the internal (global) quantum contextuality and the principle of relativity, hence, of course, it realizes the project of Parmenidean physics and engages Platonic (structural) wholes in ontology.

An equally radical and perhaps more promising tactic is the one postulating a simultaneous strong *generalization and globalization of dynamics* for **Ω**, so that relativistic kinematics and the whole geometry of spacetime would be derivative as a purely quantum effect. Classical dynamical intuitions usually require time as background, but a generalization and globalization of dynamics offers its quite new type: a kind of *quantum, atemporal 'pre-dynamics'*, which we can call 'Platonic dynamics' in QBU, which in a strictly physical sense can be considered to be equivalent to the fully adynamical interpretations as a consequence of the relativization of dynamics itself (that equivalence naturally might be troubling to a classical metaphysician). Two basic observations are of key importance here at the outset. Firstly, one of the aspects of quantum relativity in the blockworld **Ω** of 'Parmenidean physics' should be the physical-mathematical naturalness of the wholly time-symmetric formulation of QM (e.g. [Cramer, 1986], [Price, 1997], [Reznik, Aharonov, 1995], [Aharonov *et al.*, 2009], [Aharonov *et al.*, 2010], [Aharonov, Tollaksen, 2011]) – in fact nothing seems to suggest it being the opposite. We are used to treating the second law of thermodynamics as a law of nature, but we can assume that it is a certain tautology, i.e., it results from the way in which we define classical probabilities in physics from the anthropocentric perspective. The second observation concerns operator C*-algebras $\mathcal{A}$ which, as we have seen, have a certain built-in canonical, internal pre-dynamics without a temporal background, but generating purely local, *statistical* 'time flow' as derivative in QBU. It is thus possible to take into account such Lorentz invariant cosmological models of the QBU type, in which a joint violation of statistical independence $[3]_B$ and of noncontextuality $[1]_B$ is based on a certain global pre-dynamical property of **Ω** (:quantum histories) itself. Such models naturally require an appropriate generalization not only of dynamics, *but also of causality* together with the global symmetries supplied by quantum groups (a generalization of rigid Lie groups), we thus should not expect such a noncommutative quantum geometry with purely global notions to be possible to formulate on the basis of a classical topology. What we understand here as a generalization of causality could mean defining it in such a way so as to bring out the time-symmetric connection/coupling between relational quantum actions within whole histories, and not the arbitrary labelling of events on the basis of a cosmological asymmetry of 'cause' and 'effect', i.e., asymmetry of time introduced *a priori* (cf. e.g. [Heller, 2002, p. 54]). That kind of quantum time symmetry should not be viewed as some puzzling property of a deep, concealed microphysical regime of the Universe, but rather as an essential feature of physical processes, constantly manifesting itself in labs as something which Bohr and Pauli called in a Goethean spirit an 'essential/indivisible wholeness of the phenomenon'. That is, by the way, exactly what Goethe insisted on when he once announced: '*Der denkende Mensch irrt besonders, wenn er sich nach Ursach' und Wirkung erkundigt: sie beide zusammen machen das untheilbare Phänomen*'[67]. Mara Beller [1999, p. 257] has argued that 'Pauli's wholeness, as well as Bohr's, is not a crisply defined concept', but rather 'a tool for the legitimation of the Copenhagen credo', especially when one tries to transpose it to the post-Bell era, and she might be right inasmuch as Bohr's reflection was conditioned by such ideas as that of William James, or ever-recurring, misleading intuition of 'measurement disturbance', and was openly reluctant to ontology and metaphysical speculations on holism. But in a time-symmetric QM in QBU with completely relational states, there is indeed a direct connection between Bohr contextuality (as he presented it for experimental situations) and generalized causality for quantum histories if indeed 'both time boundary conditions, the pre- and post-selection, enter on equal footing' [Aharonov *et al.*, 2010, p. 32].

More radical or more conservative approaches are still possible here, depending on the ambition of the model and its mathematical realization. The former attempt to model all of universe $\Omega$ in the context of a quantization of gravity with the help of purely global (noncommutative) geometric structures, without any primary concept of locality and local dynamics (e.g. [Heller, 2002], [Lam, 2008], [Shimony, 2009, p. 489*ff.*]), which in practice might even imply the violation – in different, subtle ways – of all of Bell's conditions and the introduction of relativistic cosmology itself into a completely new mathematical world where none of its aspects recall the world of classical mechanics. A Platonic internalistic compositionist might greet such an extreme holistic cosmology with the highest approval, but less radical tactics are also interesting. What is possible first of all are internally contextual and Lorentz-invariant models with hidden variables, violating $[1]_B$ and $[3]_B$ in the framework of a time-symmetric QM with a purely temporal 'nonlocality' ([Reznik, Aharonov, 1995], [Aharonov *et al.*, 2001; 2002; 2009; 2010], [Aharonov, Tollaksen, 2011]). The pre-dynamics here is completely (timelike-) global against $[3]_B$, as the Platonic wholes are hidden quantum histories composed

67 'The thinking man errs especially when he makes an inquiry according to separate cause and effect, since they both jointly create an indivisible phenomenon' [Goethe, 1976, p. 209*f.*].

of completely relational events which are 'codetermined' jointly by past actions and by future actions (which is mirrored in the so-called Aharonov–Bergmann–Lebowitz rule). One can reduce the dynamical nature of that timelike 'nonlocality' and just postulate a global atemporal coupling of relational events *within* quantum histories, thus postulate *timelike nonseparability* for them as a structural feature of Platonic **Ω** (which makes it akin to Rovelli's and Silberstein–Cifone–Stuckey's approaches). Even John G. Cramer [1986, p. 661] himself made it clear that retrocausality of the quantum 'transactions' could be seen as a 'pedagogical convention for the purposes of description', hence, as a matter of fact, it is not a 'process' at all. Platonic dynamics in QBU, if only fully relativized and satisfying internalistic logic (without 'elements of reality' with 'dispositions'), can be indeed indiscernible from the adynamical, acausal blockworld with hidden global constraints. In any event, the extreme contextuality and relationality of the local 'states' is a striking property of such theories in full accordance with Rovelli's interpretation: internal non-relational states postulated in $[1]_B$ in such a quantum **Ω** simply *do not exist*, while the sole past of any systems never determines their actual relational 'states', and even cannot do that.

Both time-symmetric pre-dynamical QM (with a generalized global dynamics and causality) and adynamical, purely relational QBU with global constraints as relativized, antipresentist approaches with built-in Kochen–Specker contextuality (just as that of Rovelli) are able to effectively eliminate not only all the versions of the 'paradox' of the delayed choice quantum eraser, but also Hardy's paradox and all the quantum 'paradoxes' (e.g. the so-called Three-Box Paradox, Quantum Cheshire Cats or the novel Quantum Pigeonhole Paradox) based on counterfactual reasoning *without employing it*, which is a standard philosophical accusation against their formulations. As has been rightly observed [Aharonov *et al.*, 2002, p. 137], that accusation is a miss, as it is possible to prepare experimental arrangements with intermediate quantum interactions, which test those contextuality 'paradoxes' without bringing any classical counterfactual statements into play (cf. e.g. [Fedrizzi *et al.*, 2011], [George *et al.*, 2013] or [Denkmayr *et al.*, 2014]): the general formalism of projections in $\mathcal{A}$ points to a deep, consistent time-symmetric structure of quantum processes, which is still unreconcilable with externalistic compositionism. One should remember that the constructed quantum 'weak values' are just quantum multiple-time amplitudes (for generalized multiple-time 'states' treated as *timelike nonseparable, relational physical actions*) being ratios of amplitudes [Kastner, 2004], though directly physically measurable, but they are *not* any classical 'elements of reality' as some non-relational, objective 'hidden states' of systems (which even Lev Vaidman [1999, p. 866], sympathizing with the 'many-worlds' interpretation, admits). The 'weak values', calculated for arbitrary histories inside **Ω**, are as relational and contextual as ordinary 'sharp values' (being a special case of the former), and 'quantum states' in general, and they have little to do with the classical, i.e., time-asymmetric counterfactuals as such (hence the plain counterfactual interpretation of the Aharonov–Bergmann–Lebowitz rule, also that motivated by the 'many-worlds' ideology, is rather misleading). One might put it simply: *time symmetry is deeply embedded in standard QM itself*, which generates *the relational, multiple-time 'weak amplitudes' for relational events* inside nonseparable blockworld **Ω**, and the counterfactuals, loved by analytic philosophers, have nothing to do with that fact. It would be interesting to formulate time-symmetric QM for TQFT in the language of C*-algebras on boundaries $b_i$ – such topological 'states'-actions on $b_i$ might then be treated not only as relational, but *explicite* time-symmetric.

The adoption of a pre-dynamical nature of the effects connected to the violation of $[3]_B$ in **Ω** need not be a violation of the original condition of *R*-Completeness at the foundation of relational QM (seen as a principle theory), as we are still treating QM as a theory offering a *complete representation of the physical situation from the perspective of any local observer* in **Ω** (cf. e.g. the same standpoint represented by H. Price [1997, p. 311]), and standard QM formalism and its statistical algorithms for local projections in $\mathcal{A}$ do not undergo any modification 'by an iota' [Aharonov *et al.*, 2010, p. 32]. In other words, QM supplies an adequate description of the physical situation in a given local quantum frame for a Heisenberg–von Neumann cut *i* in **Ω**: the statistical algorithm is complete and offers a *maximum amount of information* a local observer can gain and one need not supplement it with a 'Ptolemean' super-quantum description, as no additional super-quantum information is locally accessible by sharp measurements. It is also not possible to turn QM into trivially extended classical mechanics, as any quantum generalizations of dynamics and causality cannot be trivial (cf. e.g. the experimental results as tests of the generically quantum contextuality/nonseparability in [Fedrizzi *et al.*, 2011] and [George *et al.*, 2013]). The constructive approach, though, postulates that statistical algorithms work effectively, because at the roots of all local quantum effects lies the ur-phenomenon of the global nonseparability of action, understood as a pre-dynamical property of **Ω** as a whole[68]. We thus uphold the thesis from Ch. 2.2 that the quantum algebra for **Ω**

68 In Ch. 3.2 we presented the *formal* nonseparability of state. The constructive and ontological approach to QM interprets that singular property of formalism as an adequate representation of the actual property of pre-dynamics of **Ω**, i.e., global, purely quantum symmetries of **Ω**.

can be interpreted as a generalized algebra of the process – fundamental *algebra of the series of nonseparable actions (i.e., hidden histories) obeying global quantum symmetries, which formally represent a violation of the condition of a timelike separability of events* $[3]_B$ *and of noncontextual realism* $[1]_B$. In accordance with the thesis from Ch. 3, the strong (local) Bohr contextuality of all the actions-events in **Ω** is a *direct consequence* of that global nonseparability, hence it is possible to introduce a more general concept of metaphysical realism consistent with KST: Local Contextual Realism *tout court*. The intrinsic contextuality of every *local* quantum action inside **Ω**, being a pillar of the old Bohrian/Heisenbergian philosophy, should be viewed as a local manifestation and direct consequence of the *global* nonseparability of **Ω**.

Good candidates for such algebras are first and foremost to be found in quantum Clifford algebras, whose fully classical variant has been launched by us as a geometric-projective toy model of the process in Ch. 2.2. Nevertheless, their introduction requires a complete transformation of the physical concepts, as it breaks away from the basic intuitions of the classical geometry and topology we work with when speaking of spacetime **M** and relativistic CBU or analyzing EPR-type correlations in the way Fig. 3 presents: that is precisely a revolution of the conceptualization the 'paradoxes' such as the delayed choice quantum eraser or Hardy's paradox demand. The very formulation of the $\lambda$-independence condition was motivated by time-asymmetric spacetime imagery of one-way causal streams with extensional compositionism of events at its roots. When thinking of its violation as an unrealistic 'causal past conspiracy' we are still guided by classical intuitions, but in the quantum geometry of **Ω** they might be completely inadequate as 'artifacts of classical geometry' (cf. [Majid, 2000, p. 3895]), hence it is justified, as I have suggested in Ch. 3.2, to abandon the correlative and troubling categories of 'locality'/'nonlocality' of the state in **M** in favour of the category of the global nonseparability in Parmenidean universe **Ω**. If we wish to construct a geometric model of the quantum structure of **Ω**, the classical 'Schopenhauer sphere' from Ch. 2.3 should be replaced with the 'quantum Parmenidean sphere': a model of the non-evolving, global series–histories in a certain noncommutative projective geometry, with the concept of locality being derivative (it might therefore be necessary to pass on to noncommutative topology). It is only then that we can expect the 'quantum eraser' or 'nonlocal correlations' (also in time) to be not some mysterious anomalies in the classical universe, but direct 'Platonic' or 'Leibnizian' traces of purely global (nonseparable) pre-dynamics in a unique **Ω**. That means that in the framework of that strategy we are conducting a conscious turnover of the ontological model of composition in quantum cosmology: we are consequently replacing externalistic compositionism with radically holistic internalistic compositionism. The very concept of a 'quantum state of the Universe' cannot be used in the same way as the concept of the 'quantum state' of the subsystem inside that Universe, which is not a standard physical object: we find the statement that 'Universe **Ω** is to be found in a quantum state Ψ' bereft of physical sense, as there does not exist any representation of that state in any Hilbert space $\mathcal{H}_{\Omega}$ construed for the action frame of any observer *O*. Let us also note that the hidden, global boundary conditions of **Ω** behave here as a fundamental law of nature with regards to its local subsystems (in that case there could be indeed no sense in a stiff, classical separation of 'boundary conditions' and 'laws' [Crane, 1994, p. 123]).

Exotic Platonic wholes are not simply a speculative excess with regards to intuitive and domesticated Aristotelian wholes, but are rather an alternative metaphysical matrix for fundamental physics in the situation when the quantum theory makes it impossible to effectively adhere to the scheme of externalistic compositionism. Perhaps one might be in the position to define that dispute between two ontological models – slightly dramatizing it for the purposes of the discussion – as a great philosophical dispute between 'radical holists' and 'naïve particularists'. It is at first astonishing to find that that has such a profound meaning for theoretical physics, but as such the situation makes us wonder less when we consider the historical fact – as we have begun to do in Ch. 1 – that ancient Greek metaphysics has up to this very day created the basic intellectual paradigms concerning the problem of the whole and parts, or of the absoluteness and relativity of cognition. When construing cosmology, wish it or not, we will be raising the fundamental topics of ancient ontology, even if were to be working solely on advanced mathematical models. The category of the 'hidden whole' of nature is in a way of a border character, more metaphysical than physical: on the one hand, that whole – as a structure – is conceivable and definable on the ground of more and more general mathematical theories (such as noncommutative topology *in statu nascendi*); on the other hand, there is no empirical sense in it when interpreted as a unique 'state of **Ω**' – it does not reveal iself otherwise than in series of local phenomena and their correlations for arbitrary internal subsystems of **Ω**. However one cannot explain them away, as e.g. the delayed choice quantum eraser effects, the Three-Box Paradox or Hardy's paradox, within an externalistic and presentist scheme populated by changing 'elements of reality', exactly as one cannot explain relativistic 'paradoxes' within the three-dimensional world with ontic becoming. Rather they are – in a Platonic spirit – physically conceivable only through a kind of internalistic logic (and the associated non-classical probability theory for internal observers) as a *structure of unique quantum history* **Ω**, or the 'time-whole', *Zeitgestalt*. In that sense, the concept of the 'ur-phenomenon' can be applied here, the *Urphänomen*, '*eine Wirkung ohne Ursache*' [Goethe, 1976, p. 209], introduced by Goethe in his natural scientific studies to denote the hidden, but effectively forming unity in the time series of the studied phenomena. (Cf. e.g. [Schad, 1989] for an exemplary illustration from the area of biology with a serious consideration of time symmetry, 'Doppelstrom der Zeit', in the theory of organic evolution). The border character of that category is probably of a fundamental character, expressed also in its link with the paradoxes of self-reference and limitative theorems. That means that that property cannot be eliminated by construing certain familiar mechanics based on absolute, non-relational 'elements of reality' (base elements) and idealized 'observers' playing the role of the absolutes of measurement interactions. Thus, fundamental physics has no choice: regression to quasi-Aristotelian schemes or to the Humean 'mosaic-world' of 'elements of reality' is not possible and it has to consistently apply the more and more deeply penetrating procedures of relativization. In Ch. 4 and 5 I have attempted to highlight the fundamental role of the relativity principle in QM by introducing the concept of quantum frames and relational states, while in Ch. 1 and 3 I have argued that the adoption of a Platonic-type ontology of internalistic compositionism is not an arbitrary step, as it is deeply grounded in the QM structure as such and in nonseparability as the ur-phenomenon of mechanics. Those two directions might seem to be divergent (the Parmenidean or Platonic type of ontology is traditionally seen as opposing any ontological relativization), but that impression is rather the effect of their perfunctory assessment and a slightly stereotypical reflection on the metaphysical tradition. Their joint acceptance in the context of cosmology leads to a certain variant of relational block ontology of the quantum universe.

In accord with some of Bohr's semi-positivistic intuitions: φύσις κϱύπτεσθαι φιλεῖ (Heraclitus, B 123). Nature loves to *hide*. Nevertheless, at the same time, in a highly nontrivial and physical sense: *In omnibus partibus relucet totum* (Nicholas of Cusa, *Dialogus de ludo globi*, I.42,7–8; [Nicolaus de Cusa, 1998, p. 47]). *The whole* is manifesting itself in all its parts. It seems that the more internally relativized physics is, the more Platonic it becomes.

# Bibliography


| | |
|---|---|
| LMS | *Leibniziens Mathematische Schriften*, hg. von C.I. Gerhardt, Bd. I–VII, Berlin – Halle 1849–63. |
| LPP | Gottfried Wilhelm Leibniz, *Philosophical Papers and Letters. A Selection Translated and Edited, with an Introduction by Leroy E. Loemker* (2nd ed.). Dordrecht – Boston – London. |
| LPS | *Die philosophischen Schriften von Gottfried Wilhelm Leibniz*, hg. von C.I. Gerhardt, Bd. I–VII, Berlin 1875–90. |
| PO | *Platonis Opera. Recognovit brevique adnotatione critica instruxit Ioannes Burnet* (*Scriptorum Classicorum Bibliotheca Oxoniensis*), tomi I–V, Oxonii e typographeo Clarendoniano, 1905–13. Φαιδων = *Phaedo* [*Ph.*]: t. I; Θεαιτητος = *Theaetetus* [*Tht.*]: t. I; Παρμενιδης = *Parmenides* [*Parm.*]: t. II; Φιληβος = *Philebus* [*Phil.*]: t. II; Τιμαιος = *Timaeus* [*Tim.*]: t. IV. |

Aharonov, Y., Bergmann, P., Lebowitz, J. (1964). Time symmetry in the quantum process of measurement. *Physical Review* **134**: 1410–6.

Aharonov, Y., Botero, A., Popescu, S., Reznik, B., Tollaksen, J. (2002). Revisiting Hardy's paradox: counterfactual statements, real measurements, entanglement and weak values. *Physics Letters* A **301**: 130–8.

Aharonov, Y., Botero, A., Scully, M. (2001). Locality or non-locality in quantum mechanics: hidden variables without 'spooky action-at-a-distance'. *Zeitschrift für Naturforschung* **56a**: 5–15.

Aharanov, Y., Kaufherr, M. (1988). Quantum frames of reference. *Physical Review* D **30**: 111–2.

Aharonov, Y., Popescu, S., Tollaksen, J. (2010). A time-symmetric formulation of quantum mechanics. *Physics Today* **63**: 27–32.

Aharonov, Y., Popescu, S., Tollaksen, J., Vaidman, L. (2009). Multiple-time states and multiple-time measurements in quantum mechanics. *Physical Review* A **79**: 052110.

Aharonov, Y., Tollaksen, J. (2011). New insights on time symmetry in quantum mechanics. In: R.Y. Chiao, M.L. Cohen, A.J. Leggett, W.D. Phillips, C.L.

Harper Jr. (eds.), *Visions of Discovery: New Light on Physics, Cosmology, and Consciousness* (pp. 105–51). Cambridge.

Aharonov, Y., Vaidman, L. (1990). Properties of a quantum system during the time interval between two measurements. *Physical Review* A **41**: 11–20.

Ajdukiewicz, K. (1977). Change and contradiction. In: K. Ajdukiewicz, *The Scientific World-Perspective and Other Essays, 1931–1963*, ed. by J. Giedymin (pp. 192–208). Dordrecht – Boston.

Ambjørn, J., Jurkiewicz, J., Loll, R. (2012). Causal dynamical triangulations and the quest for quantum gravity. In: J. Murugan, A. Weltman, G.F.R. Ellis (eds.), *Foundations of Space and Time. Reflections on Quantum Gravity* (pp. 321–37). Cambridge.

Anandan, J. (2003). Laws, symmetries, and reality. *International Journal of Theoretical Physics* **42**: 1943–55.

Ansmann, M., Wang, H., Bialczak, R.C., Hofheinz, M., Lucero, E., Neeley, M., O'Connell, A.D., Sank, D., Weides, M., Wenner, J., Cleland, A.N., Martinis, J.M. (2009). Violation of Bell's inequality in Josephson phase qubits. *Nature* **461**: 504–6.

Antonopoulos, C. (2004). Moving without being where you're not: a non-bivalent way. *Journal for General Philosophy of Science* **35**: 235–59.

Arthur, R. (1998). Cohesion, division and harmony: physical aspects of Leibniz's continuum problem (1671–1686). *Perspectives on Science* **6**: 110–35.

Aristotle (1924). *Aristotle's Metaphysics.* A revised text with introduction and commentary by W.D. Ross (2 vols). Oxford. [*Metaph.*].

Aristotle (1938). *Categories. On Interpretation. Prior Analytics*. Ed. and transl. by H.P. Cooke & H. Tredennick. Cambridge. [*Cat.*].

Aristotle (1950). *Aristotelis Physica. Recognovit brevique adnotatione critica instruxit W.D. Ross*. Oxford. [*Phys.*].

Ashtekar, A., Schilling, T.A. (1998). Geometrical formulation of quantum mechanics. In: A. Harvey (ed.), *On Einstein's Path. Essays in Honor of Engelbert Schücking* (pp. 23–65). Berlin.

Aspect, A., Dalibard, J., Roger, G. (1982). Experimental test of Bell's inequalities using time-varying analyzers. *Physical Review Letters* **49**: 1804–7.

Aspect, A., Grangier, P. Roger, G. (1982). Experimental realizaton of Einstein-Podolsky-Rosen-Bohm *Gedankenexperiment*: a new violation of Bell's inequalities. *Physical Review Letters* **49**: 91–4.

Atiyah, M. (1990). *Topological Quantum Field Theory*. Cambridge.

Auletta, G. (2005). Quantum information as a general paradigm. *Foundations of Physics* **35**: 787–815.

Bäck, A. (2014). *Aristotle's Theory of Abstraction*. Heidelberg – New York – Dordrecht – London.

Baez, J.C. (2001). Higher-dimensional algebra and Planck-scale physics. In: C. Callender, N. Huggett (eds.), *Physics Meets Philosophy at the Planck Scale: Contemporary Theories in Quantum Gravity* (pp. 172–95). Cambridge.

Baker, L.R. (1997). Why constitution is not identity. *The Journal of Philosophy* **94**: 599–621.

Ballentine, L.E. (2000). *Quantum Mechanics: A Modern Development*. Singapore.

Balme, D.M. (1941). Greek science and mechanism. II: The atomists. *The Classical Quarterly* **35**: 23–8.

Bańka, J.S. (1998). *The Unity of Being and Logos. Kant and the Presocratics* (in Polish). Poznań.

Barbour, J. (2001). *The End of Time: The Next Revolution in Our Understanding of the Universe*. London.

Barbour, J. (2003). The deep and suggestive principles of Leibnizian philosophy. *The Harvard Review of Philosophy* **XI**: 45–58.

Barrett, J.W., Crane, L. (2000). A Lorentzian signature model for quantum general relativity. *Classical and Quantum Gravity*, **17**: 3101–18.

Baugh, J., Finkelstein, D.R., Galiautdinov, A., Saller, H. (2001). Clifford algebra as quantum language. *Journal of Mathematical Physics* **42**: 1489–500.

Baugh, J., Finkelstein, D.R., Saller, H., Tang, Z. (1998). General covariance is Bose-Einstein statistics. In: A. Harvey (ed.), *On Einstein's Path: Essays in Honor of Engelbert Schucking* (pp. 67–80). New York – Berlin – Heidelberg.

Baxter, D. (1988). Many-one identity. *Philosophical Papers* **17**: 193–216.

Bell, J.S. (1987a). On the Einstein–Podolsky–Rosen paradox. In: J.S. Bell, *Speakable and Unspeakable in Quantum Mechanics* (pp. 14–21). Cambridge. (First published in: *Physics*, 1964, **1**: 195–200).

Bell, J.S. (1987b). On the impossible pilot wave. In: J.S. Bell, *Speakable and Unspeakable in Quantum Mechanics* (pp. 159–168). Cambridge. (First published in: *Foundations of Physics*, 1982, **12**: 989–99).

Beller, M. (1999). *Quantum Dialogue. The Making of a Revolution*. Chicago.

Belousek, D.W. (1999). Bell's theorem, non-separability, and spacetime individuation in quantum mechanics. *Philosophy of Science* **66** (Suppl.): S28–S46.

Belousek, D.W. (2003). Non-separability, non-supervenience and quantum ontology. *Philosophy of Science* **70**: 791–811.

Bene, G. (1997). Quantum reference systems: a new framework for quantum mechanics. *Physica* A **242**: 529–65.

Bene, G. (2000). Quantum reference systems: reconciling locality with quantum mechanics. Preprint arXiv: quant-ph/0008128v1.

Bene, G., Dieks, D. (2002). A perspectival version of the modal interpretation of quantum mechanics and the origin of the macroscopic behavior. *Foundations of Physics* **32**: 645–71.

Berkovitz, J. (1998). Aspects of quantum non-locality. I: Superluminal signalling, action-at-a-distance, non-separability and holism. *Studies in History and Philosophy of Modern Physics* **29**: 183–222.

Bigaj, J. (2003). How the hypothesis εἰ ἕν ἐστιν is to be understood (in Polish). In: M. Manikowski (ed.), *Platonic Colloquia: Παρμενίδης – Σοφιστής*, in Polish (pp. 35–47). Wrocław.

Bisognano, J.J., Wichman, E.H. (1975). On the duality condition for a Hermitian scalar field. *Journal of Mathematical Physics* **16**: 985–1007.

Blandzi, S. (2002). *Plato's Project of the First Philosophy* (in Polish). Warszawa.

Bohm, D. (1951). *Quantum Theory*. Englewood Cliffs, N.J.

Bohm, D. (1952). A suggested interpretation of quantum theory in terms of hidden variables. *Physical Review* **85**: 166–93.

Bohm, D., Aharonov, Y. (1957). Discussion of experimental proof for the paradox of Einstein, Rosen, and Podolsky. *Physical Review* **108**: 1070–6.

Bohm, D. (1965). A proposed topological formulation of the quantum theory. In: I.H. Good (ed.), *The Scientist Speculates: An Anthology of Partly-Baked Ideas* (pp. 302–14). New York.

Bohm, D., Davies, P.G., Hiley, B.J. (2006). Algebraic quantum mechanics and pregeometry [1982]. In: G. Adenier, C.A. Khrennikov, T.M. Nieuwenhuizen (eds.), *Quantum Theory: Reconsideration of Foundations 3. AIP Conf. Proc., vol. 810, 16–11 VI 2005, Växjö, Sweden* (pp. 314–24). Melville, N.Y.

Bohm, D., Hiley, B.J. (1981). On a quantum algebraic approach to a generalized phase space. *Foundations of Physics* **11**: 179–203.

Bohm, D., Hiley, B.J. (1993). *The Undivided Universe*. London – New York.

Bohr, A., Ulfbeck, O. (1995). Primary manifestation of symmetry: origin of quantal indeterminacy. *Reviews of Modern Physics* **67**: 1–35.

Bohr, A., Mottelson, B.R., Ulfbeck, O. (2004a). The principle underlying quantum mechanics. *Foundations of Physics* **34**: 405–17.

Bohr, A., Mottelson, B.R., Ulfbeck, O. (2004b). Quantum world is only smoke and mirrors. *Physics Today* **57**: 15–6.

Bohr, N. (1963). Discussion with Einstein on epistemological problems in atomic physics. In: *Philosophical Writings of Niels Bohr*. Vol. 2: *Essays 1933–1957 on Atomic Physics and Human Knowledge* (pp. 32–66). Woodbridge, CT.

Bouwmeester, D., Pan, J.-W., Daniell, M., Weinfurter, H., Zeilinger, A. (1999). Observation of three-photon Greenberger–Horne–Zeilinger entanglement. *Physical Review Letters* **82**: 1345–9.

Boyer, C.B. (1991). *A History of Mathematics* (2. ed., revised by U.C. Merzbach). New York – Chichester – Brisbane – Toronto – Singapore.

Brans, C.H. (1988). Bell's theorem does not eliminate fully causal hidden variables. *International Journal of Theoretical Physics* **27**: 219–26.

Bratteli, O., Robinson, D.W. (1987). *Operator Algebras and Quantum Statistical Mechanics*. Vol. 1: *C*- and W*-Algebras, Symmetry Groups, Decomposition of States* (2. ed.). Berlin – Heidelberg – New York.

Breuer, T. (1995). The impossibility of accurate state self-measurements. *Philosophy of Science* **62**: 197–214.

Breuer, T. (1996). Subjective decoherence in quantum measurements. *Synthese* **107**: 1–17.

Breuer, T. (1997). *Universalität und Unvollständigkeit. Eine Untersuchung über die Grenzen physikalischer Theorien*. Habilitationsschrift, Universität Salzburg, Institut für Philosophie.

Brill, D.R., Gowdy, R.H. (1970). Quantization of general relativity. *Reports on Progress in Physics*, **33**: 413–88.

Brown, H.R. (2005). *Physical Relativity: Space-Time Structure from a Dynamical Perspective*. Oxford.

Brown, H.R., Pooley, O. (2006). Minkowski space-time: a glorious non-entity. In: D. Dieks (ed.), *Ontology of Spacetime* (pp. 67–92). Amsterdam – Oxford.

Brown, M.J. (2009). Relational quantum mechanics and the determinacy problem. *British Journal for the Philosophy of Science*, **60**: 679–95.

Bruß, D. (2002). Characterizing entanglement. *Journal of Mathematical Physics* **43**: 4327–51.

Bub, J. (2005). Quantum mechanics is about quantum information. *Foundations of Physics* **35**: 541–60.

Busche, H. (1997). *Leibniz' Weg ins perspektivische Universum. Eine Harmonie im Zeitalter der Berechnung*. Hamburg.

Butterfield, J.N. (1992). Bell's theorem: what it takes. *British Journal for the Philosophy of Science* **42**: 41–83.

Butterfield, J.N. (2002). Topos theory as a framework for partial truth. In: P. Gardenfors, K. Kijania-Placek, J. Wolenski (eds.), *Invited Papers at 11th International Congress of Logic, Methodology and Philosophy of Science* (pp. 307–30). Dordrecht.

Butterfield, J.N. (2007). Stochastic Einstein locality revisited. *British Journal for the Philosophy of Science* **58**: 805–67.

Cabello, A. (1999). Quantum correlations are not contained in the initial state. *Physical Review* A **60**: 877.

Callender, C. (2008). Finding 'real' time in quantum mechanics. In: W.L. Craig, Q. Smith (eds.), *Einstein, Relativity and Absolute Simultaneity* (pp. 50–72). London – New York.

Carlip, S. (1998). *Quantum Gravity in 2+1 Dimensions*. Cambridge.

Carnap, R. (1963). Intellectual autobiography. In: P.A. Schilpp (ed.), *The Philosophy of Rudolf Carnap* (s. 1–84). LaSalle, IL.

Chibeni, S.S. (2005). A logico-conceptual analysis of the Einstein–Podolsky–Rosen argument. In: M. Pietrocola, O. Freire Jr. (eds.), *Filosofia, Ciência e História: Michel Paty com o Brasil, uma homenagem aos 40 anos de colaboração* (pp. 115–35). São Paulo.

Choquet-Bruhat, Y., Geroch, R. (1969). Global aspects of the Cauchy problem in general relativity. *Communications in Mathematical Physics* **14**: 329–35.

Cleland, C.E. (1984). Space: an abstract system of non-supervenient relations. *Philosophical Studies* **46**: 19–40.

Coles, P.J., Kaniewski, J., Wehner, S. (2014). Equivalence of wave–particle duality to entropic uncertainty. *Nature Communications* **5**: 5814.

Connes, A., Rovelli, C. (1994). Von Neumann algebra automorphisms and time-thermodynamics relation in generally covariant quantum theories. *Classical and Quantum Gravity* **11**: 2899–917.

Cramer, J. (1986). The transactional interpretation of quantum mechanics. *Review of Modern Physics* **58**: 647–87.

Crane, L. (1994). Topological field theory as the key to quantum gravity. In: J.C. Baez (ed.), *Knots and Quantum Gravity* (pp. 121–32). Oxford.

Crane, L. (1995). Clock and category: is quantum gravity algebraic? *Journal of Mathematical Physics* **36**: 6180–93.

Crane, L. (2005). Relational topology as the foundation for quantum gravity. *Modern Physics Letters* A **20**: 2017–35.

Crane, L. (2009). Categorical geometry and the mathematical foundations of Quantum Gravity. In: D. Oriti (ed.), *Approaches to Quantum Gravity: Toward a New Understanding of Space, Time and Matter* (pp. 84–98). Cambridge.

Crane, L., Yetter, D. (1993). Categorical construction of 4D topological quantum field theories, [w:] L.H. Kauffman, R.A. Baadhio (eds.), *Quantum Topology* (pp. 120–30). Singapore.

Crockett, T. (1999). Continuity in Leibniz's mature metaphysics. *Philosophical Studies* **94**: 119–38.

Dalla Chiara, M.L. (1977). Logical self reference, set theoretical paradoxes and the measurement problem in quantum mechanics. *Journal of Philosophical Logic* **6**: 331–47.

Dalla Chiara, M.L., Giuntini, R. (1989). Paraconsistent quantum logics. *Foundations of Physics* **19**: 891–904.

Dalla Chiara, M.L., Giuntini, R. (2000). Paraconsistent ideas in quantum logic. *Synthese* **125**: 55–68.

Deltete, R., Guy, R. (1991). Einstein and EPR. *Philosophy of Science* **58**: 377–97.

Dembiński, B. (2003). *Plato's Late Teaching. The Relationship between Ontology and Mathematics* (in Polish). Katowice.

Denkmayr, T., Geppert, H., Sponar, S., Lemmel, H., Matzkin, A., Tollaksen, J., Hasegawa, Y. (2014). Observation of a Quantum Cheshire Cat in a matter-wave interferometer experiment. *Nature Communications* **5**: 4492.

De Witt, B.S. (1967). Quantum theory of gravity. I: The canonical theory. *Physical Review* **160**: 1113–48.

Dickson, M. (2004). A view from nowhere: quantum reference frames and uncertainty. *Studies in History and Philosophy of Modern Physcis* **35**: 195–220.

Dirac, P.A.M. (1926). Quantum mechanics and a preliminary investigation of the hydrogen atom. *Proceedings of the Royal Society* A **110**: 561–9.

Dittrich, B. (2007). Partial and complete observables for Hamiltonian constrained systems. *General Relativity and Gravitation* **39**: 1891–927.

Dorr, C. (2005). What we disagree about when we disagree about ontology. In: M. Kalderon (ed.), *Fictionalism in Metaphysics* (pp. 234–86). Oxford.

Dürr, D. (2001). *Bohmsche Mechanik als Grundlage der Quantenmechanik.* Heidelberg.

Durt, T. (1996). Why god might play dice. *International Journal of Theoretical Physics* **35**: 2271–84.

Durt, T. (1997). Three interpretations of the violation of Bell's inequalities. *Foundations of Physics* **27**: 415–34.

Einstein, A. (1936). Physik und Realität. *The Journal of the Franklin Institute* **221**: 313–47.

Einstein, A. (1948). Quantenmechanik und Wirklichkeit. *Dialectica* **2**: 320–3.

Einstein, A. (1949a). Autobiographisches/Autobiographical notes. In: P.A. Schilpp (ed. & transl.), *Albert Einstein: Philosopher–Scientist* (pp. 1–95). La Salle, IL.

Einstein, A. (1949b). Remarks concerning the essays brought together in this co-operative volume. In: P.A. Schilpp (ed. & transl.), *Albert Einstein: Philosopher–Scientist* (pp. 665–88). La Salle, IL.

Einstein, A. (1982). *Mein Weltbild.* Frankfurt.

Einstein, A., Born, M. (1969). *Briefwechsel 1916–1955.* München.

Einstein, A., Podolsky, B., Rosen, N. (1935). Can quantum-mechanical description of physical reality be considered complete? *Physical Review* **47**: 777–80.

Elder, C.L. (2011). *Familiar Objects and Their Shadows.* Cambridge.

Elitzur, A.C., Dolev, S., Zeilinger, A. (2002). Time-reversed EPR and the choice of histories in quantum mechanics. In: I. Antoniou, V.A. Sadovnichy, H. Walther (eds.), *The Physics of Communication: Proc. of XXII Solvay Conference in Physics, Delphi and Lamia, Greece, 24–29 Nov. 2001* (pp. 452–61). Singapore.

Esfeld, M. (1999). Holism in cartesianism and in today's philosophy of physics. *Journal for General Philosophy of Science* **30**: 17–36.

Esfeld, M. (2001). *Holism in Philosophy of Mind and Philosophy of Science.* Dordrecht.

Esfeld, M. (2004). Quantum entanglement and a metaphysics of relations. *Studies in History and Philosophy of Modern Physics* **35B**: 601–17.

Fedrizzi, A., Almeida, M.P., Broome, M.A., White, A.G., Barbieri, M. (2011). Hardy's paradox and violation of a state-independent Bell-inequality in time. *Physical Review Letters* **106**: 200402.

Fine, A. (1970). Insolubility of the quantum measurement problem. *Physical Review* D **2**: 2783–7.

Fine, A. (1982). Some local models for correlation experiments. *Synthese* **50**: 279–94.

Fine, A. (1996). *The Shaky Game: Einstein, Realism and the Quantum Theory*. Chicago.

Finkelstein, D.R. (1977). The Leibniz project. *Journal of Philosophical Logic* **6**: 425–39.

Finkelstein, D.R. (1988). Finite physics. In: R. Herken (ed.), *The Universal Turing Machine. A Half-Century Survey* (pp. 349–76). Hamburg.

Finkelstein, D.R. (1989). Quantum net dynamics. *International Journal of Theoretical Physics* **28**: 441–67.

Finkelstein, D.R. (1993). Physical process and physical law. In: T.E. Eastman, H. Keeton (eds.), *Physics and Whitehead: Quantum, Process, and Experience* (pp. 180–95). Albany, N.Y.

Finkelstein, D.R. (1996). *Quantum Relativity: A Synthesis of the Ideas of Einstein and Heisenberg*. Berlin.

Finkelstein, D.R. (1999a). Action physics. *International Journal of Theoretical Physics* **38**: 447–55.

Finkelstein, D.R. (1999b). Third relativity. *International Journal of Theoretical Physics* **38**: 2937–40.

Finkelstein, D.R. (2003a). Elementary processes. In: J. Renn, L. Divarci, P. Schröter (eds.), *Revisiting the Foundations of Relativistic Physics. Festschrift in Honor of John Stachel* (pp. 479–97). Dordrecht – Boston – London.

Finkelstein, D.R. (2003b). Emptiness and relativity. In: B.A. Wallace, A.P. Lutzker (eds.), *Buddhism and Science: Breaking the New Ground* (pp. 365–85). New York – Chichester, West Sussex.

Finkelstein, D.R., Galiautdinov, A. (2001). Cliffordons. *Journal of Mathematical Physics* **42**: 3299–314.

Fraser, C.G. (1989). The calculus as algebraic analysis – some observations on mathematical-analysis in the 18$^{th}$ century. *Archive for History of Exact Sciences* **39**: 317–35.

Freire, O. (2005). Science and exile: David Bohm, the cold war, and a new interpretation of quantum mechanics. *Historical Studies in the Physical and Biological Sciences* **35**: 1–34.

French, S. (1989). Individuality, supervenience and Bell's theorem. *Philosophical Studies* **55**: 1–22.

Fry, E.S., Walther, T. (1997). On Some Aspects of an Hg Based EPR Experiment. *Zeitschrift für Naturforschung* **52a**: 20–4.

Fry, E.S., Walther, T. (2002). Atom based tests of the Bell inequalities – the legacy of John Bell continues… In: R.A. Bertlmann, A. Zeilinger (eds.), *Quantum (Un)speakables* (pp. 103–17). Berlin – Heidelberg – New York.

Fuchs, C.A. (2003). Quantum mechanics as quantum information, mostly. *Journal of Modern Optics* **50**: 987–1023.

Fulling, S.A. (1973). Nonuniqueness of canonical field quantization in Riemannian space-time. *Physical Review* D **7**: 2850–62.

Futch, M. (2002). Leibniz's non-tensed theory of time. *International Studies in the Philosophy of Science* **16**: 125–39.

Futch, M. (2008). *Leibniz's Metaphysics of Time and Space*. Dordrecht.

Gallicchio, J., Friedman, A.S., Kaiser, D.I. (2014). Testing Bell's inequality with cosmic photons: closing the setting-independence loophole. *Physical Review Letters* **112**: 110405.

Gambini, R., Porto, R.A. (2002). Relational reality in relativistic quantum mechanics. *Physics Letters* A **294**: 129–33.

Gambini, R., Porto, R.A., Pullin, J. (2004). A relational solution to the problem of time in quantum mechanics and quantum gravity: a fundamental mechanism for quantum decoherence. *New Journal of Physics* **6**: 45.

Gambini, R., Porto, R.A., Pullin, J. (2007). Fundamental decoherence from quantum gravity: a pedagogical review. *General Relativity and Gravitation* **39**: 1143–56.

George, R.E., Robledo, L.M., Maroney, O.J.E., Blok, M.S., Bernien, H., Markham, M.L., Twitchen, D.J., Morton, J.J.L., Briggs, G.A.D., Hanson, R. (2013). Opening up three quantum boxes causes classically undetectable wavefunction collapse. *PNAS* **110**: 3777–81.

Giustina, M., Mech, A., Ramelow, S., Wittmann, B., Kofler, J., Beyer, J., Lita, A., Calkins, B., Gerrits, T., Woo Nam, S., Ursin, R., Zeilinger, A. (2013). Bell violation using entangled photons without the fair-sampling assumption. *Nature* **497**: 227–30.

Gleason, A.M. (1957). Measures on the closed subspaces of a Hilbert space. *Journal of Mathematics and Mechanics* **6**: 885–93.

Gloy, K. (1996). *Das Verständnis der Natur*. Bd. II: *Die Geschichte des ganzheitlichen Denkens*. München.

Go, A. (2004). Observation of Bell inequality violation in B mesons. *Journal of Modern Optics* **51**: 991–8.

Goethe, J.W. (1976). *Maximen und Reflexionen*. Frankfurt am Main.

Goethe, J.W. (1979). *Farbenlehre* (mit Einleitungen und Kommentaren von Rudolf Steiner), Bd. 1, hg. von G. Ott und H.O. Proskauer. Stuttgart.

Grünbaum, A. (1968). *Modern Science and Zeno's Paradoxes*. London.

Grupp, J. (2006). Mereological nihilism: quantum atomism and the impossibility of material constitution. *Axiomathes* **16**: 245–386.

Haag, R., Hugenholtz, N.M., Winnik, M. (1967). On the equilibrium states in quantum statistical mechanics. *Communications in Mathematical Physics* **5**: 215–36.

Hardy, L. (1998). Spooky action at a distance in quantum mechanics. *Contemporary Physics* **39**: 419–29.

Harte, V. (2002a). *Plato on Parts and Wholes – The Metaphysics of Structure*. Oxford.

Harte, V. (2002b). Plato's problem of composition. In: J.J. Cleary, G.M. Gurtler, S.J. (eds.), *Proc. of the Boston Area Colloquium in Ancient Philosophy*, vol.XVII/2001 (pp. 1–17). Leiden – Boston – Köln.

Hasegawa, Y., Loidl, R., Badurek, G., Baron, M., Rauch, H. (2004). Violation of a Bell-like inequality in single-neutron interferometer experiments: quantum contextuality. *Journal of Modern Optics* **51**: 967–72.

Hawking, S.W., Ellis, G.F.R. (1973). *The Large Scale Structure of Space-Time*. Cambridge.

Häffner, H., Hänsel, W., Roos C.F., Benhelm, C.F., Chek-al-kar, D., Chwalla, M., Körber, T., Rapol, U.D., Riebe, M., Schmidt, P.O., Becher, C., Gühne, O., Dür, W., Blatt, R. (2005). Scalable multiparticle entanglement of trapped ions. *Nature* **438**: 643–6.

Healey, R. (2004). Gauge theories and holisms. *Studies in History and Philosophy of Modern Physics* **35**: 619–42.

Heisenberg, W. (1959). Die Kopenhager Deutung der Quantentheorie. In: W. Heisenberg, *Physik und Philosophie* (pp. 27–42). Stuttgart.

Heisenberg, W. (1969). *Der Teil und das Ganze. Gespräche im Umkreis der Atomphysik*. München.

Held, C. (1999). *Die Bohr-Einstein Debatte. Quantenmechanik und physikalische Wirklichkeit*. Paderborn.

Heller, M. (1986). The ontological involvements of contemporary physics (in Polish). *Analecta Cracoviensia* **18**: 3–20.

Heller, M. (1988). Cosmology and reality (in Polish). *Analecta Cracoviensia* **20**: 15–29.

Heller, M. (1995). 'Timaeus' – the philosophical myth about the origin and nature of the world (in Polish). In: M. Heller, *Eternity – Time – Cosmos*, in Polish (pp. 13–30). Kraków.

Heller, M. (2002). Time and physics – a noncommutative revolution. In: H. Eilstein (ed.), *A Collection of Polish Works on Philosophical Problems* (pp. 47–56). Dordrecht.

Hellmuth, T., Walther, H., Zajonc, A., Schleich, W. (1987). Delayed-choice experiments in quantum interference. *Physical Review* A **35**: 2532–41.

Herzog, T.J., Kwiat, P.G., Weinfurter, H., Zeilinger, A. (1995). Complementarity and the quantum eraser. *Physical Review Letters* **75**: 3034–7.

Hestenes, D. (1975). Observables, operators and complex numbers in the Dirac theory. *Journal of Mathematical Physics* **16**: 556–72.

Hestenes, D. (1988). Universal geometric algebra. *Simon Stevin. A Quarterly Journal of Pure and Applied Mathematics* **62**: 253–74.

Hestenes, D. (1997). Real Dirac theory. In: J. Keller, Z. Oziewicz (eds.), *Proc. of Int. Conf. on The Theory of the Electron, Cuautitlan, Universidad Nacional Autonoma de Mexico, 1995. Advances in Applied Clifford Algebras* **7** (Suppl.): 97–144.

Hestenes, D. (2003). Mysteries and insights of Dirac theory. *Annales de la Fondation Louis de Broglie* **28**: 367–90.

Hestenes, D., Sobczyk, G. (1987). *Clifford Algebra to Geometric Calculus*. Dordrecht.

Hestenes, D., Ziegler, R. (1991). Projective geometry with Clifford algebra. *Acta Applicandae Mathematicae* **23**: 25–63.

Hiley, B.J. (2001). Towards a dynamics of moments: the role of algebraic deformation and inequivalent vacuum states. In: K.G. Bowden (ed.), *Correlations. Proc. of ANPA 23. Conference, 16–21 VIII 2001, Cambridge* (pp. 104–34). London.

Hiley, B.J. (2003). Algebraic quantum mechanics, algebraic spinors and Hilbert space. In: K.G. Bowden (ed.), *Boundaries. Proc. of ANPA 24. Conference, 15–20 VIII 2002, Cambridge* (pp. 149–86). London.

Hiley, B.J. (2005). Non-commutative quantum geometry: a reappraisal of the Bohm approach to quantum theory. In: A. Elitzur, S. Dolev, N. Kolenda (eds.), *Quo Vadis Quantum Mechanics?* (pp. 299–324). Berlin – Heidelberg.

Hiley, B.J., Fernandes, M. (1997). Process and time. In: H. Atmanspacher, E. Ruhnau (eds.), *Time, Temporality and Now* (pp. 365–83). Berlin.

Hofer-Szabó, G., Rédei, M., Szabó, L. (1999). On Reicheinbach's common cause principle and Reichenbach's notion of common cause. *British Journal for the Philosophy of Science* **50**: 377–99.

Hofer-Szabó, G., Rédei, M., Szabó, L. (2002). Common-causes are not common common-causes. *Philosophy of Science* **69**: 623–33.

Hofer-Szabó, G. (2012). Separate common causal explanation and the Bell inequalities. *International Journal for the Theoretical Physics* **51**: 110–23.

Hofmann, J., Krug, M., Ortegel, N., Gérard, L., Weber, M., Rosenfeld, W., Weinfurter, H. (2012). Heralded entanglement between widely separated atoms. *Science* **337**: 72–5.

Hossack, K. (2000). Plurals and complexes. *British Journal for the Philosophy of Science* **51**: 411–43.

Howard, D. (1985). Einstein on locality and separability. *Studies in History and Philosophy of Science* **16**: 171–201.

Howard, D. (1989). Holism, separability and the metaphysical implications of the Bell experiments. In: J. Cushing, E. McMullin (eds.), *Philosophical Consequences of Quantum Theory* (pp. 224–53). Notre Dame, IN.

Howard, D. (1992). Locality, separability and the physical implications of quantum theory. In: A. van der Merwe, F. Selleri, G. Tarozzi (eds.), *Bell's Theorem and the Foundations of Modern Physics* (pp. 306–14). Singapore.

Howard, D. (1997). Space-time and separability: problems of identity and individuation in fundamental physics. In: R.S. Cohen, M. Horne, J. Stachel (eds.), *Potentiality, Entanglement and Passion-at-a-Distance: Quantum Mechanical Studies for Abner Shimony*, vol. 2 (pp. 113–41). Dordrecht.

Hurley, S.L. (1998). *Consciousness in Action*. Cambridge – London.

Isham, C. (1993). Canonical quantum gravity and the problem of time. In: L.A. Ibort, M.A. Rodriguez (eds.), *Integrable Systems, Quantum Groups and Quantum Field Theories* (pp. 157–288). Dordrecht.

Isham, C. (1997). Topos theory and consistent histories: the internal logic of the set of all consistent sets. *International Journal of Theoretical Physics* **36**: 785–814.

Isham, C., Butterfield, J. (1998). A topos perspective on the Kochen-Specker theorem (I): Quantum states as generalised valuations. *International Journal of Theoretical Physics* **37**: 2669–733.

Isham, C., Butterfield, J. (2000). Some possible roles for topos theory in quantum theory and quantum gravity. *Foundations of Physics* **30**: 1707–35.

Jacobson, T. (1995). Thermodynamics of spacetime: the Einstein equation of state. *Physical Review Letters* **75**: 1260.

Jacques, V., Wu, E., Grosshans, F., Treussart, F., Grangier, P., Aspect, A., Roch, J.-F. (2007). Experimental realization of Wheeler's delayed-choice Gedanken Experiment. *Science* **315**: 966–8.

Jarrett, J. (1984). On the physical significance of the locality conditions in the Bell arguments. *Noûs* **18**: 569–89.

Jiang, T. (2001). The problematic of whole–part and the horizon of the enlightened in Huayan Buddhism. *Journal of Chinese Philosophy* **28**: 457–75.

Johnston, M. (1992). Constitution is not identity. *Mind* **101**: 89–105.

Julsgaard, B., Kozhekin, A., Polzik, E.S. (2001). Experimental long-lived entanglement of two macroscopic objects. *Nature* **413**: 400–3.

Kaiser, F., Coudreau, T., Milman, P., Ostrowsky, D.B., Tanzilli, S. (2012). Entanglement-enabled delayed-choice experiment. *Science* **338**: 637–40.

Kaiser, G. (1981). Phase-space approach to relativistic quantum mechanics. III: Quantization, relativity, localization and gauge freedom. *Journal of Mathematical Physics* **22**: 705–14.

Karakostas, V. (2004). Forms of quantum nonseparability and related philosophical consequences. *Journal for General Philosophy of Science* **35**: 283–312.

Karimi, E., Cardano, F., Maffei, M., Lisio C. de, Marruzzi, L. Boyd, R.W., Santamato, E. (2014). Hardy's paradox tested in the spin-orbit Hilbert space of single photons. *Physical Review* A **89**: 032122.

Kastner, R. (2004). Weak values and consistent histories in quantum theory. *Studies in History and Philosophy of Modern Physics* **35**: 57–71.

Kim, Y.H., Yu, R., Kulik, S.P., Shih, Y.H., Scully, M.O. (2000). A delayed choice quantum eraser. *Physical Review Letters* **84**: 1–5.

King, L. (1979). Hua-Yen mutually interpenetrative identity and the Whiteheadean organic relation. *Journal of Chinese Philosophy* **6**: 387–410.

Kirk, G.S. (1975). *Heraclitus: The Cosmic Fragments. A Critical Study, with Introduction, Text and Translation*. Cambridge.

Kochen, S., Specker, E. (1967). The problem of hidden variables in quantum mechanics. *Journal of Mathematics and Mechanics* **17**: 59–87.

Kock, A. (1981). *Synthetic Differential Geometry*. Cambridge.

Koslicki, K. (2008). *The Structure of Objects*. Oxford – New York.

Król, Z. (2005). *Plato and the Foundations of Contemporary Mathematics. Plato's Concept of Number* (in Polish). Nowa Wieś k/Torunia.

Kuchař, K.V. (1992). Time and interpretations of quantum gravity. In: J. Kunsatter, D. Vincent, J. Williams (eds.), *Proc. of the 4. Canadian Conference on General Relativity and Astrophysics* (pp. 211–314). Singapore.

Kuchař, K.V. (1999). The problem of time in quantum geometrodynamics. In: J. Butterfield (ed.), *The Arguments of Time* (pp. 169–96). Oxford.

Kwiat, P.G., Steinberg, A.M., Chiao, R.Y. (1992). Observation of a 'quantum eraser': a revival of coherence in a two-photon interference experiment. *Physical Review* A **45**: 7729–39.

Lam, V. (2008). Structural aspects of space-time singularities. In: D. Dieks (ed.), *The Ontology of Spacetime* II (pp. 111–31). Amsterdam.

Lassere, F. (1964). *The Birth of Mathematics in the Age of Plato*. Larchmont – New York.

Laudisa, F. (1995). Einstein, Bell, and nonseparable realism. *British Journal for the Philosophy of Science* **46**: 309–29.

Laudisa, F. (2001). The EPR argument in a relational interpretation of quantum mechanics. *Foundations of Physics Letters* **14**: 119–32.

Lawson-Daku, B.J., Asimov, R., Gorceix, O., Miniatura, C., Robert, J., Baudon, J. (1996). Delayed choices in atom Stern-Gerlach interferometry. *Physical Review* A **54**: 5042–7.

Lawson, H.B., Michelsohn, M.-L. (1990). *Spin geometry*. Princeton, N.J.

Lavendhomme, R. (1996). *Basic Concepts of Synthetic Differential Geometry*. Dordrecht.

Leibniz, G.W. (1860a). Brevis demonstratio erroris memorabilis Cartesii, LMS VI, pp. 117–23 (LPP, s. 296–302).

Leibniz, G.W. (1860b). Specimen dynamicum, LMS VI, pp. 234–54 (LPP, pp. 435–52).

Leibniz, G.W. (1879a). Briefwechsel zwischen Leibniz und de Volder, 1698–1706, LPS II, pp. 139–283 (*The Leibniz–De Volder Correspondence*, transl., ed. & introd. by P. Lodge, New Haven – London 2013).

Leibniz, G.W. (1879b). Briefwechsel zwischen Leibniz und des Bosses, 1706–1716, LPS II, pp. 285–521 (*The Leibniz-Des Bosses Correspondence*, transl., ed. & introd. by B.C. Look & D. Rutheford, New Haven - London 2007).

Leibniz, G.W. (1879c). Briefwechsel zwischen Leibniz, Landgraf Ernst von Hessen-Rheinfels und Antoine Arnauld, 1686–1690, LPS II, pp. 1–138 (*The Leibniz-Arnauld Correspondence*, ed. & transl. by H. T. Mason, New York 1967).

Leibniz, G.W. (1880a). Without title ('Discourse de metaphysique'), LPS IV, pp. 427–63 (LPP, pp. 303–30).

Leibniz, G.W. (1880b). Theoria motus abstracti seu Rationes Motuum universales, a sensu et Phaenomenis independentes, LPS IV, pp. 221–40 (LPP, pp. 139–42).

Leibniz, G.W. (1880c). Systeme nouveau de la nature et de la communication des substances, aussi bien que de l'union qu'il y a entre l'ame et le corps, LPS IV, pp. 471–87 (LPP, pp. 453–61).

Leibniz, G.W. (1882). *Nouveaux essais sur l'entendement par l'auteur du systeme de l'harmonie preestablie*, LPS V, pp. 39–509 (*New Essays on Human Understanding*, ed. & transl. by P. Remnant & J. Bennett, Cambridge 1981).

Leibniz, G.W. (1885a). Without title ('Principes de la philosophie ou monadologie'), LPS VI, pp. 607–23 (LPP, pp. 643–53).

Leibniz, G.W. (1885b). Principes de la Nature et de la Grace, fondés en raison, LPS VI, pp. 598–606 (LPP, pp. 636–42).

Leibniz, G.W. (1890a). De rerum originatione radicali, LPS VII, pp. 302–8 (LPP, pp. 486–91).

Leibniz, G.W. (1890b). Without title, LPS VII, pp. 289–91.

Leibniz, G.W. (1890c). Specimen inventorum de admirandis naturae Generalis arcanis, LPS VII, pp. 309–18.

Leibniz, G.W. (1890d). Without title ('In Betreff der Seele der Thiere'), LPS VII, pp. 328–32.

Lewis, D. (1986a). *On the Plurality of Worlds*. Oxford.

Lewis, D. (1986b). *Philosophical Papers*, vol. II. Oxford.

Lewis, D. (1991). *Parts of Classes*. Oxford.

Lewis, D. (1994). Humean supervenience debugged. *Mind* **103**: 473–90. (Also in: D. Lewis, *Papers in Metaphysics and Epistemology*, Cambridge 1999, pp. 224–47).

Liu, M. (1982). The harmonious universe of Fa-tsang and Leibniz: a comparative study. *Philosophy East & West* **32**: 61–76.

Loemker, L.E. (1969). Introduction, In: LPP, pp. 1–62. Boston.

Lucretius (1917). *Titi Lucreti Cari De Rerum Natura Libri Sex* (ed. by W.A. Merill). Berkeley.

Lundeen, J.S., Steinberg, A.M. (2009). Experimental joint weak measurement on a photon pair as a probe of Hardy's paradox. *Physical Review Letters* **102**: 020404.

Lützen, J (1983). Euler's vision of a generalized partial differential calculus for a generalized kind of function. *Mathematics Magazine* **56**: 299–306.

Ma, X.-S., Kofler, J., Quarry, A., Tetik, N., Scheidl, T., Ursin, R., Ramelow, S., Herbst, T., Ratschbacher, L., Fedrizzi, A., Jennewein, T., Zeilinger, A. (2013). Quantum erasure with causally disconnected choice. *Proceedings of the National Academy of Sciences of the United States of America* **110**: 1221–6.

Majid, S. (2000). Quantum groups and noncommutative geometry. *Journal of Mathematical Physics* **41**: 3892–942.

Malin, S. (1982). Observer dependence of quantum states in relativistic quantum field theories. *Physical Review* D **26**: 1330–4.

Malin, S. (1984). Quantum states and potentialities of quantum systems. *Foundations of Physics* **16**: 1297–305.

Malin, S. (2003). *Nature Loves to Hide: Quantum Physics and the Nature of Reality, a Western Perspective*. Oxford.

Malin, S. (2006). What are quantum states? *Quantum Information Processing* **5**: 233–7.

Markopoulou, F. (2000). The internal description of a causal set: what the Universe looks like from the inside. *Communications in Mathematical Physics* **211**: 559–83.

Martinetti, P., Rovelli, C. (2003). Diamond's temperature: Unruh effect for bounded trajectories and thermal time hypothesis. *Classical and Quantum Gravity* **20**: 4919–31.

Matsukevich, D.N., Maunz, P., Moehring, D.L., Olmschenk, S., Monroe, C. (2008). Bell inequality violation with two remote atomic qubits. *Physical Review Letters* **100**: 150404.

Maudlin, T. (1994). *Quantum Non-Locality and Relativity*. Oxford.

Maudlin, T. (2007). *The Metaphysics Within Physics*. Oxford.

Maxwell, N. (1985). Are probabilism and special relativity incompatibile? *Philosophy of Science* **52**: 23–43.

Mayr, O. (1986). *Authority, Liberty and Automatic Machinery in Early Modern Europe*. Baltimore.

Maziarz, E.A., Greenwood, T. (1995). *Greek Mathematical Philosophy*. New York.

McLaughlin, W.I., Miller, S.L. (1992). An epistemological use of nonstandard analysis to answer Zeno's objections against motion. *Synthese* **92**: 371–84.

McRae, R. (1976). *Leibniz: Perception, Apperception and Thought*. Toronto.

Mercer, C. (2004). *Leibniz's Metaphysics: Its Origins and Development*. Cambridge.

Mermin, N.D. (1985). Is the moon there when nobody looks? Reality and the quantum theory. *Physics Today* **38**: 38–47.

Mermin, N.D. (1998a). The Ithaca interpretation of quantum mechanics. *Pramana* **51**: 549–65.

Mermin, N.D. (1998b). What is quantum mechanics trying to tell us? *American Journal of Physics* **66**: 753–67.

Mermin, N.D. (1999). What do these correlations know about reality? Nonlocality and the absurd. *Foundations of Physics* **29**: 571–87.

Mermin, N.D. (2006). In praise of measurement. *Quantum Information Processing* **5**: 239–60.

Miller, W.A., Wheeler, J.A. (1983). Delayed-choice experiments and Bohr's elementary quantum phenomenon. In: S. Kamefuchi, H. Ezawa, Y. Murayama, M. Namiki, S. Nomura, Y. Ohnuki, T. Yajima (eds.), *Proc. of the Int. Symp. on Foundations of Quantum Mechanics* (pp. 140–51). Tokyo.

Minkowski, H. (1909). *Raum und Zeit. Vortrag, gehalten auf der 80. Naturforscher-Versammlung zu Köln am 21. September 1908*. Leipzig – Berlin.

Mittelstrass, J. (1988). Nature and science in the Renaissance. In: R.S. Woolhouse (ed.), *Metaphysics and the Philosophy of Science in the 17th and 18th Centuries. Essays in Honour of Gerd Buchdahl* (pp. 17–43). Dordrecht.

Moehring, D.L., Madsen, M.J., Blinov, B.B., Monroe, C. (2004). Experimental Bell inequality violation with an atom and a photon. *Physical Review Letters* **93**: 090410.

Moreva, E., Brida, G., Gramegna, M., Giovannetti, V., Maccone, L., Genovese, M. (2014). Time from quantum entanglement: an experimental illustration. *Physical Review* A **89**: 052122.

Mourelatos, A.P.D. (2008). *The Route of Parmenides* (rev. and expanded ed.). Las Vegas – Zürich – Athens.

Myrvold, W.C. (2003). On some early objections to Bohm's theory. *International Studies in the Philosophy of Science* **17**: 7–24.

Newton, I. (1981). *The Mathematical Papers of Isaac Newton*, ed. by D.T. Whiteside, vol. VIII (1697–1722). Cambridge.

Nicolaus de Cusa (1998). *Dialogus de Ludo Globi. Edidit commentariisque illustravit Iohannes Gerhardus Senger* (*Nicolai de Cusa Opera Omnia*, vol. IX). Hamburg.

Norsen, T. (2007). Against 'realism'. *Foundations of Physics* **37**: 311–40.

Odin, S. (1982). *Process Metaphysics and Hua-Yen Buddhism. A Critical Study of Cumulative Penetration vs. Interpenetration*. Albany, N.Y.

Olesiński, D. (2000). Zeno of Elea and the Sophistic tradition (in Polish). In: B. Dembiński (ed.), *Within the Sphere of Classical Philosophy*, in Polish (pp. 154–63). Katowice.

Owen, G.E.L. (1986). *Logic, Science and Dialectic: Collected Papers in Greek Philosophy* (ed. by M. Nussbaum). London.

Padmanabhan, T. (2005). Gravity and the thermodynamics of horizons. *Physics Reports* **406**: 49–125.

Page, D.N., Wootters, W.K. (1983). Evolution without evolution: dynamics described by stationary observables. *Physical Review* D **27**: 2885–92.

Pan, J.-W., Bouwmeester, D., Daniell, M., Weinfurter, H., Zeilinger, A. (2000). Experimental test of quantum nonlocality in three-photon Greenberger–Horne–Zeilinger entanglement. *Nature* **403**: 515–9.

Parsons, T. (1990). True contradictions. *Canadian Journal of Philosophy* **20**: 335–53.

Paty, M. (1995). The nature of Einstein's objections to the Copenhagen interpretation of quantum mechanics. *Foundations of Physics* **25**: 183–204.

Penrose, R. (2004). *The Road to Reality: A Complete Guide to the Laws of the Universe*. London.

Peres, A. (2004). Einstein, Podolsky, Rosen, and Shannon. *Foundations of Physics* **35**: 511–4.

Peres, A., Zurek, W.H. (1982). Is quantum mechanics universally valid? *American Journal of Physics* **50**: 807–10.

Peruzzo, A., Shadbolt, P., Brunner, N., Popescu, S., O'Brien, J.L. (2012). A quantum delayed-choice experiment. *Science* **338**: 634–7.

Petkov, V. (2006). Is there an alternative to the block universe view? In: D. Dieks (ed.), *The Ontology of Spacetime* (pp. 207–28). Amsterdam.

Petkov, V. (2007). Relativity, dimensionality and existence. In: V. Petkov (ed.), *Relativity and the Dimensionality of the World* (pp. 115–35). Berlin – Heidelberg – New York.

Petkov, V. (2009). *Relativity and the Nature of Spacetime*. Berlin – Heidelberg – New York.

Petkov, V. (2013). *From Illusions to Reality: Time, Spacetime and the Nature of Reality*. Montreal.

Pittman, T.B., Strekalov, D.V., Migdall, A., Rubin, M.H., Sergienko, A.V., Shih, Y.H. (1996). Can two-photon interference be considered the interference of two photons? *Physical Review Letters* **77**: 1917–20.

Placek, T. (1997). Charting the labyrinth of Bell-type theorems. *Logic and Logical Philosophy* **5**: 93–120.

Placek, T. (2002). Partial indeterminism is enough: a branching analysis of Bell-type inequalities. In: T. Placek, J. Butterfield (eds.), *Non-Locality and Modality* (pp. 317–42), Dordrecht – Boston – London.

Poczobut, R. (2000). *The Dispute about the Non-Contradiction Principle* (in Polish). Lublin.

Popper, K.R. (1962). *Conjectures and Refutations*. New York.

Popper, K.R. (2002). *Unended Quest*. London – New York.

Poulin, D. (2006). Toy model for a relational formulation of quantum theory. *International Journal of Theoretical Physics* **45**: 1189–215.

Portmann, S., Wüthrich, A. (2007). Minimal assumption derivation of a weak Clauser-Horne inequality. *Studies in History and Philosophy of Modern Physics* **38**: 844–62.

Price, H. (1997). *Time's Arrow and Archimedes' Point: New Directions for the Physics of Time*. New York – Oxford.

Priest, G. (1979). The logic of paradox. *The Journal of Philosophical Logic* **8**: 219–41.

Priest, G. (1985). Inconsistencies in motion. *American Philosophical Quarterly* **22**: 339–46.

Priest, G. (1987). *In Contradiction: A Study of Transconsistent.* Dordrecht – Boston – Lancaster.

Priest, G. (2002). *Beyond the Limits of Thought.* Oxford.

Priest, G., Routley, R. (1984). Introduction: paraconsistent logics. *Studia Logica* **43**: 3–16.

Primas, H. (1990). Mathematical and philosophical questions in the theory of open and macroscopic quantum systems. In: A.I. Miller (ed.), *Sixty-Two Years of Uncertainty: Historical, Philosophical and Physical Inquiries into the Foundations of Quantum Mechanics* (pp. 233–57). New York.

Przełęcki, M. (2000). On the Paradoxes in Plato's 'Parmenides' (in Polish). In: M. Przełęcki, *The Platonic Readings*, in Polish (pp. 29–45). Warszawa.

Pulaczewska, H. (1999). *Aspects of Metaphor in Physics.* Tübingen.

Putnam, H. (1967). Time and physical geometry. *Journal of Philosophy* **64**: 240–7.

Reisenberger, M., Rovelli, C. (2002). Spacetime states and covariant quantum theory. *Physical Review* D **65**: 125016.

Rescher, N. (1981). *Leibniz's Metaphysics of Nature. A Group of Essays.* Dordrecht.

Rescher, N. (1991). *G. W. Leibniz' Monadology.* Pittsburgh.

Rescher, N. (1996). *Process Metaphysics: An Introduction to Process Philosophy.* Albany, N.Y.

Resnick, M. D. (1981). Mathematics as a science of patterns: ontology and reference. *Noûs* **15**: 529–50.

Resnick, M.D. (1982). Mathematics as a science of patterns: epistemology. *Noûs* **16**: 95–105.

Rey, A.-L. (2011). Action, perception, organisation. In: J.E.H. Smith, O. Nachtomy (eds.), *Machines of Nature and Corporeal Substances in Leibniz* (pp. 157–73). Dordrecht – Heidelberg – London – New York.

Reznik, B., Aharanov, Y. (1995). Time-symmetric formulation of quantum mechanics. *Physical Review* A **52**: 2538–50.

Rickles, D., French, S. (2006). Quantum gravity meets structuralism: interweaving relations in the foundations of physics. In: D. Rickles, S. French, J. Saatsi (eds.), *The Structural Foundations of Quantum Gravity* (pp. 1–39). Oxford.

Rietdijk, C.W. (1966). A rigorous proof of determinism derived from the special theory of relativity. *Philosophy of Science* **33**: 341–4.

Rovelli, C. (1991). Time in quantum gravity: an hypothesis. *Physical Review* D **43**: 442–56.

Rovelli, C. (1993). The statistical state of the universe. *Classical and Quantum Gravity* **10**: 1567–78.

Rovelli, C. (1996). Relational quantum mechanics. *International Journal of Theoretical Physics* **35**: 1637–78.

Rovelli, C. (2002a). Is there incompatibility between the ways time is treated in General Relativity and in standard Quantum Mechanics? In: A. Ashtekar, J. Stachel (eds.), *Conceptual Problems of Quantum Gravity* (pp. 126–40). New York.

Rovelli, C. (2002b). Partial observables. *Physical Review* D **65**: 124013.

Rovelli, C. (2004). *Quantum Gravity*. Cambridge.

Rovelli, C. (2006). The Disappearnace of space and time. In: D. Dieks (ed.), *Ontology of Spacetime* (pp. 25–36). Amsterdam – Oxford.

Rovelli, C. (2011). Forget time. *Foundations of Physics* **41**: 1475–90.

Rowe, M.A., Kielpinski, D., Meyer, V., Sackett, C.A., Itano, W.M., Monroe, C., Wineland, D.J. (2001). Experimental violation of a Bell's inequality with efficient detection. *Nature* **409**: 791–4.

Saunders, S. (2002). How relativity contradicts presentism. In: C. Callender (ed.), *Time, Reality and Experience* (pp. 277–92). Cambridge.

Schad, W. (1989). Die Zeitgestalt in der Evolution der *Ceratites*-Ammoniten aus dem Oberen Muschelkalk Mitteleuropas. In: W.H. Arnold (hg.), *Entwicklung. Interdisziplinäre Aspekte zur Evolutionsfrage* (pp. 99–141). Stuttgart.

Schelling, J.W.F. (1927). Einleitung zu dem Entwurf eines Systems der Naturphilosophie. In: *Schellings Werke*, nach der Originalausgabe in neuer Anordnung hg. von M. Schröter, 2. Hauptband (pp. 269–326). München.

Schopenhauer, A. (1851). *Parerga und Paralipomena: Kleine Philosophische Schriften* (Erster Band). Berlin.

Schrödinger, E. (1935a). Discussion of probability relations between separated systems. *Proc. of the Cambridge Philosophical Society* **31**: 555–63.

Schrödinger, E. (1935b). Die gegenwärtige Situation in der Quantenmechanik. *Die Naturwissenschaften* **23**: 807–12, 823–8, 844–9.

Scully, M.O., Drühl, K. (1982). Quantum eraser: a proposed photon correlation experiment concerning observation and 'delayed choice' in quantum mechanics. *Physical Review* A **25**: 2208–13.

Scully, M.O., Englert, B.-G., Walther, H. (1991). Quantum optical tests of complementarity. *Nature* **351**: 111–6.

Shapiro, S. (2000). *Philosophy of Mathematics: Structure and Ontology*. New York.

Shimony, A. (1984). Conrollable and uncontrollable non-locality. In: S. Kamefuchi, N.B. Gakkai (eds.), *Proc. of the Int. Symp. on Foundations of Quantum Mechanics in the Light of New Technology* (pp. 225–30). Tokyo.

Shimony, A. (1986). Events and processes in the quantum world. In: R. Penrose, C. Isham (eds.), *Quantum Concepts in Space and Time* (pp. 182–203). Oxford.

Shimony, A. (2009). Unfinished work: a bequest. In: W.C. Myrvold, J. Christian (eds.), *Quantum Reality, Relativistic Causality, and Closing the Epistemic Circle. Essays in Honour of Abner Shimony* (pp. 479–91). Berlin.

Sider, T. (2001). *Four-Dimensionalism: An Ontology of Persistence and Time*. Oxford.

Silberstein, M., Cifone, M., Stuckey, W.M. (2008). Why quantum mechanics favors adynamical and acausal interpretations such as relational blockworld over backwardly causal and time-symmetric rivals. *Studies in History and Philosophy of Modern Physics* **39**: 736–51.

Silberstein, M., Stuckey, W.M., Cifone, M. (2007). An argument for 4D block world from a geometric interpretation of nonrelativistic quantum mechanics. In: V. Petkov (ed.), *Relativity and the Dimensionality of the World* (pp. 197–216). Berlin – Heidelberg – New York.

Simplicius (1895). *Simplicii In Aristotelis Physicorum Libros Quattuor Posteriores Commentaria* (vol. II = *Commentaria in Aristotelem Graeca* X), hg. von H. Diels. Berlin. [*In Arist. Phys.*].

Singh, V. (2005). Einstein and the quantum. *Current Science* **89**: 2101–12.

Smerlak, M., Rovelli, C. (2007). Relational EPR. *Foundations of Physics* **37**: 427–45.

Smolenov, H. (1984). Zeno's paradoxes and temporal becoming in dialectical atomism. *Studia Logica* **43**: 169–80.

Smolin, L. (1997). *The Life of the Cosmos*. New York – Oxford.

Smolin, L. (2001). *Three Roads to Quantum Gravity*. New York.

Sorkin, R.D. (2007). Relativity theory does not imply that the future already exists: a counterexample. In: V. Petkov (ed.), *Relativity and the Dimensionality of the World* (pp. 153–61). Berlin – Heidelberg – New York.

Stein, H. (1968). On Einstein–Minkowski space-time. *Journal of Philosophy* **65**: 5–23.

Stuckey, W.M., Silberstein, M., Cifone, M. (2007). The relational blockworld interpretation of non-relativistic quantum mechanics. In: G. Adenier, C.A. Fuchs, A. Khrennikov (eds.), *Foundations of Probability and Physics 4. AIP Conf. Proc., vol. 889, 4–9 VI 2006, Växjö, Sweden* (pp. 412–21). Melville, N.Y.

Stuckey, W.M., Silberstein, M., Cifone, M. (2008). Reconciling spacetime and the quantum: relational blockworld and the quantum liar paradox. *Foundations of Physics* **38**: 348–83.

Szabó, L.E. (1995). Quantum mechanics in an entirely deterministic universe. *International Journal of Theoretical Physics* **34**: 1751–66.

Szabó, L.E. (1998). Quantum structures do not exist in reality. *International Journal of Theoretical Physics* **37**: 449–56.

Szabó, L.E. (2000a). On an attempt to resolve the EPR-Bell paradox via Reichenbachian concept of common cause. *International Journal of Theoretical Physics* **39**: 901–11.

Szabó, L.E. (2000b). Contextuality without contextuality: Fine's interpretation of quantum mechanics. In: T. Placek (ed.), *Probabilities – a New Tool for Philosophy? Reports on Philosophy* **20**: 117–30.

Teller, P. (1986). Relational holism and quantum mechanics. *British Journal for the Philosophy of Science* **37**: 71–81.

Teller, P. (1989). Relativity, relational holism and the Bell inequalities. In: J. Cushing, E. McMullin (eds.), *Philosophical Consequences of Quantum Theory: Reflections on Bell's Theorem* (pp. 208–23). Notre Dame, IN.

Thiele, R. (2007). What is function? In: R.E. Bradley, L.A. D'Antonio, C.E. Sandifer (eds.), *Euler at 300: An Appreciation. The MAA Tercentenary Euler Celebration* (pp. 63–83). Washington, DC.

Tittel, W., Brendel, J., Zbinden, H., Gisin, N. (1998). Violation of Bell inequalities by photons more than 10 km apart. *Physical Review Letters* **81**: 3563–6.

Trifonov, A. Björk, G., Söderholm, J., Tsegaye, T. (2002). Comprehensive experimental test of quantum erasure. *The European Physical Journal* D **18**: 251–8.

Vaidman, L. (1999). The meaning of elements of reality and quantum counterfactuals: reply to Kastner. *Foundations of Physics* **29**: 865–76.

Van Fraassen, B.C. (2010). Rovelli's world. *Foundations of Physics* **40**: 390–417.

Vlastos, G. (1966). A note on Zeno's arrow. *Phronesis* **11**: 3–18. (Also in: G. Vlastos, *Studies in Greek Philosophy*, vol. I, Princeton 1995, pp. 205–18).

Von Neumann, J. (1932). *Die mathematischen Grundlagen der Quantenmechanik*. Berlin.

Vorenkamp, D. (2005). Reconsidering the Whiteheadian critique of Huayan temporal symmetry in light of Fazang's views. *Journal of Chinese Philosophy* **32**: 197–210.

Wahsner, R. (1991). 'Die Theorie dient nur der Konstruktion!' Zur geometrisch-kinematischen Naturphilosophie des Thomas Hobbes. In: H. Hecht (hg.), *Gottfried Wilhelm Leibniz im philosophischen Diskurs über Geometrie und Erfahrung* (pp. 47–85). Berlin.

Walborn, S.P., Terra Cunha, M.O., Pádua, S., Monken, C.H. (2002). Double-slit quantum eraser. *Physical Review* A **65**: 033818.

Wald, R.M. (1984). *General Relativity*. Chicago – London.

Wald, R.M. (1994). *Quantum Field Theory in Curved Spacetime and Black Hole Thermodynamics*. Chicago.

Walther, P., Aspelmeyer, M., Resch, K.J., Zeilinger, A. (2005). Experimental violation of a cluster state Bell inequality. *Physical Review Letters* **95**: 020403.

Wedin, M.V. (2000). *Aristotle's Theory of Substance: The Categories and Metaphysics Zeta*. Oxford – New York.

Weihs, G. Jennewein, T. Simon, T., Weinfurter, H., Zeilinger, A. (1998). Violation of Bell's inequality under strict Einstein locality conditions. *Physical Review Letters* **81**: 5039–43.

Weingard, R. (1972). Relativity and the reality of past and future events. *British Journal for the Philosophy of Science* **23**: 119–21.

Weinstein, S. (2009). Nonlocality without nonlocality. *Foundations of Physics* **39**: 921–36.

Weisz, E., Choi, H.K., Sivan, I., Heiblum, M., Gefen, Y., Mahalu, D., Umansky, V. (2014). An electronic quantum eraser. *Science* **344**: 1363–6.

Weizsäcker, C.F. von (1982). Parmenides und die Quantentheorie. In: C.F. von Weizsäcker, *Die Einheit der Natur* (pp. 466–91). München.

Weizsäcker, C.F. von (1987). The Greek philosophy and modern physics (orig. in Polish). In: M. Heller, A. Michalik, J. Życiński (ed.), *Philosophizing in the Context of Science*, in Polish (pp. 140–51). Kraków.

Weizsäcker, C.F. von (1992). *Zeit und Wissen*. München.

Wesoły, M. (1993). Plato's concept of harmony in the light of the dialogues and unwritten doctrine (in Polish). In: A. Kijewska, E.I. Zieliński (ed.), *Plato: A New Interpretation*, in Polish (pp. 107–32). Lublin.

Weyl, H. (1931). *The Theory of Groups and Quantum Mechanics*. New York.

Weyl, H. (1952). *Space – Time – Matter.* Mineola, N.Y.

Wheeler, J.A. (1968). Superspace and the nature of quantum geometrodynamics. In: C. DeWitt, J.A. Wheeler (eds.), *Battelle Rencontres: 1967 Lectures in Mathematics and Physics* (pp. 242–307). New York – Amsterdam.

Wheeler, J.A. (1978). The 'past' and the 'delayed-choice' double-slit experiment. In: A.R. Marlow (ed.), *Mathematical Foundations of Quantum Theory* (pp. 9–47). New York.

Wheeler, J.A. (1979). Frontiers of time. In: T. di Francia (ed.), *Problems in the Foundations of Physics. Proc. of the Int. School of Physics 'Enrico Fermi', Course 72* (pp. 395–497). Amsterdam.

Whitrow, G.J. (1989). *Time in History: Views of Time from Prehistory to the Present Day.* Oxford – New York.

Wigner, E.P., Inönü, E. (1952). Representations of the Galilei group. *Il Nuovo Cimento* **9**: 705–18.

Woolhouse, R.S., ed. (1994) *Gottfried Wilhelm Leibniz: Critical Assessments.* Vol II: *Metaphysics and Its Foundations* 2 (*Substances, Their Creation, Their Complete Concepts, and Their Relations*). London – New York.

Wooters, W.K. (1990). Local accessibility of quantum states. In: W.H. Zurek (ed.), *Complexity, Entropy and the Physics of Information* (pp. 39–46). Boston.

Woszczek, M. (2011). Leibniz's series and the problem of dynamics in the quantization of gravity (in Polish). *Filozofia Nauki* **19**(2): 41–62.

Yokota, K., Yamamoto, T., Koashi, M., Imoto, N. (2009). Direct observation of Hardy's paradox by joint weak measurement with an entangled photon pair. *New Journal of Physics* **11**: 033011.

Youschkevitch, A.P. (1976/1977). The concept of function up to the middle of the $19^{th}$ century. *Archive for History of Exact Sciences* **16**: 37–85.

Zajonc, A. (1998). Light and cognition: Goethean studies as a science of the future. In: D. Seamon, A. Zajonc (eds.), *Goethe's Way of Science: A Phenomenology of Nature* (pp. 299–314). Albany, N.Y.

Zajonc, A., Wang, L.J., Zou, X.Y., Mandel, L. (1991). Quantum eraser. *Nature* **353**: 507–8.

Zhao, Z., Chen, Y.-A., Zhang, A.-N., Yang, T., Briegel, H., Pan, J.-W. (2004). Experimental demonstration of five-photon entanglement and open-destination teleportation. *Nature* **430**: 54–8.

# General Index

## D



## E

## Q



## R

**Y**



**Z**

**DIA-LOGOS**
**Schriften zu Philosophie und Sozialwissenschaften**
**Studies in Philosophy and Social Sciences**

Herausgegeben von
Edited by
Tadeusz Buksiński & Piotr W. Juchacz

"*Dia-Logos. Studies in Philosophy and Social Sciences* is a peer-reviewed book series publishing valuable monographs and edited volumes on various aspects of philosophy and social sciences. The series is intended to be an interdisciplinary forum of deliberation according to our firm belief that challenges of the contemporary world require common and multilevel research. The *Dia-Logos* series does not represent a single ideology or school of thought, but it is open to different trends and various styles of reflection, trying to understand better the contemporary world. We invite the submission of manuscripts of monographs and edited volumes from academic philosophers and social scientists."

Bd. / Vol. 1 Piotr W. Juchacz / Roman Kozłowski (Hrsg.): Freiheit und Verantwortung. Moral, Recht und Politik. 2002.

Bd. / Vol. 2 Norbert Leśniewski / Ewa Nowak-Juchacz (Hrsg.): Die Zeit Heideggers. 2002.

Bd. / Vol. 3 Marek Kwiek (ed.): The University, Globalization, Central Europe. 2003.

Bd. / Vol. 4 Ewa Czerwińska-Schupp (Hrsg.): Philosophie an der Schwelle des 21. Jahrhunderts. Geschichte der Philosophie, Philosophische Anthropologie, Ethik, Wissenschaftstheorie, Politische Philosophie. 2003.

Bd. / Vol. 5 Danuta Sobczyńska / Pawel Zeidler / Ewa Zielonacka-Lis (eds.): Chemistry in the Philosophical Melting Pot. 2004.

Bd. / Vol. 6 Marek Kwiek: Intellectuals, Power, and Knowledge. Studies in the Philosophy of Culture and Education. 2004.

Bd. / Vol. 7 Marek Kwiek: The University and the State. A Study into Global Transformations. 2006.

Bd. / Vol. 8 Andrzej Przylebski (Hrsg.): Das Erbe Gadamers. 2006.

Bd. / Vol. 9 Ewa Czerwińska-Schupp (ed.): Values and Norms in the Age of Globalization. 2007.

Bd. / Vol. 10 Tadeusz Buksiński (ed.): Democracy in Western and Post-Communist Countries. Twenty Years after the Fall of Communism. 2009.

Bd. / Vol. 11 Marek Zirk-Sadowski / Mariusz Golecki / Bartosz Wojciechowski (eds.): Multicentrism as an Emerging Paradigm in Legal Theory. 2009.

Bd. / Vol. 12 Bartosz Wojciechowski: Philosophical Approach to the Interculturality of Criminal Law. 2010.

Bd. / Vol. 13 Marek Nowak / Michał Nowosielski (eds.): (Post)transformational Migration. Inequalities, Welfare State, and Horizontal Mobility. 2011.

Bd. / Vol. 14 Tadeusz Buksiński (ed.): Religions in the Public Spheres. 2011.

Bd. / Vol. 15 Bartosz Wojciechowski / Piotr W. Juchacz / Karolina M. Cern (eds.): Legal Rules, Moral Norms and Democratic Principles. 2013.

Bd. / Vol. 16 Ewa Nowak / Dawn E. Schrader / Boris Zizek (eds.): Educating Competencies for Democracy. 2013.

Bd. / Vol. 17 Tadeusz Buksiński (ed.): Identities and Modernizations. 2013.

Bd. / Vol. 18 Karolina M. Cern: The Counterfactual Yardstick. Normativity, Self-Constitutionalisation and the Public Sphere. 2014.

Bd. / Vol. 19 Vito Breda / Lidia Rodak (eds.): Objectivity in Judicial Narratives. Theories and Critiques. 2015.

Bd. / Vol. 20 Marek Woszczek: Platonic Wholes and Quantum Ontology. Translated by Katarzyna Kretkowska. 2015.

www.peterlang.com

www.ingramcontent.com/pod-product-compliance
Ingram Content Group UK Ltd.
Pitfield, Milton Keynes, MK11 3LW, UK
UKHW041826100926
8556IPUK00002B/43

9 783631 666296